3 · HILLSLOPE FORM AND PROCESS

CAMBRIDGE GEOGRAPHICAL STUDIES

HILLSLOPE
FORM AND PROCESS

by M. A. CARSON
Associate Professor of Geography, McGill University

and M. J. KIRKBY
Lecturer in Geography, Bristol University

CAMBRIDGE UNIVERSITY PRESS

CAMBRIDGE UNIVERSITY PRESS
Cambridge, New York, Melbourne, Madrid, Cape Town, Singapore, São Paulo, Delhi

Cambridge University Press
The Edinburgh Building, Cambridge CB2 8RU, UK

Published in the United States of America by Cambridge University Press, New York

www.cambridge.org
Information on this title: www.cambridge.org/9780521109116

First published 1972
Reprinted 1975
This digitally printed version 2009

A catalogue record for this publication is available from the British Library

Library of Congress Catalogue Card Number: 74–163061

ISBN 978-0-521-08234-1 hardback
ISBN 978-0-521-10911-6 paperback

CONTENTS

Contents

PREFACE

The study of hillslopes has attracted geomorphologists for a long time and it is viewed by many as the most central single theme in the subject. Much research has been directed to the topic over the past two decades, but, to date, no attempt has been made to produce a comprehensive survey of it in the way that Leopold, Wolman and Miller's *Fluvial processes in geomorphology* did for stream channel form and process. A similar survey for hillslopes has been our aim in writing this book. It is directed at the graduate level, but it should also be readily understood by any final-year undergraduate with a basic science background. We have clearly drawn heavily on the work of others, often geomorphologists or pedologists, but equally engineers working in the fields of rock mechanics, soil mechanics or hydraulics. We feel strongly that slope studies and geomorphology generally must be approached from an inter-disciplinary point of view.

Although we have both done our best to give a balanced account of the topics on which we have written, there remain minor differences in viewpoint between the sections which the two of us have written. In a subject where few, if any, conclusions are final, we have felt it honest to retain such differences and leave the reader free to accept either or neither version. At the same time it may assist him to distinguish these intended differences if we list our responsibilities for individual chapters, as follows:

M. A. Carson: Chapters 2, 4, 6, 7, 11, 12, and 15;
M. J. Kirkby: Chapters 1, 3, 5, 8, 9, 10, 13, 14, and 16.

However, we would like to affirm that we consider this book to be a joint venture, with responsibility equally shared between us: our names appear in the title in alphabetical order according to bibliographical practice.

Parts of the manuscript have been critically reviewed by our colleagues, especially J. B. Bird, R. J. Chorley, D. Ingle Smith and Eiju Yatsu, to all of whom we owe a large debt. Similarly we would like to thank the Editorial Board of the Cambridge Geographical Studies for their advice and continued patience in the preparation of the text. Lastly our thanks go to our wives for their constant assistance in reviewing, proof-reading and living with the manuscript during the last two years.

April 1971   M. A. CARSON   M. J. KIRKBY

ACKNOWLEDGEMENTS

Thanks are due to the following authors, publishers and learned bodies for permission to reproduce figures and passages from books and journals.

Addison-Wesley (fig. 7.10); Allyn and Bacon (fig. 7.7); American Geophysical Union (figs. 4.1, 8.15); American Journal of Science (figs. 6.5, 6.9, 6.11, 9.4, 9.17, 11.8, 13.6); American Society of Civil Engineers (figs. 7.3, 10.7); The Arctic Institute of North America (fig. 12.3); The Association of American Geographers (fig. 12.1); Biuletyn Peryglacjalny (fig. 12.6); Butterworths (fig. 7.8c); Cambridge University Press (figs. 2.4a, b); R. J. Chorley; Civil Engineering (fig. 3.6); D. R. Currey; Die Erde (fig. 15.5); Dover Publications; Geografiska Annaler (figs. 10.9, 12.4); Geographical Analysis (figs. 7.16, 7.17); The Geological Society of America (fig. 15.1); K. J. Gregory; Institute of British Geographers (figs. 11.7, 16.8); The Council of the Institution of Civil Engineers, London (figs. 4.10, 4.19, 7.12); International Association for Scientific Hydrology (fig. 16.21a); H. Jenny; L. C. King; D. Koons; Koninklijke Nederlandse Akademie Van Wetenschappen; W. B. Langbein; Longman (fig. 11.3); Macmillan & Co. Ltd (London) (figs. 2.1, 2.2, 15.3, 15.4); McGraw-Hill Book Co. (fig. 4.12); M. A. Melton; Methuen & Co. Ltd (figs. 3.13, 7.5, 8.9); National Research Council – Highway Research Board, (U.S.A.) (fig. 4.13); Nature Norwegian Geotechnical Institute (figs. 6.8, 7.13; table 4.1); Office of Naval Research, (U.S.A.) (fig. 16.20); J. C. Pugh; A. Rapp; Royal Geographical Society (Geographical Journal) (fig. 12.2); Royal Society of London (fig. 3.8 (inset)); S. A. Schumm; A. N. Strahler; Soil Science (figs. 3.4, 3.10); University of Chicago Press (fig. 15.2); University of Toronto Press (fig. 4.7); U.S. Department of Agriculture (figs. 3.12, 8.13); U.S. Geological Survey (figs. 9.2, 9.3, 16.21b); Vierteljahsschrift der Naturforschenden Gesellschaft (figs. 6.13, 6.14, 6.15); D. E. Walling; A. Young; E. G. Youngs; Zeitschrift fur Geomorphologie (figs. 6.10, 12.5).

INTRODUCTION:

CHAPTER 1
GEOMORPHIC SYSTEMS AND MODELS

Hillslopes have the distinction of being the commonest and, at the same time, least studied geomorphic features, especially in terms of the processes acting on them. In some ways it is their very ubiquity which is responsible for this neglect because researchers have preferred to study unique or restricted features instead of facing the massive sampling problem which is involved in characterizing slopes. In addition, hillslopes present difficult research problems because their forms change either slowly or infrequently, particularly when compared to rivers.

The study of slopes is essential not only to an understanding of natural landscapes, but also as a practical means for controlling erosion and sedimentation which result when man modifies the landscape through agriculture, engineering construction or dumping operations. Cultivation and grazing have long accelerated natural rates of surface wash and have led to the initiation of gullies; and this aspect of soil erosion has been studied intensively particularly since the 1920s. Stripping of vegetation for agriculture has also induced more rapid wind erosion in some areas. Engineering construction often produces artificially steep slopes in cut or fill material, and the aim of design procedures is to reduce the tendency of such slopes to fail suddenly in a landslide, especially in the short term.

Geomorphologists, in common with other environmental scientists, study complex systems which contain within them many interactions and feedbacks. The two main ways of beginning to understand a system are, first inductively, through measuring landscape variables at sample points and analysing the results using statistical, usually multivariate, methods, and second deductively, building models on established physical laws whenever possible. These two approaches are inevitably interlocked, the inductive approach helping to generate some deductive models and testing others. In this book our emphasis is on landforms and the processes which form them, and consequently on deductive process–response models rather than on statistical analysis of data which are necessary to verify them.

Different workers have approached geomorphic problems from dif-

1

ferent standpoints which have, often unintentionally, limited the range of problems which could be tackled and the types of solution which were acceptable. There is not one correct paradigm of this sort: each is the product of its time and stimulates a period of fruitful research before being rejected. Geomorphic fashions have swung from the qualitative statements of early observers to the highly mathematical treatments of Bakker and Scheidegger, and back again to inductive studies derived from field measurement (Carson, 1969*b*). The approach of this book, with its emphasis on process–response models, is no less a part of current fashion but seems to represent a paradigm which is not yet exhausted. To examine this approach, we must delve a little into systems and model methodology.

SYSTEMS

Harvey (1969, p. 451) has described a system as containing:

'(1) A set of elements identified with some variable attribute of objects.
(2) A set of relationships between the attributes of objects.
(3) A set of relationships between those attributes of objects and the environment.'

In geomorphology, the objects are normally landforms, and their attributes consist of the topographic, soil, vegetation, etc., properties of the landforms. The relationships between these attributes consist of exchanges of energy, debris or water so that there are causal or other links between the attributes. The choice of elements which belong to the system, as opposed to the environment, varies with the problem and with scale, as will be discussed below.

An important distinction which clarifies the different approaches of geomorphologists, is that between open and closed systems. A closed system is one in which there are no relationships between the system elements and the environment. While it is trivially true that no geomorphic system is closed, much of classical physics and chemistry has developed from closed system models and Chorley (1962) has pointed out the extent to which closed system *thinking* has influenced geomorphic thinking. Closed systems typically consist of near-frictionless perpetual motion machines, for example the solar system; or else of systems in which initial variations in mechanical and thermal energy are progressively smoothed out to produce a totally uniform, constant-temperature equilibrium. The former type of equilibrium is somewhat analogous to a geomorphic 'steady state' in which the elements of the system are simply constant through time and unaffected by the environment over the time scale considered. The latter type of equilibrium is

2

analogous to a peneplain type of equilibrium in which the landscape system is progressively running down towards a dead level uniform plain. This is not to suggest that these geomorphic equilibria are supposed to occur within a closed system, only that they are types of equilibrium which arise from closed system *thinking*.

In a closed system, the sum of mechanical and thermal energies is conserved, but only in a frictionless system is each *separately* conserved. Otherwise frictional losses convert mechanical energy into heat energy and the system shows a progressive increase in entropy, which is itself defined in two ways which can be shown to be equivalent. If an amount of heat energy ΔH is associated with material at absolute temperature θ, then the entropy of the system, S is given by:

$$S = \sum_\theta \frac{\Delta H}{\theta}. \tag{1.1}$$

Alternatively the molecules of the system can be considered to be in discrete energy states 1, 2, 3 . . ., n, a proportion p_i of the molecules being in state i. On this definition, the entropy of the system,

$$S \propto -\sum_{i=1}^{n} p_i \log p_i. \tag{1.2}$$

It can be shown that at equilibrium, when all mechanical energy has been transformed to heat, the entropy is at a maximum, corresponding to a condition of constant temperature, and, in the unconstrained case, to a condition in which all the p_is of equation 1.2 are equal. This condition is one in which there is a minimum of organization or differentiation between the various states; or a condition for which the *a priori* knowledge of the system is at a minimum.

In an open system, which has links with the external environment, the energy of the system is no longer conserved, and equilibrium positions and criteria for equilibria are more difficult to define. Open systems may run down towards a static equilibrium, but many maintain a dynamic equilibrium (Hack, 1960) which responds to variations in the external environment, and is maintained by negative feedbacks which react against an environmental stimulus so that the system partly accommodates the stimulus and partly arranges itself to resist the stimulus. This self-regulation mechanism has been formalized, initially in chemistry, as Le Châtelier's principle, which states:

Any system in . . . equilibrium undergoes, as a result of a variation in one of the factors governing the equilibrium, a compensating change in a direction such that, had this change occurred alone it would have produced a variation of the factor considered in the opposite direction. (Prigogine and Defay, 1954, p. 262.)

3

This statement is almost identical to the definition proposed by Mackin (1948) for 'grade' of a river; but its possible geomorphic implications have not been fully explored.

An alternative criterion for equilibrium in an open system which has been proposed for organisms (Denbigh, 1951) is that the *rate of production* of entropy is a minimum. If entropy is seen as a measure of differentiation, it can be seen that the survival of an organism is dependent on maintaining its internal differentiation, and that a breakdown in differentiation is fatal to it. In a system where temperature differences are unimportant, this condition is equivalent to one of minimum work; a concept which has been widely used as a criterion in otherwise indeterminate mechanical systems, and has been applied to river meanders (Bagnold, 1960). The analogy between a geomorphic system and an organism is an attractive one, because both exhibit a steady progression over a long time span, in conjunction with a near-equilibrium in response to short-term fluctuations. It is therefore reasonable to hypothesize that geomorphic features measured over decades can be treated as being in a true dynamic equilibrium like that of an organism, and not merely as showing a trend which is obscured by large amounts of random noise in short measuring periods.

The concept of minimum entropy production has been used in a somewhat different manner by Leopold and Langbein (1962). An analogy was made between heat energy and mechanical energy, and between absolute temperature and elevation above base level, to define entropy as in equation 1.1. As with other analogies it must be tested within the appropriate system. The thermodynamic argument only shows that a quantity (entropy) which is conserved in a frictionless (reversible) cycle must have the form:

$$\Delta S = \frac{\Delta H}{f(\theta)}, \qquad (1.3)$$

where f can be any function of θ; and the particular function, f, must be chosen with reference to the particular system. For a river system where θ is elevation, the function can only be a constant, so that minimum entropy production corresponds to minimum work. In the same paper Leopold and Langbein attempt to estimate geomorphic distributions in terms of most probable states, an approach which is linked to the concept of entropy in its information theory context, and which has since led to the development of a theory of minimum variance (Langbein and Leopold, 1966) for meanders and other features of channel geometry.

The discussion of landscapes as systems leads naturally to a comparison with other systems, which provide a fruitful source of analogies and allow hypotheses to be set up for the geomorphic system. This procedure has been formalized as General Systems Theory (von Bertalanffy, 1951) and has been discussed at length in the geomorphic context by Chorley (1962), but it has not reached a level at which conclusions for *general* systems can be applied to particular systems without re-testing in the particular system.

TIME AND SPACE SCALES

Schumm and Lichty (1965) have considered the conflict between short-term equilibrium and long-term evolution for landscapes, referred to

TABLE 1.1. *Systems for the study of drainage basin variables* (modified from Schumm and Lichty, 1965, table 1)

Variables	Cyclic	Graded (a)	Graded (b)	Steady
Climate	E	E	E	E
Geology	E	E	E	E
Regional relief	S	E	E	E
Slope forms	S	S	E	E
Soil properties	S	S	E	E
Vegetation properties	S	S	S	E
Drainage density	S	S	S	E
Sediment discharge	S	S	S	S
Channel geometry and micro-morphology	I	S	S	E
Water discharge	I	I	S	S

E = environmental variable.
S = element of system studied.
I = irrelevant or meaningless variable in system studied.

above, in terms of the choice of a suitable system. If we wish to examine geomorphic change over a particular time span, then features which change only in much longer periods may be considered as fixed, that is as belonging to the external environment. Landscape features which change in the chosen time period form elements of the system to be studied; and features which change very rapidly within the chosen time span can be considered as having average values which are a part of the system, and random deviations which are irrelevant to the system.

Schumm and Lichty distinguish cyclic, graded and steady time spans.

Cyclic time ... refers to a time span encompassing an erosion cycle ... A fluvial system when viewed from this perspective is an open system undergoing continued change ... (p. 113).

The *graded time* span refers to a short span of cyclic time during which a graded condition or dynamic equilibrium exists.... When viewed from this perspective

5

one sees a continual adjustment between elements of the system, for events occur in which negative feedback (self-regulation) dominates (p. 114).

During a *steady time* span a true steady state may exist in contrast to the dynamic equilibria of graded time. These brief periods of time are referred to as a steady time span because in hydraulics steady flow occurs when none of the variables involved in a section change with time (p. 115).

The periods implied by these definitions depend on the rate at which a particular landscape is developing, but might, under normal conditions, refer to periods of about 10^6, 10^2 and 10^{-2} years respectively. Table 1.1 (adapted from Schumm and Lichty, table 1) indicates systems for study which may be appropriate to the time scale we are concerned with, but close scrutiny shows that almost all the variables listed may be split into components which are relevant to each time span. For example, long-term average climate and climatic change are relevant to *cyclic time*; frequency distributions of present climate are relevant to *graded time*; and instantaneous rainfall and temperature values are relevant to *steady time*. The *graded time* span has been divided into two in table 1.1, because *graded time* spans mean different things in slope and river studies. *Graded time* (*b*) corresponds more closely to Schumm and Lichty's usage, and appears to correspond to a time period in which the channel geometry and channel extension reach equilibrium, say ten- to one-hundred-year periods. *Graded time* (*a*) refers to a longer period, say 1,000 to 100,000 years, in which slope profile forms reach an equilibrium to which the word grade has also been applied.

For the longer time spans of *cyclic time* and *graded time* (*a*), the system is essentially a sediment system, and the system can be defined in terms of sediment distribution and flows, and their variation over time and space. In this system the flows of water are incidental, merely providing an agent for the transportation of the debris. For the short time periods of *graded time* (*b*) and *steady time*, however, the overall topography is fixed and the system describes the flows of water over this surface. Sediment discharge remains a system variable, but only 5 per cent or less of the available flow energy is absorbed in debris transportation. This water system is entirely relevant to hydrologic studies and forms most of the system for considering equilibrium hydraulic geometry and drainage density, but it is of only marginal relevance to slope studies. Since all measurements are made in *steady time* spans and even long periods of record apply only to *graded time* (*b*) for typical systems, slope studies are limited by the problem of extrapolating to long time periods. It is clear that the critical variable must be sediment discharge, since only this is a systems variable in all time spans; but it is also clear that extrapolations must be rather crude over long time spans.

Although *cyclic* and *steady time* spans refer to very long and short time periods respectively, the establishment of *graded time* implies that dynamic equilibria exist, and that the *graded time* span is not merely an artifact of the period of accurate records. If a dynamic equilibrium exists, then a *graded time* system can be examined in which the fluctuations of the system variables need not be referred to events in longer time spans. All that has been said about time scales applies equally to space scales, and we may study systems which refer to a single point or quadrat; to the whole world; or to intermediate areas which show maximum inter-action and minimal linkage with other areas. For slope studies, the appropriate spatial systems appear to be the drainage basin and the slope profile. Because slope profiles, defined as lines of greatest slope, may migrate laterally during their evolution, they are less satisfactory than drainage basins in minimizing external links, but their greater simplicity outweighs this disadvantage, particularly if we consider pro-files on which the contours remain straight so that lateral migration is minimized.

On a slope profile sediment is transferred from point to point, with little or no transfer across the profile, so that hillslopes can be simpli-fied to two-dimensional spatial systems with a relatively small loss of information and considerable gain in analytical simplicity. In this book, therefore, we have concentrated on the slope profile system over long periods of time; that is, we view it as a sediment system. In chapter 16, we will return to drainage basins, to examine the extent to which the third dimension modifies the slope models considered in this book, but it is as well here to anticipate a little and state that the modifications entailed are relatively slight.

MODELS OF HILLSLOPE SYSTEMS

Many early workers, with the notable exception of Gilbert, were pre-occupied with the way in which the external environment changes through cyclic time, but it seems that we can only examine this after we have constructed sound models linking process and form within the system (Carson, 1969*b*). Theory construction proceeds, as always, by an alternation of deductive and inductive steps (fig. 1.1). For example, a study may set out to test a model; and will do this by (1) making deductions which are testable, (2) designing an experiment to collect suitable data, (3) analysing this data to a form which is comparable with the deductions from the model and (4) testing. If the test is satisfactory, then the model is not disproved and may be checked against other areas: otherwise the model must be modified using fresh concepts to refine it.

Davis (1899) is chiefly responsible for the historical bias of geomorphology towards genetic, evolutionary models, but although his general evolutionary framework remains valid (see chapter 2), the neglect of both process and slope geometry by Davis and his successors limits the explanatory power of their deductions. In this book we are seeking to examine processes and profiles in a deductive framework which is compatible with established physical laws. The basic models are causal, although this term is taken to include causal feedback loops.

On the process level, the causal model is of a mechanical equilibrium between *forces* which tend to transport material and *resistances* which

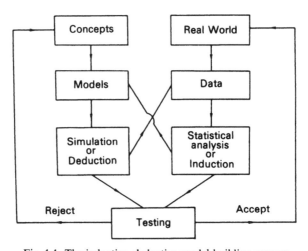

Fig. 1.1. The inductive–deductive model building process.

oppose the movement. Analysis is directed to an understanding of the factors responsible for the movement, and also to an understanding of temporal and spatial variations in its rate of action. On the landform level, the model seeks to link spatial variations in transport rate to sequences of slope profile development; that is, a process–response model. Formally this may be done as a differential equation (see chapters 5 and 15) which relates the temporal process rate to the spatial change in hillslope forms. Despite Hack's (1960) paper on dynamic equilibrium, little has been done on the application of equilibrium models to hillslope forms, although the criteria for open system equilibria appear to be applicable. It might, however, be noted that equilibrium models have been most successful in river studies at the micro- or mesoscale, and least successful in predicting overall long profiles.

At an early stage in an investigation it may be preferable to aim at generating models after collecting and analysing data which are considered relevant to a range of possible models. This inductive approach relies heavily on multivariate statistical methods, notably multiple regression and factor analyses. However, the inherent difficulty in making large numbers of reliable measurements, especially of process rates, so limits sample sizes that there are severe practical problems in verifying or generating models. The best that can be done is to choose sample points which are contiguous and so minimize extraneous differences, but which at the same time retain a full range of internal variation due to the operation of the system – in other words to select sample points from within slope profiles or basins, which thus form suitable sample clusters as well as suitable spatial systems. A more serious problem which cannot usually be overcome is the impossibility of adequate sampling over time spans relevant to slope profile development.

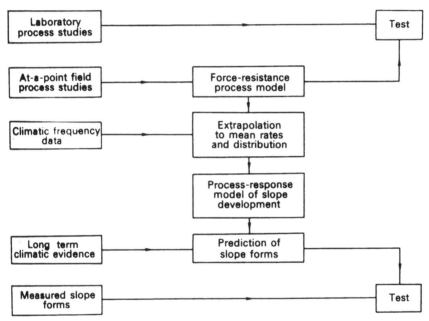

Fig. 1.2. The pattern of hillslope studies.

The most elegant solution to this problem is to find locations where space may be substituted for time (Savigear, 1952); but this is rarely possible in a simple way. Alternatively, we can invoke the ergodic hypothesis (Harvey, 1969, p. 128) and assume that the spatial distribution of a variable is the same as its temporal distribution. On a world scale, this is equivalent to an assumption of strict uniformitarianism – a rather doubtful hypothesis; and within a basin it implies that all profiles follow the same sequence, but at different rates – a statement which appears, on the available evidence, to be false.

Introduction

The difficulties inherent in slope studies mean that the most effective studies are of slope *process* at a point over short periods, and of slope *forms* over an area. The process studies must be linked to a physical force-resistance model, which can be independently tested in some cases against laboratory experiments, and then extrapolated using climatic frequency data to produce long-term rates, and which then form the basis for a process–response model (fig. 1.2). The process–response model may be combined with long-term climatic data to produce predictions of slope forms which may be compared with measured forms, and this is the only other point at which an independent test can be made. If the test rejects the predicted forms, then the error may be derived from any of the stages, and not only from the original process measurements.

Although there is no unique way to examine slope forms and processes, the relationship between data and models shown in fig. 1.2 typifies most recent work, although it clearly excludes Davis' approach. This book follows the same pattern, although reflecting the preponderance of quantitative work on *process* rather than *form*: an imbalance in recent work which is perhaps a reaction to an earlier over-emphasis on slope forms, interpreted in the context of denudation chronology.

CHAPTER 2

STARTING POINTS: SYSTEMS OF REFERENCE

In the previous chapter, a distinction was made between the formulation of a model and the construction of a system of reference for it. The cycle of erosion of William Morris Davis (Davis, 1899) contains aspects of both a model and a system. It is a system of reference in the sense that it maintains that the landscape evolves through time; indeed, it has been referred to by Chorley (1962) as the classic geomorphic example of closed system thinking. The particular *mode* of landscape development envisaged by Davis constitutes one possible model within this evolutionary system.

Many geomorphologists have attacked the specific model suggested by Davis, whilst still accepting the notion of an evolutionary system. In the introduction of his *Canons of Landscape Development*, for instance, Lester King (1953) writes: 'Some authorities have rejected the cyclic concept altogether; others, with whom we align ourselves, have accepted the general concept of a cycle of landforms developed under erosion, while considering that the detailed forms and sequences depart considerably from those visualized and adduced by Davis' (p. 723). In this text, we also accept the system of reference provided by Davis, whilst regarding his actual model of slope development as one of dubious utility.

THE DAVISIAN SYSTEM

The essence of the Davisian system, built on Powell's (1875) concept of base level, is very simple and may be summarized as follows. The emergence of a landmass above sea level, due to uplift, takes place rapidly on a geologic time scale. Uplift occurs so rapidly, relative to denudation processes which begin to act on the landmass, that very little modification of the landscape occurs during uplift. It is not until uplift suddenly comes to a halt, in fact, that sub-aerial denudation becomes important.

Initially the major process of denudation is stream incision into the landmass and concomitant development of narrow valleys. Stream downcutting cannot continue indefinitely, however, since streams can-

11

not cut down below the level of the sea, which is now assumed to be constant relative to the landmass. Actually the limit of downcutting depends upon distance from the sea; although it corresponds to sea level at the mouths of streams, there must exist some slope to allow water to flow to stream mouths. The locus of points which represent the limit of active downcutting is termed the graded stream profile. Streams are assumed to attain the lower limit of active downcutting first of all at the mouth, and then progressively upstream. Once the stream is graded, undercutting of side slopes ceases, except for occasional lateral undercutting, and slopes are moulded only by processes acting over the whole length, and develop independently of the stream.

There are two key assumptions here. One is that uplift and denudation cannot occur concurrently, that is, denudation can only assume any importance once the landmass is tectonically stable. The second is the assumption that streams undergo two phases of activity: rapid incision initially and then virtual dormancy once grade is attained. Criticism of the Davisian system has been levelled at both of these assumptions.

The concept of grade, as outlined by Davis based on Gilbert's (1877, 1914) ideas, has recently been challenged by engineering geologists in the United States, e.g. Leopold and Maddock (1953), who claim that most streams attain grade relatively early in the process of downcutting. In other words, the attainment of grade in no way relates to the change in stream activity from rapid downcutting to floodplain development as Davis suggested. These new ideas seem as plausible, at least, as those of Davis, but they probably do not invalidate the cycle of erosion. Streams with floodplains clearly cannot lower their valley floors at the same rate as streams which need only to excavate material from their channel beds. Whatever the worth of Davis' ideas on grade, it still seems probable, assuming that the landmass is tectonically stable, that subsequent to a phase of rapid stream downcutting, a period follows when floodplain development dominates, and after this time slopes develop relatively independently of streams.

The other assumption in the Davisian scheme, that uplift and denudation act successively on a landmass, is the one which attracted the earliest criticism particularly from Walther Penck (1923, 1924, 1925). In his *Die Morphologische Analyse* he admits that Davis 'kept well in mind the importance of concurrent uplift and denudation' (p. 12), although 'this was a notion which he scarcely ever used, and his followers never'. Penck's ideas on landscape development and tectonic history are summarized in fig. 2.1. The Davisian sequence of uplift succeeded by denudation is clearly one extreme case; the other extreme is uplift and denudation proceeding concurrently at constant rates. Between these two extremes exist an infinite number of possible courses in which uplift is initially most important and then becomes subsidiary to denudation. In all these intermediate cases, however, neither uplift nor denudation is completely insignificant, although the relative importance is continually changing. Now, although Penck believed that large parts of the world are tectonically stable, he claimed that many areas have undergone continual diastrophism for very long periods of time, and that the landforms of these areas could not, therefore, be understood in terms of a cycle of erosion based on rapid uplift followed by still-stand.

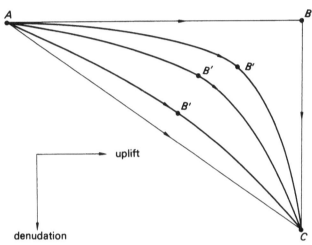

Fig. 2.1. Possible systems of upheaval and denudation according to Walther Penck (based on Penck, 1953).
ABC Davisian system of successive uplift and denudation.
AB'C Systems of concurrent uplift and denudation with eventual dominance of denudation.
AC Extreme case of concurrent uplift and denudation at constant rates.

THE IDEAS OF WALTHER PENCK

In order to fully appreciate Penck's beliefs, it is necessary to separate two distinct, although related, components of his thesis. We shall, therefore, first of all leave on one side the issue of crustal instability versus crustal stability, and focus our attention on the relationship between slope development and the intensity of stream erosion. We shall *then* note how Penck believed that stream erosion, under certain circumstances, might depend upon the rate of tectonic uplift, so that slope profiles might also reflect the diastrophic history of an area.

Penck's ideas on the relationship between the intensity of stream erosion and the development of side slopes are summarized in pp. 177–80 of *Morphological Analysis of Landforms*. Envisage a stream beginning to incise itself into a landmass, and think of this downcutting in terms of

13

infinitesimally small periods of geologic time. In each period of time, a slope unit with a specific gradient is produced. The gradient depends on the amount of incision and the amount of denudation on the developing slope. On a slope cut into homogeneous material, the amount of weathering and the amount of denudation should be constant in each small period of time. As a result, slope units produced by rapid incision will be steeper than slope units resulting from slow downcutting. Once a slope unit has been formed, subsequent denudation is assumed to produce a retreat of this unit without changing its angle of slope. As an illustration, a profile of the side slopes developed under conditions of increasing erosional intensity of the stream is given in fig. 2.2; the increasing rate of

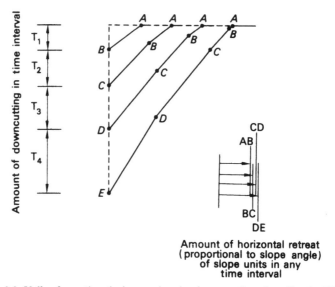

Fig. 2.2. Valley formation during waxing development (based on Penck, 1953).

stream downcutting is reflected in a convex slope profile. It can be seen from the diagram that, not only do slope units retreat over time, but they also move upslope and eventually disappear completely. A period of increasing erosional intensity, as in fig. 2.2, was termed *aufsteigende Entwicklung*, that is, waxing development; a period of decreasing erosional intensity was labelled *absteigende Entwicklung*, translated as waning development. Penck argues that 'this succession, one above the other, of slope units with different gradients provides a sensitive means of following up the erosional intensity (of the stream) at a definite place: a convex form is proof of an increase in erosional intensity, a concave form is proof of decrease' (Penck, 1953, p. 179). By the same reasoning, a straight slope is indicative of a phase of downcutting with constant erosional intensity and termed *gleichformige Entwicklung* or uniform development.

Penck did recognize that there were limits to both waxing and waning development, particularly the first of these. He emphasized that there exists a maximum angle of stability for any rock mass, and, once stream erosion had steepened side slopes to this critical angle, there would be no further steepening of the slope. A straight slope at the critical angle of stability would then develop irrespective of any subsequent increase in erosional intensity. In some ways this admission, usually ignored by Penck's critics, points to the futility of the whole concept, since it is probable that most streams cut down sufficiently rapidly that they attain this critical angle of slope at a fairly early stage in the dissection of a landmass. Some evidence for this is provided in chapters 6 and 7.

With this relationship in mind, we can now briefly examine Penck's ideas on the relationship between river erosion and crustal movements and, in turn, between crustal movements and the development of slope profiles. The case of uniform rate of uplift (Penck, 1925, p. 89) affords the simplest illustration. Assume that the mouths of rivers are fixed in location during uplift, and that streams are cutting down into a fairly homogeneous mass. It can be shown, as Penck claimed, that the gradient of a stream at each point becomes so adjusted that the rate of downcutting is everywhere the same and also equal to the rate of upheaval. Subdivide a river into numerous very short reaches of constant length, and denote the lowest reach by s_1, the reach next upstream from it by s_2, and so on up the course of the river. Compare the rate of downcutting in reach s_1 with the rate of upheaval of the landmass. In the case where downcutting is more rapid than uplift, the reach will lower its position relative to base level. Since the position of the river mouth is assumed to be fixed in a horizontal sense, the net result must be a lowering of the gradient in the s_1 reach. This decrease of gradient must produce a decrease in the rate of downcutting, and eventually this will be reduced until it equals the rate of uplift. In the case in which downcutting is initially slower than the upheaval rate, the reach s_1 must rise to a higher position relative to base level, increasing the stream gradient of the reach, and increasing the rate of downcutting there also. Eventually the gradient will become sufficiently steep that, again, uplift and downcutting balance each other. Now, if uplift and downcutting are balanced on reach s_1, the upper end of this reach is fixed in altitude. The upstream end of this stretch of the river is, however, also the downstream end of reach s_2 and acts as base level for it in just the same way that the sea did for reach s_1. The upstream end of s_2 is thus uplifted relative to the down-

15

stream end. Again, whether the upheaval rate is initially in excess or smaller than the downcutting rate, eventually the two will come into equilibrium for reasons outlined in the case of reach s_1. Successively then each stretch of the river must come into equilibrium and eventually the whole river will cut down at a rate equal to the upheaval rate. On the basis of this reasoning, a uniform rate of uplift, provided that it lasts long enough, will produce a pattern of uniform development in terms of downcutting, and, as a result, straight valley-side slopes.

It is debatable whether Penck intended his ideas to be of any more than theoretical importance, and whether he really believed that waxing, waning and uniform stream development should always be attributed to particular patterns of crustal instability. Admittedly the impression exists that Penck viewed slope profiles as a direct reflection of upheaval patterns, but this is due more to the writings of Davis than Penck himself. As pointed out by Simons (1962), in attempting to define Penck's terms waxing and waning development, Davis unfortunately linked them to *upheaval patterns* rather than modes of *stream behaviour* as Penck intended. The term waxing development, *as translated by Davis*, applies to 'forms developed during upheaval of increasing rate and hence characterized by convex valley walls'; the term waning development 'during later upheaval of decreasing rate and hence characterized by valley-side slopes which are, near their base at least, concave' (Davis, 1932, p. 427). Subsequently, one of the most noted of Davis' disciples, Douglas Johnson, repeated the same mistranslation describing Penck's (supposed) views as 'one of the most fantastic errors ever introduced into geomorphology' (Johnson, 1940, p. 23). The error was, in fact, Johnson's, following his teacher, and not Walther Penck's.

Nevertheless Penck did apparently believe that, in certain cases, non-uniform upheaval of landmasses did occur, contrary to the Davisian thesis, and that under these circumstances landforms would inevitably reflect the pattern of diastrophy. Some of his evidence used to support this view was quite impressive, and his argument that the upheaval pattern of a continental landmass should be reflected in the pattern of sediments surrounding it, could prove to be a very valuable approach. In addition, his attempts to relate uniform *development* to uniform *upheaval* of the landmass, anticipated by over three decades similar ideas put forward by Hack (1960) in a more general attack on the closed system framework of the Davisian system.

It is easy, however, to overstress Penck's ideas on crustal instability. Notwithstanding the theoretical importance of the ideas summarized above, he did recognize that many large parts of the earth's crust are stable today, and admitted that, in these areas, the landscape would evolve systematically through time, although in a manner rather different to that envisaged by Davis. His ideas on the cycle of erosion in tectonically stable areas are discussed in chapter 15.

THE ROLE OF ISOSTATIC READJUSTMENT

The concept of rapid intermittent periods of upheaval of a landmass separated by long periods of stability in which denudation dominates, has recently been supported by Lester King (1955) and J. C. Pugh (1955) working in Africa. It is ironic, however, that only through rejecting the particular Davisian *model* of slope development, could they support the Davisian *system* of successive upheaval and denudation. Since these ideas are not widely cited in the literature of geomorphology they are briefly summarized below.

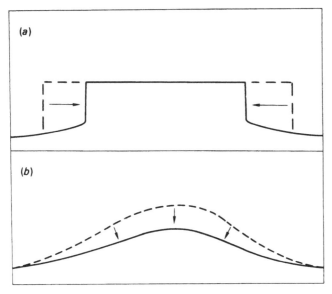

Fig. 2.3. Idealized contrast between down-wearing and slope retreat.

The Pugh–King model assumes that initially there is a stable upraised continental block, bounded by steep sides along the coast and incised by stream canyons. Denudation of the landmass takes place through parallel retreat of the canyon walls and the coastal scarps, rather than through flattening as in the Davis cycle. The retreat of these scarps results in the emergence of basal pediments eventually coalescing to produce wide, sweeping pediplains.

These authors argue that removal of this surface material through

17

scarp recession must result in some form of isostatic uplift as compensation. Admittedly, if this is true for *pediplanation* it should also be true for *peneplanation*, but the difference between the two erosional processes would result in different modes of isostatic readjustment. In the cycle of erosion put forward by Davis, denudation proceeds through a vertical lowering of the land surface throughout an extensive area; in the pediplanation model, material is moved laterally from the edges of an upland block. This difference (fig. 2.3) has important results. According to Pugh (1955, p. 373–4), whereas there is an almost immediate isostatic response to *vertical* lowering of a landscape through erosion, there is a distinct threshold of *lateral* denudation before isostatic readjustment occurs. In the Davisian cycle, then, denudation and isostatic response should be virtually synchronous (this point is challenged by Schumm, 1963) and, as a result, isostatic compensation must considerably delay the full development of a peneplain. Assuming pediplanation, in contrast, isostatic compensation occurs only when a threshold amount of denudation has taken place, and is thus an intermittent event. Once isostatic readjustment has occurred, a new coastal scarp and cycle of canyons is created, and in the following period of crustal stability, this in turn will be subject to further retreat and development of a new pediplain surface.

Gunn (1949) calculated that this lateral threshold distance was about 480 km. Once this has been exceeded and uplift takes place, the actual amount of uplift is given by the equation:

$$h = \frac{B}{A}.r \tag{2.1}$$

where h is the height of isostatic compensation, B is the specific gravity of the surface rocks removed, A is the specific gravity of the deep-seated (sima) inflow material, and r is the thickness of the surface layer removed. Holmes (1944) suggested values for A and B of 3·4 and 2·6 respectively, whilst du Toit (1937) preferred slightly higher values of 3·5 and 2·65. In both cases the ratio B/A approximates to 0·76.

These ideas were applied to different parts of the African continent separately by the two workers. Although there were slight differences in detail between the two approaches, a brief summary of Pugh's work in west Africa will be sufficient to illustrate the major points.

The dominant feature of the landscape of west Africa, at least at the regional scale, is the succession of high-level erosion surfaces separated from each other by steep scarps. The two highest, at 1,080–1,350 m and at 600–750 m, are particularly well developed. The highest is termed the Gondwana surface (Pugh and King, 1952) and the one immediately beneath it, the post-Gondwana surface. An impressive step of about

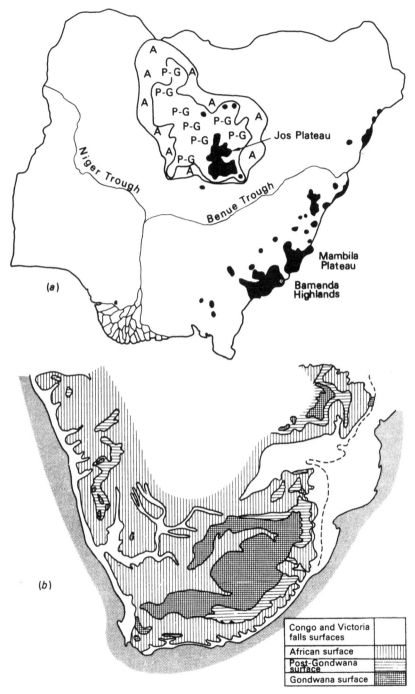

Fig. 2.4. Surfaces of pediplanation in (*a*) northern Nigeria (from Pugh, 1955), and (*b*) southern Africa (from King, 1951).

19

120 m also separates the post-Gondwana surface in northern Nigeria from another below it, the African surface. The Gondwana surface derives its name from the former super-continent, Gondwanaland, which according to King (1950, 1951) broke up in the Cretaceous and, through continental drift, separated into the component southern hemisphere continents of today. Prior to the tectonic instability that resulted in the break-up of the old continent, a very extensive erosional surface at about 600 m had been developed. The Gondwana surface of west Africa is believed to be a remnant of this surface. The break-up of Gondwanaland produced steep coastal scarps around the edges of the newly formed landmass, and pronounced rejuvenation of the existing drainage net. The retreat of these scarps and canyon walls produced another pediplain surface, the post-Gondwana; the high-level surface in west Africa just below the Gondwana surface is believed to have originated in this way. Towards the end of the Cretaceous, retreat of the scarps that had cut into the Gondwana surface was disturbed by upheaval of the landmass, and a further cycle, the African, was initiated. According to Pugh and King, this upheaval stems from isostatic readjustment to the pediplanation process. In a similar way the African cycle continued through the Tertiary until the Oligocene/Miocene when further uplift, again attributed to isostatic compensation, occurred. Similar cycles initiated by this uplift and comparable to the Victoria Falls and Congo cycles of southern Africa (fig. 2.4), may have been produced by uplift following the African cycle, but there is no clear evidence of this in west Africa to date.

Now, if we assume that the two isostatic uplifts between the Gondwana, post-Gondwana and African cycles represent the total upheaval of west Africa since the Cretaceous, successive substitution of appropriate values in equation 2.1 should enable prediction of the present-day levels of the upper surfaces. The first upheaval should be about 456 m (0.76×600), and the second upheaval about 345 m (0.76×456). This would give elevations for the present-day Gondwana, post-Gondwana and African surfaces of 1,401 m ($600 + 456 + 345$), 801 m ($456 + 345$) and 345 m respectively. The agreement of the values for the two highest surfaces with the actual upper limits of these surfaces in west Africa is good, although the correlation is by no means perfect and many points remain unexplained. These authors admit, however, that this simple pattern of isostatic compensation is bound to be complicated by local uplifts, warping and tilting of the surfaces, and other non-isostatic tectonic forces. Whatever the merits of the *detailed* hypotheses of these workers, their investigations certainly appear to support the idea that rapid intermittent upheaval of landmasses is a fairly common phenomenon.

EMPIRICAL EVIDENCE

The controversy concerning whether upheaval and denudation can occur together without one dwarfing the other in magnitude can be approached in many ways. One of the most attractive, assuming that the principle of uniformitarianism is applicable to *rates* of processes as well as to *types* of processes, is through the comparison of present-day rates of orogeny with modern denudation rates. An interesting contribution towards this has been made by Schumm (1963) in an article summarizing data from numerous different sources. In view of the potential importance of this material, it is discussed briefly below.

Early attempts to calculate rates of denudation for the major river systems of the United States, by Dole and Stabler (1909), indicated average values ranging from 0·027–0·057 m per thousand years. These values are probably low for a number of reasons. One is that they were based on only suspended and dissolved stream loads and neglected bed load; the other is that they refer to very large drainage basins. Sediment removal from large basins takes much longer than in small basins because much of it is redeposited and retransported several times before reaching the basin mouth. Since our purpose is to compare denudation rates with orogenic rates, we should use only denudation rates relevant to the size of basin found in typical mountain ranges; Schumm suggests a figure of 3,900 km² (1,500 mile²). Measurement of suspended and dissolved loads at gauging stations in the United States with catchment areas of about this size (Langbein and Schumm, 1958) suggest slightly higher rates (0·03–0·1 m per thousand years) than Dole and Stabler's data. Attempts to include bed-load, by using measurements of debris trapped behind dams, produce still higher values (0·057–0·22 m per thousand years), although the drainage basins examined are much smaller in this case. The highest sediment-yield rate recorded by the Federal Inter-Agency River Basin Committee (1953) is for a very small catchment in the Loess Hills of Iowa, and amounts to a rate of 12·6 m per thousand years. Using the relationship between sediment yield and catchment size determined by Brune (1948), the Iowa rate would correspond to a hypothetical figure of about 3 m per thousand years, assuming a basin of 3,900 km². This is perhaps the upper limit of present day denudation rates in medium-size catchments. Actual measurements in the large mountain areas are scarce, but rates of 0·6 m per thousand years and 0·96 m per thousand years have been reported by Wegman (1957) in the northern Alps and by Khosla (1953) in parts of the Himalayas respectively. Similar values have been reported by Corbel (1959). These values, intermediate between those for very small basins in areas of weak rock and those in very large river basins, are probably typical of high-mountain areas.

21

TABLE 2.1. *Estimates of rates of denudation for drainage basins of different size* (after Schumm, 1963)

Drainage basin area in 1,000 km²	Denudation rate in metres per 1,000 years	Source	Denudation rate in feet per 1,000 years	Drainage basin area in 1,000 mile²
37–3,280	0·03–0·06	Dole and Stabler, 1909	0·09–0·19	14·3–1,265
3·9	0·03–0·10	Langbein and Schumm, 1958	0·10–0·34	1·5
0·08	0·06–0·22	Langbein and Schumm, 1958	0·19–0·65	0·03
0·003	2·55	Flaxman and High, 1955	8·5	0·001
0·0003	12·6	Federal Inter-Agency River Basin Commission, 1953	42	0·0001

One of the most interesting features of sediment-yield data, often concealed among the apparently random variation among different basins, is the relationship between sediment yield and the relief of the catchment. This was examined by Schumm and Hadley (1961) in basins of about 2·6 km² in areas underlain by sandstone and shale in semi-arid regions of the western United States. They obtained the following relation:

$$\log S = 27{\cdot}35\,R - 1{\cdot}187 \qquad (2.2)$$

where S is sediment yield in acre-ft/mile² and R is the relief–length ratio. The latter parameter is the average vertical relief of a basin divided by the drainage basin length. A plot of this curve, converted to denudation rates for catchments of 3,900 km² is shown in fig. 2.5.

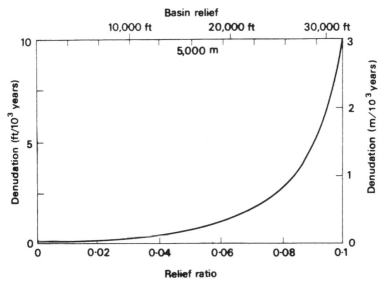

Fig. 2.5. The decrease in the rate of denudation with decreasing relief and relief ratio (after Schumm, 1963).

Assuming that these data are valid for a single catchment of fixed area and length, during the progressive change in relief with denudation, it can also be viewed as a pattern of the change in denudation rate during the actual cycle of erosion. The curve again appears to support the previous data emphasizing the relatively low rates, and indicating that only with vertical relief of about 7·5 km would denudation rates get as high as 0·6–1 m per thousand years.

Modern rates of orogeny in tectonically active areas, based on repeated precise levelling, appear to be much higher than erosion rates.

Some of these data are shown in table 2.2. On average it appears that modern rates of orogeny approximate about 7·5 m per thousand years, a figure comparable to measurements of present-day rates of upheaval associated with isostatic adjustment in areas depressed by Pleistocene ice sheets. These figures are markedly higher than the data on epeirogenic uplift along sea coasts presented by Cailleux (1952), which average about 0·9 m per thousand years.

These data clearly indicate that, in tectonically active areas today, modern rates of orogenic upheaval are significantly greater than denudation rates. On the basis of these data, it seems very unlikely that

TABLE 2.2. *Estimates of upheaval rates under (a) orogenic, (b) isostatic and (c) epeirogenic conditions* (based on data from Schumm, 1963)

	Location	Uplift per 1,000 years m	ft	Source
(a)	California	4·8–12·6	16–42	Gilluly, 1949
	Southern California	3·9–6·0	13–20	Stone, 1961
	Japan	0·8–75	3–250	Tsuboi, 1933
	Persian Gulf	3·0–9·9	10–33	Lees, 1955
(b)	Fennoscandia	10·8	36	Gutenberg, 1941
	Southern Ontario	4·8	16	Gutenberg, 1941
(c)		0·1–3·6	0·3–12	Cailleux, 1952

time-independent landforms as envisaged by Penck (1924) and Hack (1960) could be produced. Indeed this disparity between modern rates of orogeny and denudation strongly supports the Davisian system of rapid upheaval of mountain chains with little modification by erosion until orogeny has ceased.

The ultimate significance of Schumm's evidence hinges on the relevance of modern rates of orogeny and denudation to rates of these processes during the geologic past. On this point, it is very difficult to comment. It might be expected that modern rates of erosion have been increased a great deal by human activity, but Gilluly's (1949) estimate of 1·5 m per thousand years for the denudation of the Rocky Mountains during the Cretaceous is comparable to present-day rates in mountain areas. In addition, there is no obvious reason why orogenic rates should have differed significantly in the geologic past.

CONCLUSIONS

Notwithstanding the ideas of Penck, it seems from evidence so far available, that the much simpler uplift–denudation system of W. M.

24

Davis is probably more appropriate in the majority of cases. Whether actual denudation proceeds via the peneplanation model itself is a much more dubious issue and this will be examined in chapter 15. Even those who accept the peneplanation concept must face the problem of whether periods of tectonic stability have lasted long enough for the development of a peneplain. Schumm (1963), on the basis of modern erosion rates, and assuming no isostatic adjustment, argued that denudation of about 1·5 km of material would probably take between 3 and 110 million years. This is surprisingly close to Davis' (1925) own estimate of 20–200 million years for the planation of fault-block mountains in Utah. In view of these estimates, it is interesting that de Sitter (1956, p. 471) has suggested that periods, of about 200 million years, of relative quiescence do occur between shorter periods of diastrophic activity. Together, this evidence led Schumm to conclude that 'ample time exists between orogenic periods for the development of peneplains' (Schumm, 1963, Abstract). Much of this is clearly speculative, and not all geomorphologists (e.g. Thornbury, 1966, pp. 184–5) share the view that periods of still-stand of landmasses are long enough for the development of a peneplain. The peneplanation model may thus be challenged on a number of grounds, but, despite this, it seems that the Davisian *system* of successive upheaval and denudation of the landmass is probably the most appropriate framework for the study of slope forms. This is the system of reference used in this book.

PART ONE: FORCE AND RESISTANCE

CHAPTER 3

FORCE: SOURCES OF ENERGY FOR DEBRIS TRANSPORT

INTRODUCTION

A landslide or rockfall is a dramatic single expression of force over-coming resistance, but all hillslopes are examples of systems in which force and resistance are continually opposed. For any system in equilibrium, the forces tending to promote movement are exactly balanced by the resistances opposing it; and such an equilibrium system may be either in a state of rest or undergoing movement at a uniform rate. Acceleration of the system, or a part of it, is the inevitable result of the forces becoming greater than the resistances. To put it into a slope context: equilibrium usually means a state of rest so that any acceleration of the system results in the beginning of debris movement. When the forces within a moving debris mass become less than the resistances to movement of the mass, then the material will slow down and eventually stop. In order to understand hillslope systems, the forces promoting movement of debris will be examined in this chapter, and the next chapter (4) will study the nature of resistance on slopes.

Force requires energy, and all energy in geomorphic systems is ultimately derived from either gravity or climate. The force provided by gravity is simply that of the weight of each debris particle, which will move it downwards if it is not resisted. Climate, through its control on temperature and available water, provides energy for the most important forces on hillslopes, those of static and moving water. Climate also directly controls the forces of water freezing and thermal expansion of rock minerals, and indirectly affects biological and chemical forces such as those produced by plant growth and evaporite formation.

These forces vary not only in the gross amount of work done (magnitude) but also in the net result of downslope movement of material that they achieve (efficiency). The orders of magnitude and efficiency of some of the main forces acting are listed in table 3.1. It shows that although very large amounts of thermal and biochemical energy are applied to the landscape, only minute amounts of these forces are actually effective in transporting debris. It is mechanical and hydraulic forces that move most hillslope debris because these operate at relatively high efficiencies of 0·01–100 per cent. Of these forces, water flow is by far the most important both in total energy expended and in total debris transport achieved. Water flow itself has two components; hillside flow and stream flow. Hillside flow accounts for the greater part of the total work done by flowing water but is so inefficient compared with streamflow that it actually moves only a fraction of the debris that streams can transport.

This chapter introduces the most important forces on hillslopes; those of gravity, water tension and pressure, water flow, rainfall impact, expansion forces due to changes in moisture, freezing and temperature changes. Lastly, small random forces are considered together as diffusion forces.

TABLE 3.1. *Orders of magnitude of geomorphic forces and their efficiency in transporting debris*

Force	Total work done by force (J/m² year)	Total work done in downward debris transport (J/m² year)	Efficiency (%)
Gravity:			
downward transport by slides, falls, etc.	1–100	1–100	100
Water flow:			
dissolved and suspended load, mainly in rivers	10^5–10^6	10–100	0·01
	2,000	0·02 (grass)	0·002
Rainfall impact	500	0·5 (bare)	0·1
Heaves:			
(a) Moisture	2,000	0·4	0·02
(b) Frost	5×10^7	0·1	$2·5 \times 10^{-7}$
(c) Temperature	2×10^8	0·02	$1·5 \times 10^{-9}$

GRAVITY FORCES

The force of gravity on any object is expressed in terms of the weight of that object, and this force acts vertically downwards. On a slope, the weight of an object can be resolved into two components; the first is a downslope force which tends to move the object downhill parallel to the surface; and the second is a force perpendicular to the surface which acts to hold the material on to the slope. Because slope movements are generally along or parallel to the surface, the downslope component of weight is the more important in promoting debris transport. However, some slope movements are attributable to the perpendicular force, for example movements such as contraction after frost heave and soil consolidation. But the extent to which the perpendicular weight component actually promotes debris transport is minimal compared to its role as a direct resistance to hydraulic lift forces and in determining frictional resistances (chapter 4).

The gravity force depends ultimately upon differences in elevation, and it is the downslope component of weight which transmits this force. This component is equal to the weight force multiplied by the sine of the slope gradient angle, so that it is the sequence of slope gradients, or in other words the slope profile which communicates the information about elevation differences between points in the landscape. The slope or river longitudinal profile thus serves a function as a telephone line which is passing the message about conditions at the foot of the slope or river up towards the divides. As well as acting as a communications link the gradient may also be thought of as the means whereby the total available energy inherent in the relief is distributed over the landscape, so that the gravity forces are most effective on the steepest slopes, because it is there that the downslope weight component is the greatest. For example river incision or lateral migration is able to redistribute the effective gravity forces considerably in a landscape, by forming steep slopes (and thus concentrating gravity forces) near the rivers or by moving the position of the steepest slopes from one hillside to the opposite one, and can achieve this redistribution without altering the total available relief.

Because the force of gravity depends on differences in elevation, the principal source of energy for gravity is tectonic uplift of the landscape. The lowering of base level by eustatic changes in sea level produces a similar effect but does so because below base level the hydraulic forces dramatically change their character (although they do not stop acting because the force of gravity remains unchanged). On a very much smaller scale the gravity force is constantly being renewed by climate-derived forces which are lifting material; for example, in frost and moisture heaving, in the upward translocation of salts in the soil, in plant growth, and in the formation of wormcasts, anthills, etc.

The gravity forces, though similar, are slightly different in the cases of (1) a particle resting on a surface of inclination β (fig. 3.1a), and (2) a possible slip plane within the soil but parallel to the surface (fig. 3.1b). In the first case, the force on the isolated particle is simply its own weight acting through its centre of gravity. If it has mass m, then the vertical weight is equal to $m.g$, where g is the acceleration due to gravity. The downslope component, which is tending to promote movement along the surface equals $m.g.\sin \beta$; and the perpendicular component, which is tending to promote consolidation into the surface equals $m.g.\cos \beta$.

In the second case, of a possible slip plane parallel to the surface (fig. 3.1b) in a uniform medium, the relevant gravity force acting on the slip plane is the pressure exerted by the weight of the material resting on that plane. This overburden pressure also acts vertically downwards, and

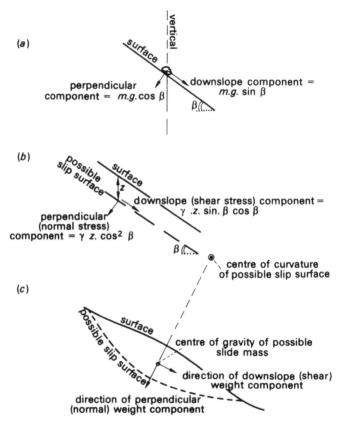

Fig. 3.1. Gravity forces (a) acting on a particle of mass m resting on a surface of slope β, (b) acting at a possible slip surface at depth z below the surface of a medium having unit weight γ and (c) acting on a soil mass above an irregular slip surface.

equals γz, where γ is the unit weight of the soil and z is the vertical distance between the surface and the slip plane. As in the first case, the components in the two directions are:

downslope component = shear stress = $\gamma.z.\sin \beta.\cos \beta$:
perpendicular component = normal stress = $\gamma.z.\cos^2 \beta$.

In more complex situations, where the possible slip surface is not a plane (fig. 3.1c), then the gravity force is the weight of the corresponding slide mass acting through the centre of gravity of the mass; and that part of the force which is promoting movement is the component acting in the direction in which the centre of gravity of the mass will begin to move if a slide occurs. The tendency for the mass to rotate can also be calculated as the moment of the gravity forces about the centre of curvature of the slip surface.

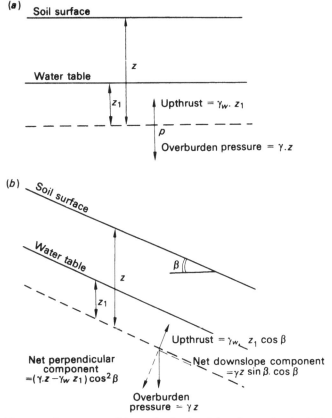

Fig. 3.2. (*a*) Pressure forces on a soil beneath a horizontal surface and water table. (*b*) Pressure forces beneath an inclined surface and a water table parallel to it.

WATER PRESSURE AND TENSION FORCES

Below the surface of a static body of water, gravity produces a pressure which is equal to the overlying weight of water per unit area, and this pressure increases with depth below the surface. Any object or material immersed in the water is subjected to an upthrust or relief of weight, which is equal to the weight of water which its bulk displaces. These

31

two statements are simply alternative expressions of the same properties, for the upthrust on a submerged body is obtained because of the difference in pressure between its top and bottom.

Below the water table, all open pores are saturated (except perhaps in clay soils), and the same physical laws apply. That is to say that at a point below a horizontal water table, there is a hydraulic upthrust or relief of weight equal to the water pressure at that point (Terzaghi, 1936). That is; at point P in fig. 3.2a, there is (1) a water pressure of $\gamma_w.z_1$, where γ_w is the unit weight of water, and z_1 is the depth below the

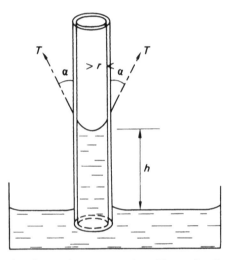

Fig. 3.3. The capillary rise of water in a narrow tube, of internal radius r. (Diameter of tube is exaggerated in the diagram.)

water table; (2) a gravity overburden pressure of $\gamma.z$, and (3) a hydraulic upthrust or relief of weight, which is equal to the pressure difference between the top (at atmospheric pressure) and bottom ($\gamma_w.z_1$ in excess of atmospheric pressure), and therefore to the total hydraulic pressure of $\gamma_w.z_1$.

Where the water table is sloping, then isolines of pressure are parallel to the water table, and hence the upthrust experienced is perpendicular to the water table. In the rather simple case where the water table slopes, and is parallel to the land surface (fig. 3.2b), then the upthrust is also perpendicular to the land surface, and has no component in the downslope direction. The hydraulic forces therefore tend not to reduce the downslope gravity forces which encourage downslope movement, while they are effective in relieving the perpendicular gravity forces which help to hold the material down on the slope.

Above the water table, there may still be some moisture in the soil, but in an unsaturated state. If an open-ended narrow tube is placed with its bottom end below the water surface, then water will rise up the tube (fig. 3.3). The force which lifts the water up the tube is a surface tension or capillary force which acts around the circumference of the tube, and pulls upwards at a slight angle to the glass, called the angle of contact (α in fig. 3.3). The lifting force is therefore equal to the component of the surface tension in a vertical direction multiplied by the circumference of the tube: that is

$$\text{lifting force} = T.\cos \alpha.2\pi r,$$

where T is the surface tension and r is the radius of the tube. The height of rise is determined by the weight of water which this force can support. If the capillary rise is h, then the weight of the lifted column of water is $\gamma_w.h.\pi r^2$. Equating the lifting force to the weight lifted, the height of capillary rise is,

$$h = \frac{2T.\cos \alpha}{\gamma_w.r}. \tag{3.1}$$

In this formula, it is clear that the height of rise is inversely proportional to the radius of the tube, and a tube of a given diameter can be said to exert a suction or tension on the main body of water, equal to h.

The pore spaces in a soil mass can be considered as an interconnecting network of tubes of differing diameters, and if a block of soil is lowered into water, there will be a corresponding capillary rise in the soil, but the rise will be least in the largest pores and greatest in the smallest pores. Above the water table there is therefore a zone of diminishing water content in which only progressively smaller and smaller pores contain water as one proceeds upwards, and this zone is called the capillary fringe (Baver, 1956, chapter 6). Water is also present in soil considerably above the capillary fringe and is derived mainly from rainwater which is percolating downwards. At any point in the unsaturated zone the water exerts a suction. With a simple tube the suction is always constant, but in the soil the suction is variable and depends on the largest pores which contain capillary water. Wetter and wetter soil has water in larger and larger pores and can exert progressively less moisture tension. Experimental curves relating moisture content to tension (fig. 3.4) show a hysteresis, with different relationships during wetting and drying of the soil. For shrinking soils the wetting–drying loops are non-repeatable, and depend on the previous history of the soil. Marshall (1959) contains a fuller account of these soil moisture relationships.

Below the water table there is a positive upthrust relieving the overburden weight of the soil, and in a similar way at levels above the water table there is a hydraulic force which increases the effective weight of the soil by an amount equal to the tension force acting on it and this hydraulic force is called the capillary cohesion. Because moisture tends to move downwards from the surface, isolines of equal tension tend to be roughly parallel to the surface, so that the capillary cohesive force tends to be entirely perpendicular to the surface, and to have no component parallel to the surface modifying the effective downslope gravity

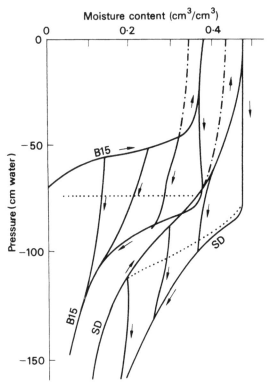

Fig. 3.4. The relationship between soil moisture and capillary tension for two experimental soils (from Youngs, 1957).

force. Although the water table and isolines of equal tension are usually roughly parallel to the slope surface, they are not always so and when they are not, hydraulic tension or pressure forces contribute directly to promote or discourage downslope movement. Increased capillary cohesion during drying of the soil also provides the forces which produce shrinkage in soils with high clay contents, and the shrinkage takes place both perpendicular to and parallel to the surface; the latter causing mud-cracks to form.

34

WATER FLOW FORCES

Flows over the surface

Surface water flow is of vital importance in the understanding of stream-flow processes, but it is also relevant to the movement of material on hillslopes. Water may flow in a uniform layer over the surface, but more usually breaks up into threads of greater and lesser flow depth or even into distinct small channels or rills. In these circumstances sediment transport obeys very similar laws to those operating in rivers, but there are a number of differences. First of all, the flow depths are usually small and the elements which contribute roughness to the flow are

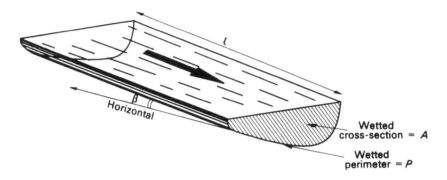

Fig. 3.5. Uniform flow in an open channel.

relatively large and may even protrude above the surface. Secondly the flows are usually present only during rainstorms, during which rain-splash impact both impedes the flow and also assists in throwing material into suspension. Lastly the flows are usually very ephemeral, so that specialized channel features such as ripples or dunes, meanders or indeed well defined channels are seldom formed.

The rate of flow in open channels can be calculated by balancing the down-channel component of the weight of water in the channel against the frictional forces acting on its bed and banks (Chow, 1959). Fig. 3.5 represents a section of channel of cross-section A, wetted perimeter P, and of length l, the whole unit being inclined at slope angle β. A and P are measured in a *vertical* cross-section. The weight of this section is $\rho_w.g.A.l \cos \beta$, where ρ_w is the density of water. The downslope component of this weight is $\rho_w.g.A.l.\sin \beta.\cos \beta$. This force must be balanced against the frictional forces, which are acting over a total area of $P.l$ (assuming a relatively wide, shallow channel). The frictional force per unit area is empirically related to the velocity, v, by a power law, with an exponent which increases as the flow becomes more turbulent. For turbulent tranquil (or streaming) flow, the most normal state, the frictional force is proportional to the square of the flow velocity, and is usually expressed in the form: $\frac{1}{2}\rho_w.f.v^2$, where f is the dimensionless friction factor. The total friction force is therefore $\frac{1}{2}\rho_w.f.v^2.P.l$. Equating this to the weight component downstream for a steady uniform flow, and expressing velocity in terms of the remaining quantities:

$$v = \left(\frac{2g}{f}\cdot\frac{A}{P}\cdot\sin\beta.\cos\beta\right)^{\frac{1}{2}} \tag{3.2}$$

In this equation the centre term, A/P is called the hydraulic radius, r, and is a mean depth. For a uniform thin flow the hydraulic radius is equal to the depth. The friction factor, f, is related to the size of the grains forming the roughness by a relation of the form:

$$\left(\frac{1}{f}\right)^{\frac{1}{2}} = 4\cdot07\log_{10}\left(\frac{r}{d_{84}}\right)+\text{constant}, \tag{3.3}$$

where the constant has values of 2·0 (Wolman, 1955) to 4·24 (Nikuradse, 1932) and where d_{84} is the diameter which is equal to or larger than

TABLE 3.2. *Values of exponents m, n in the flow relationship* $v \propto r^m.(\sin\beta)^n$ *as the type of flow varies* (after Horton, 1945)

Flow type	m	n
Laminar	2·00	1·00
Turbulent	0·5–0·67	0·50
Subdivided	$\geqslant 0\cdot0$	$\leqslant 0\cdot5$

84 per cent of the bed particles. The friction force therefore depends strongly on the size of the largest particles on the bed of the flow. As the type of flow varies, the flow velocity can be expressed in the more general form:

$$v \propto r^m.(\sin\beta)^n; \tag{3.4}$$

(Horton, 1945) where the exponents m, n have the values in table 3.2.

The type of flow which occurs depends on the relative magnitudes of viscous, inertial and gravity forces in the flow. Reynolds' number expresses this in the dimensionless ratio:

$$\frac{\text{Inertia force}}{\text{Viscous force}} = \frac{\bar{v}.r.\rho_w}{\mu}, \tag{3.5}$$

where $\bar{v}$ = mean flow velocity and μ = dynamic molecular viscosity.

When viscous forces predominate, the flow is *laminar* and Reynolds' number has values of less than about 500. In laminar flow layers of fluid move over one another with no mixing between the layers, as this is damped out by the relatively large viscous forces. When the Reynolds' number is greater than 2,000, then the flow is *turbulent* with considerable mixing between the layers, so that the path of a water particle is no longer a straight line parallel to the bed but is instead extremely sinuous. The sinuosity is random, so that the mean path of water particles is still parallel to the bed, but the random variations are large, even close to the bed.

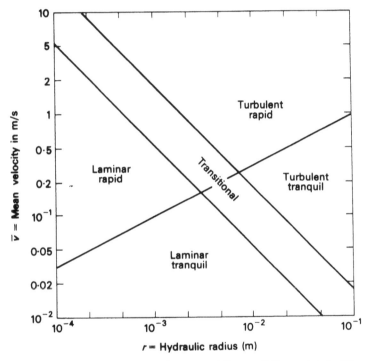

Fig. 3.6. Types of water flow (from Robertson and Rouse, 1941, reprinted with permission from the March 1941 issue of *Civil Engineering – ASCE*, official monthly publication of the American Society of Civil Engineers).

Water flow commonly occurs as one of two types; tranquil or streaming flow at lower speeds, and rapid or shooting flow at higher speeds. The point at which this changeover occurs depends on the ratio between inertial and gravity forces, usually expressed in the dimensionless Froude number:

$$\left(\frac{\text{Inertia force}}{\text{Gravity force}}\right)^{\frac{1}{2}} = \frac{\bar{v}}{(g.r)^{\frac{1}{2}}} \qquad (3.6)$$

37

Tranquil flow is the more usual but rapid flow is common in weirs and may occur during river flood flows. It is also relatively common in thin flows such as occur over smooth hillsides. Rapid flow is associated with free surface instability which is characterized by roll waves and patterns of intersecting diagonal ripples on the surface of the flow (Koloseus and Davidian, 1966). In thin flows, laminar rapid flow is a fairly common condition (fig. 3.6) where the surface is smooth and fairly steep. Where the surface is very rough, with stones or grass protruding from the surface of the flow, then although the flow is slow enough to be laminar, the roughness elements themselves divert the flow so that water particles follow sinuous paths typical of turbulent flow: this type of flow has been called subdivided flow (Horton, 1945).

Several forces act on particles at the base of a flow. First, the speed of the current exerts a dynamic pressure on the particle because there is a momentum difference between the upstream and downstream faces of the particle. Second, the difference in mean downstream velocity between the top and bottom of the particle produces an aerodynamic lift in accordance with Bernoulli's principle. Third, in turbulent flows the eddy velocities provide randomly directed forces which will lift some bed particles into the flow. These forces are resisted by the weight of the particles, friction and interlocking between them, and by cohesive forces if the particles are embedded in the soil. Once in the flow, lift forces are balanced by the particle weight which tends to produce settling. In near-bed movement particles may also strike other moving particles providing an additional lift force, called the dispersive grain stress (Bagnold, 1954).

The friction force which resists the flow of the water is transmitted to the bed (and bank) materials as a tractive force which tends to move them downstream. As described above, the friction force per unit area in turbulent flow is:

$$\tfrac{1}{2}\rho_w . f . v^2 = \rho_w . g . r \sin_\iota . \cos \beta. \tag{3.7}$$

This quantity is the tractive shear stress. A part of this tractive stress acts on the surface area of a particle to produce a propelling force (which is proportional to the square of its diameter), but the resistance to motion of the particle depends on its weight (which is proportional to the cube of its diameter), so that the critical diameter which can be moved is directly proportional to the tractive force (Rubey, 1938). If particles are closer together than about eight diameters, then they shield one another from the flow so that each is less likely to move: and once moved each is more likely to come to rest against another (Leopold, Emmett and Myrick, 1966). The effective tractive force is reduced by a factor of $(1-c)$, where c is the particle concentration on the bed (Bagnold, 1956).

Bernoulli effect lift is in practice less than the lift forces derived from turbulent eddies, as they have velocities up to twice the horizontal velocity in the region near the bed. On average the upward and downward turbulent fluctuations are equal, but in the interchange of particles between a higher and a lower point in the flow, a higher proportion move down than move up because of the greater intensity of turbulence at higher levels (but still near the bed) and also because of the overall tendency to fall under gravity. In equilibrium, however, these tendencies are balanced by a difference in concentration between the two levels; the lower level having a denser concentration of particles, so that an equal total number of particles are exchanged between the levels. For each grain size therefore, there is a concentration gradient from high at the bed to progressively lower away from the bed. Because of the greater influence of the gravity force for larger sizes, this gradient is steeper than for fine material.

In flows where large concentrations of coarse material are moving, the normal distinction between bed-load and suspended load breaks down as the bed is no longer clearly defined. Particles follow a zig-zag path between collisions with other particles. These impacts help to maintain debris in the flow, and transmit part of its weight to the bed through a chain of impacts. Under these circumstances bed-load can be defined as that part of the load for which the weight is carried by the solid bed; and the suspended load as that part for which the weight is carried by the water, and which increases the effective density of the water (Bagnold, 1956).

Formulae for sediment transport are all to some extent empirical, but are related either to the tractive force direct, or else to the dimensionless ratio of total tractive force on a particle of diameter d to the gravity forces acting on it (Engelund & Hansen, 1967). This ratio is denoted by

$$\theta = \frac{\rho_w . g . r . \sin \beta . \cos \beta . d^2}{(\rho - \rho_w) . g . d^3}, \tag{3.8}$$

where ρ, is the density of sediment. Writing the ratio of submerged sediment density to water density,

$$\frac{\rho - \rho_w}{\rho_w} = \Delta, \tag{3.9}$$

and replacing the slope, $\sin \beta \cos \beta$, by the symbol s, then the dimensionless shear,

$$\theta = \frac{r . s}{\Delta . d}. \tag{3.10}$$

At a critical value of the dimensionless shear stress (entrainment function), then material will begin to move, and this critical value has been shown (Shields, 1936) to be a function of the Reynolds' number of the flow at the bed, which is equal to

$$\frac{(g.r.s)^{\frac{1}{2}}}{v}.d,$$

with the same notation as above, and where v is the kinematic molecular viscosity (μ/ρ_w) of water ($0\cdot013$ cm^2/s at $10\,°$C; $0\cdot008$ cm^2/s at $30\,°$C). The experimental relationship is shown in fig. 3.7. For fully turbulent flow a constant value of $\theta_c = 0\cdot06$ is a satisfactory approximation.

Expressions for sediment transport are generally derived for a dune bed, but appear to have much wider applicability in practice. Theoretical justification and good empirical fit is given by the expressions:

$$f.\Phi = 0\cdot1\,\theta^{5/2} \tag{3.11}$$

or

$$f.\Phi = 0\cdot077\,\theta^2(\theta^2+0\cdot15)^{\frac{1}{2}} \tag{3.12}$$

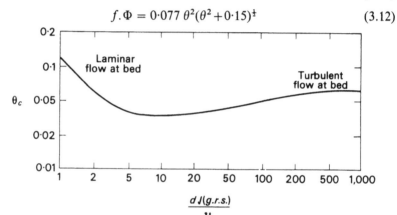

Fig. 3.7. Critical (dimensionless) tractive force at which particles on the bed begin to move (after Shields, 1936).

over the range $0\cdot07 < \theta < 10$; $10^{-4} < f.\Phi < 10$ (Engelund and Hansen, 1967, pp. 48–9); where the dimensionless sediment transport

$$\Phi = \frac{S}{(g.\Delta.d^3)^{\frac{1}{2}}} \tag{3.13}$$

and S is the sediment transport per unit width, expressed as a volume per unit time. Perhaps more relevant are formulae for bed-load transport, many of which consider the movement of fairly coarse material. Formulae vary in ease of application and in theoretical justification (e.g. Straub, 1934; Einstein, 1950; Kalinske, 1947; Engelund and Hansen, 1967), but in practice yield comparable results over a range of values of dimensionless stress. For simplicity the Meyer-Peter bed-load formula has much to recommend it (Meyer-Peter and Müller, 1948):

$$\Phi = 8(\theta - 0\cdot047)^{3/2}. \tag{3.14}$$

The relationship proposed by Bagnold (1956) is very similar, but has several refinements which allow it to be used under a wider range of conditions: it makes some estimate of suspended load in addition to the bed-load when this becomes important; it allows for variations in the critical tractive stress with the type of flow, using the data shown in fig. 3.7; it allows for variations in the angle of friction between water and grains with grain size; and it allows for variations in concentration and their influence on the effective tractive stress.

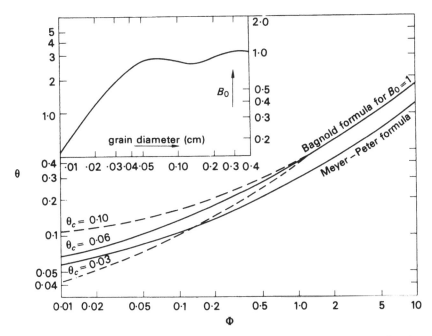

Fig. 3.8. Comparison between Meyer-Peter and Bagnold bed-load transport function. Bagnold function is proportional to B_0; main curve is for $B_0 = 1.0$ and inset shows values of B_0 for transport in a liquid as a function of grain diameter. (Plotted from expressions in Bagnold, 1956 and Meyer-Peter and Muller, 1948. Inset curve after Bagnold, 1956.)

Fig. 3.8 compares the Meyer-Peter and Bagnold expressions for bed-load transport. Bagnold's basic form for bed-load transport is:

$$\Phi = B_0(1-c)^{3/2} \cdot (\theta - \theta_c) \cdot \theta^{\frac{1}{2}}, \qquad (3.15)$$

where θ_c is the critical bed shear obtained from fig. 3.7;

$$c = \frac{\text{grain occupied space}}{\text{whole space}} = \text{concentration of sediment; and } B_0 \text{ is an}$$

41

empirical function of the solid–fluid friction angle, obtained from the inset in fig. 3.8. These and similar expressions have not, however, been obtained specifically for thin flow conditions, and so should be used with some caution for hillslope flows.

Flows through the soil

Most subsurface flow occurs at slow rates and through small pore spaces in the soil, so that the flow is almost always laminar. Two forces are acting on subsurface water to promote flow: the water flows under a gravity force which is the weight of the water, and a hydraulic force produced by the pressure or tension gradient. These forces produce flows under saturated and unsaturated conditions, and the rate of flow is given by Darcy's law (1856):

$$q_z = -K . \text{gradient} (\phi), \tag{3.16}$$

where q_z is the mean discharge flowing through the soil per unit cross-section, K is the capillary conductivity of the soil, and ϕ is the total gravitational and hydraulic potential.

It should be noted that although discharge per unit area, q_z, has the dimensions of velocity, it is *not* the velocity, v at which water is travelling through the soil, but an average through the whole soil cross-section. Thus

$$q_z = \varepsilon . v . \frac{m}{m_{\text{SAT}}}, \tag{3.16a}$$

where m is the soil moisture content, m_{SAT} the saturated content, ε is the soil porosity in cm^3/cm^3 and v is the velocity of water movement.

The basis for equation 3.16 lies in Poiseuille's formula for the laminar flow of a fluid through a tube of radius r:

$$v = \frac{-r^2 . \text{gradient} (\phi)}{8\mu}, \tag{3.17}$$

where μ is the dynamic viscosity of the fluid.

For saturated flow, this relationship can be used to deduce the permeability of a non-shrinking soil with a given pore-space distribution by calculating from the experimental curve of moisture content against moisture tension for the soil, leading to the relationship (Marshall, 1958):

$$K = 270 \frac{\varepsilon^2}{n^2} [h_1^{-2} + 3h_2^{-2} + 5h_3^{-2} + \ldots + (2n-1)h_n^{-2}] \text{ cm/s}; \tag{3.18}$$

for water at 20 °C, and where the moisture–tension curve is divided into n equal segments on the moisture scale between dry and saturated so that h_i is the water tension at the ith of these divisions, in cm of water. This equation also shows one source of variation of the conductivity with moisture content; for in unsaturated conditions the porosity, ε, should be considered as the volumetric moisture content of the soil rather than the total pore space. However, the conductivity increases more rapidly with moisture content than is indicated by a square-power law (fig. 3.9), for it appears that at low moisture contents only a part of the pore space is accessible to the flowing water.

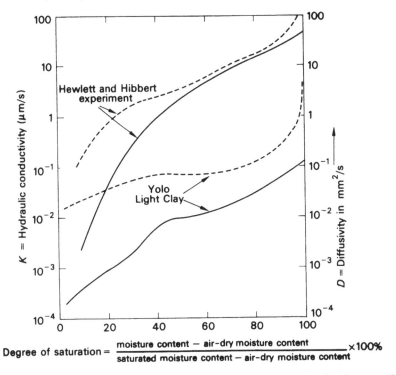

Fig. 3.9. Values of hydraulic conductivity (solid line) and diffusivity (broken line) for two soils (data from Hewlett and Hibbert experiment, calculated in Kirkby and Chorley, 1967; and from Philip, 1957).

For a simple model of flow along a layer inclined at angle β to the horizontal, the potential gradient is made up of a gravity term, $(-\sin\beta)$ (measured in the downslope direction), and a tension term, $\partial\psi/\partial x$, where ψ is the soil moisture tension and x is distance measured downslope along the layer. Darcy's law can therefore be re-written in the form:

$$q_z = K . \sin \beta - K . \frac{\partial \psi}{\partial x}$$

$$= K . \sin \beta - K . \frac{\partial \psi}{\partial m} . \frac{\partial m}{\partial x}$$

$$= K . \sin \beta - D . \frac{\partial m}{\partial x}, \tag{3.19}$$

where m is the soil moisture content, and the quantity D, $= K\partial\psi/\partial m$, is called the diffusivity, which, like the conductivity, varies with the moisture content for each soil (fig. 3.9). The two properties, conduc-

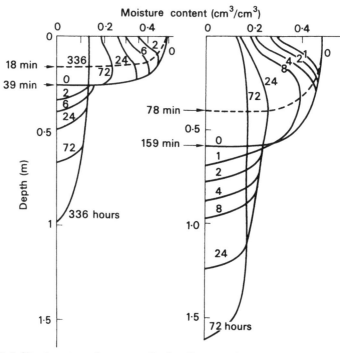

Fig. 3.10. Infiltration at maximum rate (broken lines and times in minutes at left) and redistribution after infiltration (solid lines and figures showing times in hours elapsed from end of infiltration) in an experimental soil (*SD* in fig. 3.4) (after Youngs, 1957 and 1958).

tivity and diffusivity, control the unsaturated flow of water through a soil. Their relative values are such that, for vertical and downslope flows, the conductivity term is numerically the more important at moderate moisture contents, but the diffusivity term may have a larger influence in both very dry and near-saturated conditions.

Flow during the entry of water into the soil, or infiltration, is of particular importance to hillslope processes. Most work has been done for the case where water is supplied to the surface as fast as it can enter the soil, so as to establish the maximum rates of infiltration. During this process, the upper layers of soil become saturated (fig. 3.10), and the saturated levels are separated from the dry layers below by a sharp wetting front, 1–5 cm deep, in which the moisture content drops abruptly. A model of how water flows into the soil can be made by considering the two components of the flow equation. The conductivity flow under gravity provides a flow, A, at a steady rate under a constant potential gradient. The diffusivity flow, however, operates with a declining average gradient as the wetting front advances into the soil, and to a first approximation the rate of diffusivity flow is inversely proportional to the square root of the time elapsed. A simple model of the total rate of infiltration is therefore:

$$f = A + B \cdot t^{-\frac{1}{2}}, \quad \text{or} \quad F = At + 2Bt^{\frac{1}{2}} \tag{3.20}$$

where f is the rate of infiltration, F is the total amount infiltrated after time t, t is time elapsed from the beginning of infiltration, and A, B are constants for the soil and its previous moisture content. This equation (Philip, 1957) has a sound physical basis, and explains well the initial very rapid decline in infiltration rate, but it is less satisfactory as a model after long infiltration times.

An alternative semi-empirical infiltration equation, which is a better fit at long time periods and retains the same form for very short periods, has also been proposed by Philip (1954):

$$t = A\left[F - Z \cdot \log_e\left(1 + \frac{F}{Z}\right)\right], \tag{3.21}$$

with the same notation as before, and the constant $Z = \frac{1}{2}A \cdot B^2$, to give the same response for small time periods. A better known empirical equation is (Horton, 1933):

$$f = f_c + (f_0 - f_c) \cdot e^{-k \cdot t}, \tag{3.22}$$

which gives the rate of instantaneous infiltration, f, at time t after the beginning of infiltration, where f_c, f_0 and k are constants. Comparing this equation with the others, the constant f_c is the same as A, namely the steady minimum rate of infiltration, but the other constants cannot be identified with physical quantities as in the Philip equation, and the Horton equation is seriously in error as a model for infiltration at very brief time periods.

After the end of infiltration, water continues to penetrate downwards, though at a moisture content below saturation (fig. 3.10). For many soils this moisture content remains fairly uniform, and can be identified with the concept of field capacity, which is defined as the amount of water held in the soil after excess gravitational water has drained away and the rate of downward movement has materially decreased (Veihmeyer and Hendrickson, 1931).

The tractive force on soil particles which tends to move them with subsurface water flows has received little attention, but may be compared with a model of diffusion under an external applied stress, discussed by Culling (1963) in the context of soil creep (see below p. 61). Soil particles are considered to be free to move in random directions into neighbouring pores. These movements alone will produce a net diffusion from regions of low porosity to regions of higher porosity. In an applied force field, such as that applied by a moving fluid, an additional net drift will occur in the direction of the force. The rate of movement of debris should therefore be of the form:

$$S = S_0 + D . \text{gradient } c, \tag{3.23}$$

where S_0 is the rate of movement under the applied force, D is a diffusivity of particle movement, and c is the concentration of pore space. Diffusivity will vary with particle size, according to the concentration of each size fraction in the soil and the distribution of pore sizes. The rate of movement under an applied force will also depend on these factors together with the magnitude of the fluid forces involved. A small particle of diameter d in a laminar flow of velocity v is subjected to a force of $3 . \pi . \mu . d . v$, where μ is the viscosity of the fluid. This is Stokes' law, which is usually applied to the determination of settling velocities. The resistances to this force may depend on surface activity, for clay size particles, and be proportional to d^2, or else to the mass of the particle, proportional to d^3.

The critical diameter carried is therefore proportional to either v or else to $v^{\frac{1}{2}}$. As with sediment transport in a surface flow, the dimensionless sediment transport, Φ, is perhaps empirically related to the dimensionless ratio of tractive force to resistance, θ, by a power law of the type:

$$\Phi = A(\theta - \theta_c)^n, \tag{3.24}$$

for some constants A, θ_c, n, which themselves depend on the soil properties.

DISTRIBUTION AND MOVEMENT OF HILLSLOPE WATER

As it reaches the ground, falling rain water is split into a number of components (fig. 3.11). A part of the water is intercepted by vegetation and, of this, some stays on the foliage surfaces until it is evaporated, some drips through to the ground after a delay and some is diverted down the plant to emerge as stemflow around its base. A part of the rainwater reaches the ground surface directly or indirectly: some of it fills small surface irregularities, puddles, plough furrows, etc., as depression storage; some percolates into the soil and some is unable to percolate, but becomes diverted downslope as overland flow across the land surface. Here, as elsewhere in this book, the term 'overland flow' is used for flow physically over the hillslope surface. The term 'runoff' is used only for streamflows and is not associated with any particular hillslope flow component. Water which has percolated into the soil may remain to increase soil moisture content, or may continue to percolate downwards towards the water table; but because lower soil horizons and bed-

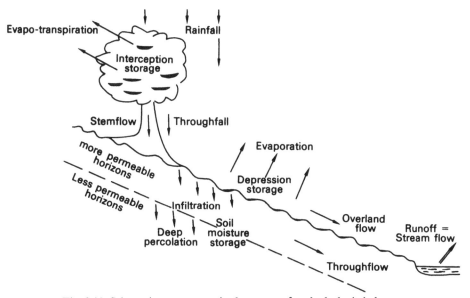

Fig. 3.11. Schematic components in the near-surface hydrologic balance.

rock are commonly less permeable than the surface layers, a part of the water percolating is diverted laterally as throughflow within the upper soil layers. Lastly some of the water which is on the soil surface or in the uppermost layers of the soil may evaporate. Over a period, the inflows and outflows must balance each other, but instantaneous balances for the system or a part of it must always be of the form:

$$\text{inflow} - \text{outflow} = \text{increase of storage.} \qquad (3.25)$$

During a storm, evaporation from plants and from the ground surface can usually be discounted, since evaporation rates depend on vapour pressure differences which are usually low while rain is falling. Interception varies with the type and density of vegetation, but a working estimate for a full tree cover is that the first 1·0 mm and 20 per cent of the subsequent rainfall (Horton, 1919) is intercepted and *stored* on the

47

foliage, from where it is mainly evaporated after the storm. The remaining rainfall does reach the soil, including 1–5 per cent concentrated as stemflow and the rest which comes through the canopy as throughfall. Interception by a grass cover is known to be similar, but there are no reliable measurements of stemflow in grass. Depression storage is able to absorb about 2–5 mm of water but the actual value varies very considerably with land use, and roughly inversely to surface slope (Linsley, Kohler and Paulhus, 1949, p. 269).

Infiltration rates decrease very rapidly with time from the beginning of rainfall, and most comparisons have been based on the steady rates

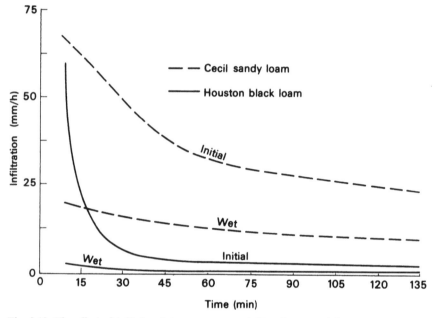

Fig. 3.12. The effect of initial moisture content on infiltration rates (after Free, Browning and Musgrave, 1940).

which are achieved after several hours of infiltration (the constants A or f_c in the infiltration equations (3.20–3.22) above). This minimum rate varies from zero to over 100 mm per hour for normal soils. Factors which affect the rate include soil grain size (or more strictly, pore size); soil moisture content at the beginning of infiltration; and the ground vegetation cover, which influences infiltration via soil structure and via raindrop impact protection.

For an initially saturated soil, the minimum rate of infiltration is correlated most closely with the organic content of the soil and its non-capillary porosity, that is the open spaces which are a part of its structure rather than of its grain size distribution (Free, Browning and Musgrave, 1940). As a rough estimate for an initially wet soil, the minimum infiltration rate,

$$A = 320 \, (\varepsilon_* - 0\cdot15) \text{ mm/h},\tag{3.26}$$

where ε_* is the non-capillary porosity in cm^3/cm^3. The influence of initial soil moisture on the whole form of the infiltration curve is illustrated in fig. 3.12, which shows that the initial infiltration rate is

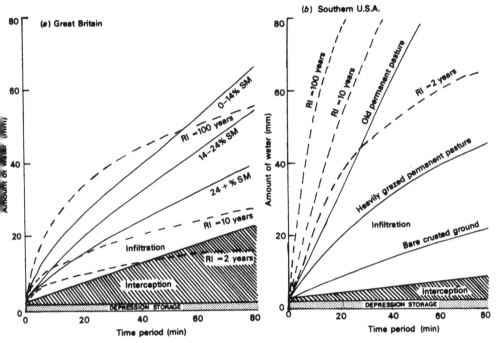

Fig. 3.13. Representative components of hillslope hydrology and relative frequencies of overland flow (from Kirkby, 1969).
RI = Recurrence interval; SM = Soil moisture.

drastically reduced for a wet soil, and that there is a lesser reduction in the minimum infiltration rate, to about one half or one-third of its dry value. Holtan and Kirkpatrick (1950) report a reduction in the infiltration rate of six to ten times in comparing old permanent pasture with bare crusted ground, as an illustration of the influence of the vegetation cover.

The variations in infiltration rate, interception and depression storage must be balanced against expected rainfall intensities in order to assess the likelihood that rain will be forced to flow over the surface because it cannot be absorbed or stored as fast as it is falling. In fig. 3.13, some

49

comparisons are made for representative conditions in Great Britain and in the southern United States. For Britain, the interception and infiltration rates are based on a total vegetation cover, and variations in infiltration rate correspond to differences in initial soil moisture. It can be seen that the relatively low rainfall intensities in Britain are rarely able to exceed the storage capacities of the soil, except when the soil is extremely wet previously. For the southern United States, interception is based on a 50 per cent vegetation cover; the soil is considered initially dry, and infiltration variations correspond to differences in vegetation cover. The greatest single factor in determining when surface storage capacity will be exceeded is high rainfall intensity, but sparse vegetation cover is also very important.

Overland flow

Overland flow which is produced by exceeding the surface depression storage and infiltration capacities is called *Horton* overland flow (Horton, 1945). Overland flow can also, however, occur when the soil is saturated so that no more water is able to enter, even though the infiltration capacity has not been exceeded. Rainfalls of long duration and low intensity are able to infiltrate into the soil until all pore spaces are saturated. If the rainfall intensity is greater than the increase of through-flow across a section of the hillslope, then the excess rainfall cannot enter the soil, but must flow over its surface. Overland flow of this type is called *saturation* overland flow (Kirkby and Chorley, 1967).

The hydrograph of overland flow can be calculated from the flow equation, which depends on the type of flow and surface roughness; and on the amount of rainfall excess above the requirements of infiltration, depression storage, etc. The continuity equation is: rainfall excess = increase in depth of flow + increase in overland flow discharge; or

$$i_* = \frac{\partial r}{\partial t} + \frac{\partial q_s}{\partial x}, \tag{3.27}$$

where i_* is the intensity of rainfall excess; r is the depth of flow; q_s is the overland flow discharge per unit width; and x is the distance downslope (measured horizontally). The flow equation is of the form:

$$q_s = k.r^m.s^n, \quad \text{see above (p. 36)} \tag{3.28}$$

where s is the slope gradient, and k,m,n are constants of the surface.
These equations are difficult to solve in general, but two aspects of their solution can be seen readily. First for constant rainfall excess, i_*, and for uniform slope, s; then the beginning of the overland flow hydrograph is approximately:

$$r = i_*.t \quad \text{and} \quad q_s = k.(i_*.t)^m.s^n. \tag{3.29}$$

If rainfall excess continues for a long time, at a rate which is constant over time but may vary along the hillslope, then the flow will ultimately achieve a steady state of flow, given by the equation:

$$q_s = \int_0^x i_*.dx; \qquad (3.30)$$

that is a discharge equal to the total rainfall excess upslope from the measurement point. Typical overland flow velocities are of the order of 1–100 cm/s, so that the times required to achieve a steady state are usually a matter of minutes only and this equation is a reasonable estimate of peak overland flow discharges.

Horton overland flow results from rainfall intensities in excess of the infiltration capacity of the soil, a parameter which varies relatively little over a small area. When Horton overland flow occurs therefore, it will occur widely, so that the rainfall excess is more or less constant over the hillside. Under these circumstances, the steady state overland flow discharge increases linearly with distance downslope from the divide. Saturation overland flow, however, is more local in its occurrence, appearing preferentially where soils are more readily saturated, especially in areas of thin or less permeable soils, in areas of flow concentration produced by profile concavity or contour curvature, and in areas adjacent to flowing streams, where soils tend to be wettest (Kirkby and Chorley, 1967). The rainfall excess will therefore only exist in certain parts of a hillside, and the area of rainfall excess will shrink or grow with changes in the rainfall intensity and antecedent soil moisture. Under these circumstances, which will apply in all but the most intense storms, overland flow is concentrated near streams, but its discharge is relatively difficult to estimate because of the great areal variations in the rainfall excess.

Throughflow

Water which has percolated into the soil may be diverted laterally to flow downslope as throughflow within the soil, and this can occur in any zone or above any interface where the downwards permeability is decreasing with increasing depth. A reduction of this sort is most common at the base of organic and 'A' horizons in the soil, but may occur at any level. In such a reducing permeability zone, water percolating from above begins to back up during a rainstorm and the increasing moisture content will allow an increasing lateral flow downslope. Throughflow is usually unsaturated except close to flowing streams and at the base of each permeable horizon, so that in many ways the permeability differences in the soil produce a water system which could be described as a perched water table.

If the permeable upper layers of soil are considered as a single unit in which the moisture content is constant with depth, but varies along the slope, then the flow model can be simplified. Flow must obey the continuity equation, which in this case is:

$$p = z_1 \cdot \frac{\partial m}{\partial t} + \frac{\partial q_T}{\partial x}, \tag{3.31}$$

where p is the net rate of percolation into the soil (incoming − outgoing); m is the moisture content of the soil; z_1 is the soil thickness through which flow is taking place; and q_T is the throughflow discharge per unit width (summed through the soil profile). The rate of flow, q_T is also governed by the flow equation 3.16. This model can be solved numeri-

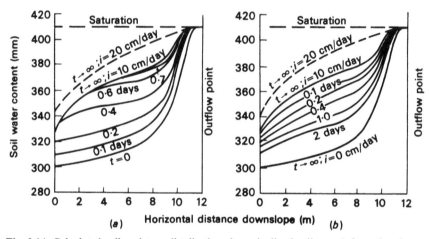

Fig. 3.14. Calculated soil moisture distributions in an inclined soil trough from data in an experiment by Hewlett and Hibbert (1963), ignoring hysteresis effects. (*a*) During infiltration at a constant rate of 100 mm per day after a previous long period of drainage; (*b*) during drainage without rainfall or evaporation, after a previous long period of infiltration at 100 mm per day.

cally for given soil characteristics and percolation rates, and gives a moisture gradient from saturated at the outflow point by the stream to progressively dryer conditions upslope (fig. 3.14). Increasing flows give less steep moisture gradients, and a longer near-saturated zone at the base of the slope. Decreasing flows produce steeper moisture gradients; and under conditions of net evapo-transpiration loss the flows may be reversed and the river supply water to the hillslopes. In the simplest steady state, with no net percolation, then the moisture conditions satisfy the relationship:

$$\text{capillary potential} = \text{elevation above river}, \qquad (3.32)$$

exactly as they would in a vertical column of soil. In practice, a steady state of this sort is never achieved, because of the extremely slow movement of water through the smallest pore spaces, and because the existence of a free surface and especially evaporation from it, tend to break the capillary films which are needed to transmit moisture tension gra-

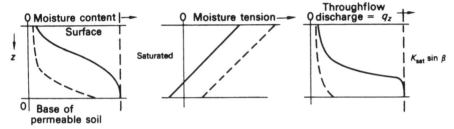

(a) Conditions of steady throughflow.

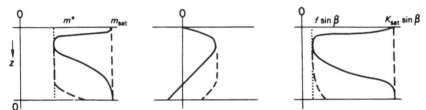

(b) Throughflow during infiltration at intensity $f = K(m_*)$ which is less than maximum infiltration capacity.

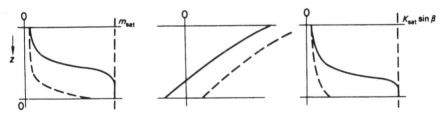

(c) Throughflow with evaporation, which is assumed in this diagram to come equally from all levels.

Fig. 3.15. Moisture distribution in the soil during throughflow, ignoring hydraulic gradients in the downslope direction. Solid lines show an example with some saturated throughflow: broken lines show only unsaturated throughflow.

dients. In a dry period therefore, a progressively smaller and smaller area adjacent to the stream is the only part of the hillside able to transmit hydraulic gradients. After rain, water films in the soil are reconnected and the connected hydraulic system expands upslope from the stream once again.

In practice there are differences of soil moisture not only along the

53

slope but also within the soil profile at a point. For the simplest example, of a soil layer of uniform properties resting on an impermeable base, some notions of the flow and moisture distribution can be obtained. For a steady throughflow with no percolation down from the surface, and with a flow q_z at elevation z above the impermeable base (fig. 3.15a), there is no flow at right angles to the surface, so that approximately

$$\psi(z) = \psi(0) + z, \tag{3.33}$$

where $\psi(z)$ is the capillary tension at z. If conditions do not vary downslope (i.e. throughflow is constant), then the flow is solely under gravity forces, at rate:

$$q_z = K(m).\sin \beta, \tag{3.34}$$

where $K(m)$ is the permeability at moisture content m which corresponds to the capillary potential ψ; and β is the slope angle. Greater throughflow discharges are obtained with an increased basal moisture potential, $\psi(0)$. Once the base of the layer has become saturated ($\psi(0) \leqslant 0$), further increases in throughflow occur if the depth of saturated soil is increased, although the throughflow component, q_z, remains constant within the saturated zone. The maximum throughflow discharge is obtained when the soil is completely saturated, and is equal to the saturated permeability, multiplied by the slope (sin β) and by the thickness of the soil layer. Further increase in flow can only be obtained with an overland flow component.

During rainfall at less than the infiltration rate, there is saturation at the surface, and a moisture gradient down to a steady value, m_* (fig. 3.15b). Percolation into a dry soil will show a wetting front below the constant moisture transmission zone, but in a thin permeable layer, this wetting front grades into a throughflow profile, with a less steep moisture gradient corresponding to the downward flow of percolation water which is contributing to the throughflow. Similarly during evaporation (fig. 3.15c), there is a steeper moisture gradient allowing an upward movement of water in addition to the throughflow. Under uniform percolation or evaporation conditions, the upper part of the moisture profile will be similar in all unsaturated parts of the slope, and the downslope moisture gradient will be mainly confined to the lower part of the profile which is producing the throughflow.

Rainfall produces increased throughflow with a hydrograph form very similar to stream hydrographs (Whipkey, 1965), and at a rate fast enough to mainly form the flood hydrograph of small streams (Weyman, 1970). Downslope discharges become very much larger when a part of the throughflow is saturated, because of the higher permeabilities (and

54

hence velocities) associated with saturation. As the saturated layer becomes thicker, the delay between rainfall and throughflow discharge also becomes less, as shorter periods are needed for water to percolate from the surface to the saturated soil. Recession flow, after rainfall, is continued for a matter of weeks at measurable levels, and is of the right order of magnitude to account for base flows in upland basins on impermeable bedrock (Hewlett, 1961; Weyman, 1970).

RAINFALL IMPACT FORCES

During a sudden impact momentum is conserved, although energy is lost. Raindrops striking a surface exert a net pressure on the surface equal to the rate of loss of momentum per unit area by the raindrops, which is the same as the rate of gain of momentum per unit area by the surface. On a sloping surface, this force has two components; a consolidation force acting *into* the surface and a shear force acting *along* the surface. The average magnitude of these overall forces is rather small. Data from Laws and Parsons (1943) give a normal pressure of:

$$\sigma = 1 \cdot 3 \times 10^{-4} \, i_*^{1 \cdot 1}, \tag{3.35}$$

where σ is the pressure in kgf/m^2, and i_* is the rainfall intensity in mm per hour. For rather intense rainfall at 100 mm per hour, this pressure amounts to only 0·02 kgf/m^2. The consolidation and shear forces acting on the soil considered as a mass are therefore almost negligible.

Looking at individual particles however, the forces involved appear rather more effective. A five millimetre raindrop has a terminal velocity of 9·5 m/s, and a momentum of 62 g cm/s. If the downslope component of this momentum is all transferred to a 5 mm quartz grain on a 10° slope, then the grain will begin to move at 0·61 m/s. In fact the momentum is not transferred very efficiently between raindrop and debris particle, but even so, each raindrop concentrates the momentum in a way which is more efficient in moving debris than the net forces would suggest. A surface can absorb forces acting perpendicular to it by consolidation, or by the packing of *surface* particles down into the soil pore spaces, or by the formation of small craters. If there is water on the surface, the impact forces can be absorbed in the production of greater turbulence in the flow. Some of the impact momentum is not absorbed but is reflected, carrying water droplets and some soil upwards. The downslope component of the impact forces is partly used up in producing an asymmetric splashback of water droplets, but mainly acts as a shear force along the surface. If there is no water on the surface, these forces can act most effectively on detached debris particles. If there is water flow on the surface, then the shear is applied to the fluid surface. It is not, however, clear if the flow is accelerated in this way because the accompanying increase in turbulence tends to slow the flow.

EXPANSION FORCES

Expansions within a soil tend to produce upward heaves of the surface, and also form polygonal patterns when acting laterally. Although expansions and contractions do not directly move debris downslope, they play an important role in overcoming soil resistances, so that gravity is able to modify the directions of expansion and contraction movements in a way which does promote downslope movement. The most important causes of expansion are changes in moisture content, changes in temperature, freezing and thawing of soil water; and to a lesser extent chemical changes during weathering and plant root growth.

Expansion due to changes in moisture content are a result of changes in capillary cohesion (p. 76). During rainfall, the tension forces are concentrated at the wetting front, but during drying they are distributed throughout the soil profile. Drying may occur as a result of downward drainage, especially on steep slopes and where rainfall is greater than evaporation; or it may occur as a result of evapo-transpiration. In the latter case, the presence or absence of vegetation and differences in rooting depth, make considerable differences to the total water lost through evaporation, and hence to the total ground movement. Ward (1953) has shown that the vertical contraction of the surface during drying amounts to between one third and one sixth of the change in water deficit, and has an upper limit of 15–30 mm under grass in Britain.

Thermal expansion of solid rock is able to exert very great forces which may shatter the rock if temperatures change by at least 300 °C (Griggs, 1936). Such temperature changes are not normally encountered except during forest fires or lightning strikes, but lesser temperature changes caused by solar heating may accelerate chemical weathering in the presence of moisture (Griggs, 1936). Temperature changes may also cause movement amongst stones as each expands and contracts, and can cause downslope movement directly (Moseley, 1869). Temperature expansion also takes place within the soil to produce a heave, but the coefficient of expansion is low (about 0·001 per cent per deg C) so that the total lift produced is very slight (1–2 mm). Temperature changes are conveyed into the soil principally by heat conduction and by circulation of water. The amplitude of the temperature wave produced by conduction declines exponentially with depth, so that at a depth of one wavelength (0·2–1·0 m), it is reduced to 1/535 of its amplitude at the surface (Geiger, 1965, chapter 1).

Expansion during freezing is more complex because (1) the presence of a freezing interface alters the temperature gradients in the soil; (2) flows of water within the soil are of major importance in both the growth of ice and the conveyance of heat; and (3) moisture tensions themselves produce a lowering of the freezing point. The simplest model is of cooling by conduction only, of a soil of constant water content, m, without allowing movement of water during the freezing process. The quantities of heat which must be removed to freeze the water are so much larger than the changes in heat elsewhere in the soil that the latter may be ignored, leading to Casagrande's (1931) approximate equation, which states that the depth of freezing is proportional to the square root of the number of day-degrees below freezing point:

$$z^2 = \left\{ 1 \cdot 73 \times 10^5 \frac{\kappa}{m \rho_w L} \right\} . H, \qquad (3.36)$$

where z is the depth of penetration of freezing into the soil (cm), κ is the thermal conductivity of the wet soil (cal/deg C cm s), ρ_w is the density of water (g/cm^3), m is the volumetric moisture and ice content of the soil (cm^3/cm^3), L is the latent heat of fusion of ice (80 cal/g), and H is the number of accumulated days-deg C below the freezing point of the water. Values of the bracketed numerical constant range between about 0·1 and 10. The force exerted during freezing can be measured by the pressure which must be applied in order to lower the freezing point enough to prevent it. This pressure is $1 \cdot 4 \times 10^6$ kgf/m^2 for each degree Centigrade of freezing point depression; so that if the air temperature is $-5°C$, then a pressure of five times the above value is required to prevent any freezing. Lesser pressures will reduce the depth of freezing by lowering the freezing point to a small extent, and so altering the number of day-degrees below the freezing point. As with other soil pressures, the effective stress, composed of the overburden pressure augmented by moisture tension forces (p. 33), is the pressure which contributes to freezing point depression.

The expansion produced by freezing of water is about 9 per cent of its volume. In a saturated soil, of moisture content m_s, which freezes without drawing in additional water, the whole expansion must take place vertically, and amounts to $9 . m_s$ per cent expansion. In unsaturated soils, part of the ice expansion can take place into air-filled pore spaces, so that the vertical expansion should be less than $9 \times$ (moisture content) per cent. Another factor which reduces the total expansion is that not all water in the soil is normally frozen: as freezing progresses, water is left in smaller and smaller pores, which eventually exert strong enough tension forces to prevent further freezing. However, very much *larger* expansions are quite usual because water is drawn in by tension forces to the freezing front, leading to complete lifting of the overlying soil and the formation of ice lenses or more commonly needle-ice. For this' segregated ice to grow, the heat flow from the freezing front must be balanced by a flow of water for freezing from below drawn up by tension gradients. Extending Casagrande's model, the volume of water frozen per second per square centimetre of freezing front;

$$q_z = \frac{\kappa}{L \cdot z}(\theta_0 - \theta), \tag{3.37}$$

where the notation is as before, with the addition of θ as the surface temperature, and θ_0 as the freezing point temperature. This rate of flow must be supplied by the hydraulic gradient at rate (equation 3.19 above):

$$q_z = D \cdot \frac{\partial m}{\partial z} - K, \tag{3.19a}$$

where D, K are the diffusivity and permeability at moisture content m. Equating these two rates, the moisture gradient,

$$\frac{\partial m}{\partial z} = \frac{\kappa(\theta_0 - \theta)/L.z + K}{D}. \tag{3.38}$$

For coarse-grained soils, the flow velocity required to feed the ice growth can easily be maintained provided that there is moisture film continuity to transmit hydraulic gradients, but this is often absent. For fine-grained soils, the flow velocity requires very high hydraulic gradients which can only be maintained in close proximity to a water table. For these reasons needle ice and ice lenses are best developed in fairly fine-grained silts which are fine enough to preserve a continuous moisture film but not so fine that a very high hydraulic gradient is necessary for flow (Taber, 1930; Beskow, 1935). Near the surface, the flow velocity required to form segregated ice is very large, so that ice growth usually begins a short distance below the surface, and will continue to form as long as conditions are cold enough and adequate water flow can be maintained by the hydraulic gradient.

Frost heaving can be produced by either daily or annual frost cycles. The former, though of low amplitude, are repeated often enough to produce a larger amount of accumulated up and down movement in temperate to sub-arctic areas than do the single high amplitude heaves produced annually in very cold areas, and are therefore more influential in producing downslope movement which results from the heaving process. However, the thick ice layers produced in arctic conditions form an impermeable layer during spring thaws, which produces a surface soil layer which has an extremely high water content, and on which gravity forces are able to promote downslope movement of the solifluction type. Similarly autumn freezing is able to sandwich a melted layer of high water content between surface frozen soil and deeper permafrost soil, again promoting relatively rapid movement.

Chemical precipitation of soluble salt crystals in arid climates can lead to heave forces similar to those produced by the formation of ice, and with the same cumulative effect, as the crystal formation may be reversed by solution. Other chemical processes, however, especially those concerned in weathering, although they produce net expansion forces, are less readily reversible, so that their cumulative effect in promoting downslope debris movement is less.

DIFFUSION FORCES

In many geomorphic contexts, movement is made up of a large number of small steps, which are random in direction, although often with a net drift in the direction of an applied force, usually gravity. Movement processes of this type have been implicit in the discussion of (1) suspended sediment concentration (p. 39), in which fluid turbulence applies small random forces to debris particles falling under gravity; (2) in the equation for unsaturated flow through a porous medium (p. 44), in which water moving through randomly arranged pore spaces in the medium follows random paths but with an overall downward trend under gravity; and (3) in the discussion of particle transport by a fluid through a porous medium (p. 46), in which small random forces were considered to be superimposed on a steady tractive stress applied by the fluid. Other debris movements within the soil may have a largely random pattern, produced by the movement of plant roots, by the movement and burrowing of small organisms in the soil and by micro-seisms, among other causes. A simple random walk, composed of a large number of small steps, is able to act as model for many of these processes, and allows a physical significance to be attached to the constants in diffusion-type equations which result from this model and in the cases discussed above, among others.

A one-dimensional random walk is constructed to show the movement of each particle. At each time interval, the particle may take a step forward, with probability p or backward with probability q. These probabilities are considered to be independent of position, and the step (if taken) is of constant length δ (fig. 3.16a). Many particles are considered to be executing similar random walks, so that there are $c(x)$ particles at position x at time t, and the concentration $c(x)$ may vary with both position and time. If r steps are taken in unit time, then, in fig. 3.16b;

Rate of movement from $(x-\delta)$ and $(x+\delta)$ to x, minus
Rate of movement from x to $(x-\delta)$ and $(x+\delta)$
= Rate of increase of number at x.

In symbols this may be written as:

$$r\{[p.c(x-\delta)+q.c(x+\delta)]-[p.c(x)+q.c(x)]\} = \frac{\partial c}{\partial t}. \qquad (3.39)$$

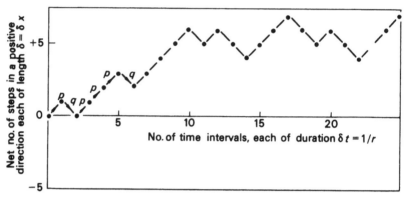

(a) Random walk model with probabilities p of taking a step in a positive direction; and q of taking a step in a negative direction. One step of length $\delta x = \delta$ is taken in each time interval δt; a total of r steps in unit time.

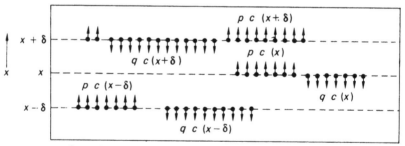

Lateral dimension along which no exchange occurs $= y \longrightarrow$

(b) Schematic exchange of particles in one time interval between levels $x-\delta$, x and $x+\delta$ with concentrations of particles per unit distance in the y direction of $c(x-\delta)$, $c(x)$, $c(x+\delta)$.

Fig. 3.16. Random walk diffusion model.

Expanding the concentration terms using Taylor's theorem, and grouping similar terms:

$$-(p-q).\delta.r.\frac{\partial c}{\partial x}+\tfrac{1}{2}(p+q).\delta^2.r.\frac{\partial^2 c}{\partial x^2}+ \ldots = \frac{\partial c}{\partial t}. \qquad (3.40)$$

In the limiting case of many small steps, $\delta \to 0$; $r \to \infty$; and $(p-q) \to 0$ in such a way that:

$$(p-q).\delta.r \to v \quad \text{and} \quad (p+q).\delta^2.r \to D, \tag{3.41}$$

where v, D are constants. In the limit the higher order differentials vanish and the exact equation of the diffusion process is:

$$-v.\frac{\partial c}{\partial x} + \tfrac{1}{2}D.\frac{\partial^2 c}{\partial x^2} = \frac{\partial c}{\partial t}. \tag{3.42}$$

A fuller discussion of this equation and variants of it may be found in, for example, Feller (1950, chapter 14); and in Culling (1965) in a geomorphic context. The concentration c is most commonly interpreted as the volumetric concentration of debris in the soil mass. The diffusivity, D, is equal to the mean-square displacement of a particle in unit time, which is the same as the variance for the position of a particle after unit time. The drift velocity, v, is proportional to the ratio of steady forces to randomly oriented forces, which must be small if no overall failure is to occur. In many situations, these quantities are directly measurable in the field.

In practice, the diffusivity and drift velocities vary with both concentration and with particle size. Variations with concentration lead to a modification of the diffusion equation to:

$$-\frac{\partial}{\partial x}(v.c) + \tfrac{1}{2}\frac{\partial}{\partial x}\left(D.\frac{\partial c}{\partial x}\right) = \frac{\partial c}{\partial t}. \tag{3.43}$$

In equilibrium, the steady force is balanced by a concentration gradient:

$$\frac{\partial c}{\partial x} = 2c.\frac{v}{D}, \tag{3.44}$$

For a constant ratio v/D, the concentration will increase exponentially with increasing distance in the direction of the steady force (e.g. downwards for gravity). Variations of the ratio with concentration will modify the form to some extent. Increasing grain size tends to reduce both v and D, and probably decrease the ratio v/D, so that within a soil fine material will tend to increase more rapidly with depth than coarse, and will also reach an equilibrium concentration very much more rapidly.

CHAPTER 4

RESISTANCE

It is well known that water, when subjected only to the force of gravity, tends to assume a level surface. The same gravitational force also acts on the mineral material which makes up the continental land masses; the fact that the land surface is not a widespread plain, but contains hills and mountains, is explained ultimately by the presence of strength in the rock masses acting against the force of gravity. Similarly the ground surface offers a resistance to other processes or forces, not just the gravity force, that act to transport debris. This resistance that opposes the shear stresses imposed by gravity and moving fluids is termed *shear strength*.

The character of the ground surface may influence denudation, not only through the intrinsic shear strength of the ground material, but also indirectly through encouraging some processes and impeding others. It is widely believed, for example, that limestone usually forms upstanding terrain, not because limestone is a particularly strong rock, but because such rock masses tend to absorb most of the surface water and thus reduce the strength of the various surface erosional forces. In this chapter an attempt will be made to treat the indirect and direct components of the resistance of the ground surface to denudational forces separately; but, because various indirect facets of resistance have been discussed in chapter 3, the former are treated rather briefly here.

MITIGATION OF FORCE

The *indirect* resistance offered by the ground surface to denudation applies primarily to those forces associated with rain drop impact and the flow of water over and through the ground. There is no mechanism by which the ground can substantially alter the shear stresses imposed upon it by gravity, although, as shown in the next section, the shear strength that a hillside can muster against the gravity force varies considerably with the character of the rock mass. The three main factors that affect the magnitude of these other forces are the transmissibility characteristics of the hillmass, the moisture content of the slope material, and the amount and type of vegetation cover. These are discussed below in general terms; the details of this interaction depend intimately on the type of denudational process operating and they are deferred to those chapters (part two) dealing with processes individually.

Transmissibility

The strength of the overland flow force, as shown in chapter 3, is determined by the *amount* of overland flow; this in turn, is inversely correlated with the ability of the hillside to absorb water. In the Horton (1945) model, overland flow occurs when the rainfall intensity exceeds the infiltration capacity of the soil mantle. The surface infiltration capacity depends partly on the antecedent moisture content, discussed below, and, more particularly, on the size of the pores in the soil. Infiltration should increase as the porosity increases, although it is usually the volume of super-capillary size pores, rather than total porosity, that is most important here. Sandy soils, therefore, tend to absorb water most easily and clays least, and, assuming that there is little difference in the shear strength, sandy soils should, accordingly, be less erodible than clay soils. The infiltration capacity of surface soils is not, however, a time-invariant property; in the dry state, clay soils can be very permeable owing to the temporary development of desiccation cracks.

Hydrologists today are beginning to question the validity of the Horton model and, as noted in the previous chapter, many now regard overland flow as an expanded form of throughflow. It is thought that most rainfall, at least outside semi-arid areas, is absorbed by the soil mantle and percolates down until it reaches an impeding layer of lower permeability. Saturated throughflow forms above this layer and, if the storm lasts long enough, may thicken sufficiently to seep out along the ground surface. In this model, the transmissibility of the lower soil horizons and the underlying bedrock is more important than the infiltration capacity of the ground surface. Soil profiles with well-developed illuviated horizons, along with soils over impermeable rock, should be expected to produce large amounts of throughflow and, possibly, surface runoff.

The throughflow model may thus explain the small amount of surface runoff that occurs in limestone areas; soils there are usually clayey but, because the underlying rock is heavily fractured and jointed, and therefore highly permeable, little surface water occurs. An important element in the throughflow model is the thickness of the soil mantle above the impermeable contact; shallow soils will quickly become saturated, whereas thicker mantles may absorb rainfall throughout the longest storms. The significance of this point led Hewlett (1961, Hewlett and Hibbert, 1967) to coin the term *hydrologic depth* to describe the runoff (throughflow and overland flow) potential of a hillside. The large runoff amounts in semi-arid areas, for instance, may be attributed partly to the shallowness of soils above the underlying rock. Soil mantles in these areas simply do not have sufficient hydrologic depth to absorb all the rainfall that is shed to them.

Soil moisture

In the Horton model, soil moisture plays a vital role in determining the relative balance between overland flow and the water which infiltrates into the ground. Soils that already contain much water have low infiltration capacities, whereas drier soils can absorb water at much faster rates. Among other things, this is due to the reduced capillary conductivity, stemming from the smaller tension forces in moist soils. Overland flow is, as a consequence, much more likely to develop during storms that occur at times when the ground surface is already moist. This conclusion holds for the throughflow model, but for different reasons. In this case, the influence of antecedent moisture content on the infiltration capacity of the soil is not unimportant; the critical point is the amount of water that the mantle can hold. When the existing moisture content is high, there is, clearly, little space available for more moisture; the throughflow layer will thicken rapidly and there is a much greater probability that water will break out at the ground surface and appear as overland flow. Whatever the exact mechanics of the process, then, overland flow is more probable at times when the existing moisture content of the soil is high.

Vegetation

A well-developed vegetation cover may influence the intensity of denudational processes a great deal, not only through indirectly controlling the size of the surface forces, but also through adding to the shear strength of the soil mantle. Indeed, under certain circumstances (chapter 7), the extra shear strength imparted by root penetration into a shallow soil mantle may control both the type and rate of surface instability. In this section, attention is confined to the more general role of vegetation in reducing the strength of surface water forces. Two components of the vegetation cover need to be considered here. One is the canopy and the other is the litter zone on the ground surface. Both can be very effective in absorbing the energy of surface water processes.

The vegetation canopy, in intercepting rainfall at some height above the ground surface, serves a dual purpose. In the first place, some of the rainfall is lost completely through evaporation from the canopy, and the potential amount of runoff is reduced accordingly. Such interception losses vary markedly with the type as well as the density of the vegetation, although, overall, they tend to be greater in wooded areas (Ward, 1967) than on grass-covered surfaces. Secondly, the canopy tends to protect the ground surface from the full impact of raindrops. Once intercepted by the canopy, much of the water trickles down stems to the ground surface and, in this way, reduces the total impact on the soil. Admittedly some of the water forms as new drops on the tips of leaves, but, except with very high canopies, the impact of these newly-formed droplets is usually much less than the impact that would have occurred in the absence of the canopy.

The presence of litter on the ground surface also serves in a dual capacity. It cushions the ground surface against the impact of raindrops and, probably more effectively than the canopy, acts to reduce the shear forces of any overland flow. It achieves this, not only through reducing the actual contact between moving water and the soil surface, but also through increasing the roughness of the ground, slowing down the flow of water and, thus, allowing more time for infiltration.

The combined effect of canopy and litter on soil splash by raindrop impact may be quite remarkable. It was noted by Ellison (1948) that on ground covered with about 325 kg of forage and litter per hectare, the measured soil splash was 11·2 tonnes per hectare, whereas, on the same soil, with a forage–litter cover of about 720 kg per hectare, the comparable splash figure was only 0·4 tonnes per hectare. Comparable data showing the effect of the vegetation cover solely on soil loss due to overland flow are more difficult to obtain, but there is little doubt that, again, vegetation is a major element in cutting down soil loss. This, of course, is well known and forms the basis of many soil conservation practices.

The geomorphic implications of this are interesting. The average tractive force associated with fluvial processes on slopes should be greater in wet areas than in dry areas simply because runoff volumes should be greater. On the other hand, the mitigating effect of a vegetal cover should also increase in moving from dry to moist areas because the density of the vegetation mantle is largely a function of precipitation. According to the data provided by Langbein and Schumm (1958) (fig. 4.1), these two factors combine to produce maximum sediment yield from drainage basins in areas with an effective precipitation of about 350–400 mm. Outside this range (grass-covered terrain), sediment yield decreases either due to the thicker forest cover in wetter areas or the smaller amount of overland flow in more arid areas.

The vegetation cloak on the ground surface exerts an influence on

65

denudation in other ways besides its control over surface erosion processes. And, indeed, particularly in relation to weathering, a well-developed vegetal cover may accelerate the rate of landscape destruction. One obvious mechanism is the physical breakdown of rock masses through root penetration into joints and bedding planes. More subtle is the effect on the chemical disintegration of rocks through the influence of the vegetal cover on the chemical quality of subsurface water. It is, however, beyond the purpose of this chapter to pursue these topics here. Some reference is made to them in subsequent chapters, and for more detailed treatment the reader is referred to Oostig (1958), Eyre (1963) and Greig-Smith (1964) and other biogeographic texts. The

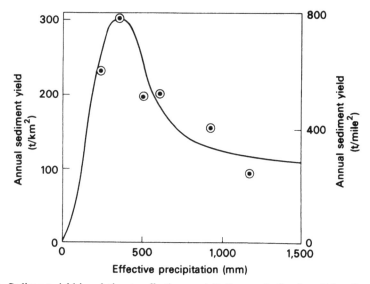

Fig. 4.1. Sediment yield in relation to effective precipitation: an indication of the effect of vegetation (after Langbein and Schumm, 1958).

rest of this chapter is devoted to the other aspect of resistance offered by the ground surface to denudational processes, that is, the intrinsic shear strength of hillside material.

SHEAR STRENGTH

The strength mustered by solid material against a stress which is acting to produce internal deformation, or shear, may be derived from a number of components. One of these is the *plane friction* produced when one grain attempts to slide past another; a second element is the *interlocking* of the various particles that constitute the material. These two components are often grouped together as the *internal friction* of the material; the overall frictional characteristics of a mineral assemblage are usually expressed by the parameter ϕ, the *angle of internal friction* (or *angle of shearing resistance*), or, alternatively, by the tangent of this angle, as the coefficient of internal friction. The total force developed by this frictional strength is the product of the internal friction coefficient of the material and the force that acts to push the particles together. This force is termed the *effective normal force*, or, as a stress, *the effective normal stress*. In addition to this frictional strength, there is a force which acts to pull tiny particles together, making them stick or cohere, and, appropriately, this source of strength is called *cohesion*.

It is fairly easy to appreciate the breakdown of shear strength into these four components. It is not so easy to picture the *exact manner* in which they combine. At this stage, it is sufficient merely to state that

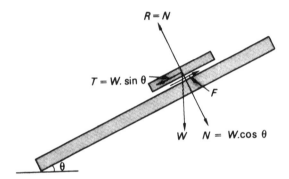

Fig. 4.2. Demonstration of angle(s) of plane friction.

the most popular, and the simplest, interpretation of shear strength is still that provided long ago by Coulomb (1776):

$$s = c + \sigma . \tan \phi \qquad (4.1)$$

where s, in units of stress, such as kgf/cm^2, is the shear strength, c (also in units of stress) is cohesion, σ is the normal stress and ϕ is the angle of internal friction. In the discussion of the individual components of shear strength below, some mention will be made of problems in the Coulomb interpretation.

Plane friction

Imagine the situation (fig. 4.2) in which two dry slabs of sandstone, with smooth faces, rest in contact, and the lower slab is tilted until the upper slab just begins to move. The forces acting on the upper slab are depicted in the diagram. The weight of the upper slab is subdivided into $W.\sin\theta$ acting along the slope and $W.\cos\theta$ acting normal to the slope. The latter force is counteracted by a force R acting in an opposite direction and equal to it in magnitude. The shear force (T) is balanced by the friction (F) acting upslope between the two slabs. The shear strength, in this case the friction between the two slabs, depends upon the coefficient of (plane) friction μ and the force acting to press the two slabs together. In general terms we have:

$$F \leqslant N.\mu; \tag{4.2}$$

the two sides of the expression are equal at the time that the upper slab is just about to slide. At that time, the shear force and the shear strength are, also, just balanced, and we have:

$$T = F, \tag{4.3}$$
$$W.\sin\theta = N.\mu,$$
$$W.\sin\theta = W.\cos\theta.\mu,$$
giving
$$\tan\theta = \mu. \tag{4.4}$$

Tilting the two slabs until the upper one moves by sliding is thus one way in which to determine the coefficient of friction of a surface. Strictly speaking, this coefficient refers only to the *starting* or *static* friction; once sliding has begun, the friction decreases slightly to a new value determined by the coefficient of *sliding, dynamic* or *kinetic* friction. This could be demonstrated in the above experiment by slowly decreasing the tilt of the slabs beneath the initial angle. The upper slab would at first continue to slide and, indeed, accelerate; eventually, at a lower angle, acceleration would cease and the slab would slide at a constant velocity. At this time, assuming no air resistance, the shear force and the friction force are again just balanced, and application of Newton's Second Law of Motion yields, once again, equation 4.4. Since the angle θ is now lower than at the start of sliding, it follows that the angle of sliding friction is lower than the angle of static friction.

The difference between the static and sliding coefficients of friction may be illustrated in a more practical way. Envisage the situation (fig. 4.3) in which one slab of sandstone rests upon another, on a horizontal surface, and the lower slab is pushed continuously forward. The friction between the two slabs must result in the upper slab being pushed forward as well. Now suppose that the upper slab is in contact with a proving ring; as it moves forward it must compress the proving ring and thus produce a force which pushes back on the upper slab. Initially the friction force will exceed the force in the proving ring, and the upper slab will move with the lower one. As the compression of the proving ring increases, however, the force within the proving ring also increases, until eventually it equals the friction force between the two

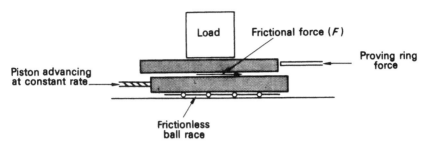

Fig. 4.3. Apparatus to determine angle(s) of plane friction.

slabs. At this point there is no further movement forward of the upper slab despite the continuing movement forward of the lower one. The force in the proving ring is a measure, therefore, of the friction force between the slabs. In turn, the friction force is given by the product of μ and the effective normal force (weight of the load and upper slab). It is thus a fairly simple matter to compute the coefficient of static friction.

Once displacement of the lower slab relative to the upper slab, that is, sliding, occurs, the upper slab moves fractionally away from the proving ring, and, then, apart from minor fluctuations, remains stationary, despite the continuing movement of the lower sandstone slab. The slight movement of the upper slab away from the proving ring indicates that once sliding begins there is a slight decrease in the friction force. The excess of the proving ring force over the new friction force results in the slight backward movement of the upper slab until, with the reduced compression of the proving ring, the force within the proving ring is reduced to precisely that value of the new friction force.

A plot of the proving ring force, obtained from a dial gauge indicat-

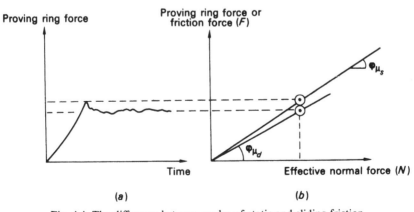

Fig. 4.4. The difference between angles of static and sliding friction.

ing the compression of the proving ring during the course of the test, would follow the pattern indicated in fig. 4.4*a*, and this may be used to give the friction force. Since the total friction force may be determined from the effective normal force and the appropriate coefficient of friction, the latter is easily determined, for both the starting and sliding conditions, as shown in fig. 4.4*b*. It should be emphasized that the actual friction coefficients between two surfaces depend considerably on the smoothness of the surfaces, the presence or absence of moisture on the contact surfaces, and many other factors. These have been discussed in some detail by Horn and Deere (1962). In subsequent discussions, unqualified reference to ϕ_μ should be taken as indicating the dynamic rather than the static coefficient.

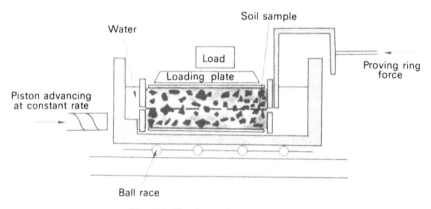

Fig. 4.5. The direct-shear apparatus.

In the outline above the term *force* rather than *stress* has been employed. The argument is in no way changed by replacing the former term by the latter. A stress is simply the force per unit area; in the previous examples, the shear stress and the effective normal stress would be obtained from the comparable forces by dividing by the area of contact between the two slabs. In fig. 4.4*b*, the two axes could have been expressed in terms of stresses instead of forces without affecting the angles of true friction.

Interlocking

Tests similar to those just described for the determination of the true friction coefficients of flat mineral surfaces could also be employed to measure the shear strength of an assemblage of mineral particles. The apparatus in which these tests would be made is illustrated in fig. 4.5; it is termed direct-shear apparatus and is widely used by civil engineers.

The difference between this type of test and the one depicted in fig. 4.3 is that, in the present situation, the resistance provided against shearing is composed of interlocking among the mineral particles in addition to plane friction along mineral faces. In order that the upper box and its contents may slide relative to the lower frame of the shear apparatus and its particles, it is necessary that some particles move upwards over others. The amount of work necessary to perform this will increase with the magnitude of the load that is pressing the upper and lower parts of the sample together. Interlocking strength, as well as friction, thus increases with the effective normal stress.

Some typical test results are shown in fig. 4.6a and b. The development of a peak strength and then a decrease to an ultimate residual

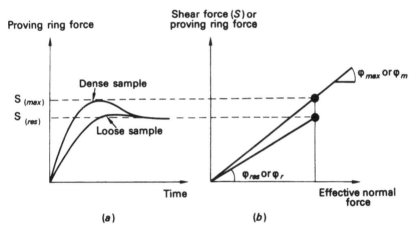

Fig. 4.6. Typical direct-shear test results: (a) change in strength of one sample during a single test; (b) summary of many tests at different effective normal forces.

value appears to resemble the replacement of a peak static friction by a lower sliding friction in the case of the tests shown in fig. 4.3. The analogy is, however, misleading; by the time the peak shear strength of a mineral assemblage has been attained it is very probable that true inter-particle friction has already dropped to the dynamic value. The relationships between ϕ_m, ϕ_r (as defined in fig. 4.6(b)) and ϕ_μ are still only partially understood by engineers; many conflicting opinions still exist. The discussion below is based primarily on the work of Rowe (1962, 1963; Rowe, Barden and Lee, 1964). The problem has also been approached by Bishop and Eldin (1953), Newland and Allely (1957), Roscoe, Schofield and Wroth (1958) and Poorooshasb and Roscoe (1961). At the outset it should be emphasized that both ϕ_m and ϕ_r are greater than ϕ_μ for the same minerals.

The reason why ϕ_r is greater than ϕ_μ appears to be due to the extra energy that must be overcome when interlocked particles move past each other. In the simplest case, consider one spherical particle moving around the circumference of another particle for a distance of half the periphery; contrast this with the same particle moving over the straight length of the other particle's diameter. The energy loss in the former situation is now increased by a factor $\frac{1}{2}\pi$ relative to the plane friction case, and, as shown by Caquot (1934),

$$\tan \phi_r = \tfrac{1}{2}\pi \tan \phi_\mu = 1.57 \tan \phi_\mu. \qquad (4.5)$$

Data provided by Rowe (1962) tend to support this interpretation. It now appears, however, that the correcting factor may vary anywhere

TABLE 4.1. *The residual shear strength of different assemblages of minerals* (from Kenney, 1967b)

Mineral	Grain size (% < 2μm)	ϕ_r°	tan ϕ_r
Quartz	0	35·0	0·69
	100	35·0	0·70
Feldspars (albite and	0	35·0	0·69
microcline)	100	35·0	0·69
Calcite	0	35·0	0·69
Muscovite	0	20·0	0·36
	100	24·5	0·45
Na-hydrous mica	100	16–26	0·29–0·49
K-hydrous mica	100	21–25	0·39–0·46
Ca-hydrous mica	100	24–25	0·44–0·47
Na-muscovite	100	17–19	0·31–0·34
Kaolinite	72	15·0	0·27
Ca-montmorillonite	100	10·0	0·17–0·18
Na-montmorillonite	100	4–10	0·07–0·18

between 1·0 and 1·57 depending on the angularity of the particles in the material. Since ϕ_μ varies with mineralogy, it is not surprising that ϕ_r has also been shown to depend to a large extent on mineral composition. Data presented by Kenney (1967b) are shown in table 4.1.

By the time a sample has been sheared to its ultimate or residual strength, inter-particle movement has become completely random and there is no preferred upward motion of the particles contributing to volume expansion. Prior to this condition, it is common for volume expansion to accompany the shearing process; this is especially true for densely packed samples. This expansion of the sample involves particles moving upwards against the normal stress and, inevitably, extra energy is lost. Using elementary mechanics, Rowe, Barden and Lee (1964) have shown that this would lead to the following increase in strength:

$$\phi_m = \phi_r + \theta \tag{4.6}$$

where θ represents the rate at which the sample dilates during the shear process. Since this rate, expressed in terms of distance of shear rather than time, will depend on the angularity of the particles and the density of packing, it should be possible to relate θ to the angle of interlock of the particles. This task, in fact, was previously attempted by Newland

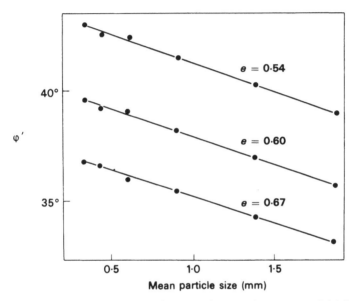

Fig. 4.7. Effect of void ratio on angle of internal friction of sandy material (after Kirkpatrick, 1965).

and Allely (1957), and more recently applied by Patton (1966) to interlocking joint surfaces in rock masses. Since the angle of interlock is strongly dependent upon the density of packing of a particulate mass, it follows that the development of a peak in shear strength, above the ultimate value, should also be affected by the initial sample density. This is borne out by test data. Fig. 4.6*a* contrasts the behaviour of a dense and a loose sample; fig. 4.7 shows the importance of density of packing, or conversely the void ratio (*e*), in controlling the internal friction developed by a sample.

At this stage it is perhaps appropriate to attempt to show how the *angle of repose* of particulate material relates to these other strength

73

parameters. The angle of repose (fig. 4.8*a*) is usually defined as that angle at which a cone develops when grains are poured from a low height onto a flat surface. It seems likely, as suggested by Skempton (1945), that the angle of repose of sand grains approximates to the angle of internal friction of the same material in the loosest state. It should thus be similar to ϕ_r as defined previously. In a rather controversial paper, Metcalf (1966) attempted to argue against this point of view and certainly the issue is still open to debate. One should note, in any case, that the angle of repose of loose particles is not the maximum angle at which they will stand. As shown in fig. 4.8*b*, it is possible to tilt

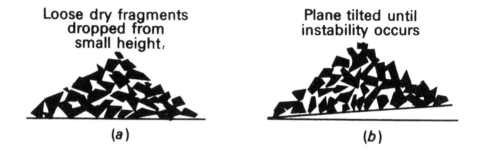

Loose dry fragments dropped from small height,

Plane tilted until instability occurs

(a)   *(b)*

Fig. 4.8. Angle of (*a*) repose and (*b*) maximum slope for loose material.

the platform on which the cone stands, up to as much as 10°, before failure occurs. It seems likely that this *angle of maximum slope* is steeper than the angle of repose due to its dependence on static, rather than dynamic, friction, although more work is needed to substantiate this point.

Effective normal stress

It has been shown in the two previous sections that the amount of both frictional and interlocking strength mustered by an assemblage of mineral particles depends on the stress that acts to push the particles together. At this stage, it is probably advisable to enlarge upon the role of this stress. Initially this will be approached with reference to the direct shear test illustrated in fig. 4.5. It will then be expanded to natural situations involving hillside slopes.

The normal stress acting on the shear surface in fig. 4.5 is defined as the force acting normal to the shear surface per unit area of that surface. Strictly speaking this should be termed the *total* normal stress. This stress is absorbed by the lower sample in two different ways. Some of it is taken up by solid-to-solid contact across the shear zone. Indeed, all of it would be absorbed by this mechanism if solid-to-solid contact existed throughout the shear surface. In the situation in which discrete particles are packed together, as in fig. 4.5, only part of the shear surface exists as solid contact between particles; the remainder is pore space that is occupied by air or water. Naturally there is pressure in this pore space. This pressure must, to some extent, support the total normal stress, just as the solid-to-solid contact supports it. The total normal stress σ is thus approximately given by:

$$\sigma = \sigma' + u \tag{4.7}$$

where σ' is the effective normal stress (the inter-particle force per unit area of the shear surface) and u is the pore pressure. In the development of frictional resistance, it is clearly the *effective* normal stress, and not the *total stress*, that is important. In situations in which the effective normal stress is employed in equation 4.1, it is convention to symbolize cohesion and internal friction by c' and ϕ' respectively. The reason why equation 4.7 is only *approximately* true has been summarized by Wu (1966, 79–80); the discrepancy is so slight that it is ignored here.

As indicated in equation 4.7, therefore, the pore pressure, through its control over the effective normal stress, is extremely important in determining the amount of internal friction developed. In dry soils the pores are fully occupied by air; the pressure of the air relative to atmospheric is zero. In equation 4.7 u is therefore zero, and the effective and total normal stresses are identical in value. It has already been shown in chapter 3 that, in soils which are only partially saturated with water, the pore pressure is negative relative to atmospheric, and that in soils saturated with freely draining water, the pore pressure exceeds the atmospheric value. The effect on the *effective* normal stress is given by substituting these statements into equation 4.7: the total stress is *supplemented* by the capillary suction in the case of partial saturation, but *decreased* by the positive pressures when pores are fully saturated. The physical significance of this is easy to appreciate. Capillary suction acts to pull the particles together and thus adds to the total stress in building up the grain-to-grain contact force; positive pore pressure tends to act against the total stress and must, therefore, reduce the solid-to-solid contact stress. Skempton (1960) provides a detailed treatment of the role of pore pressures in the determination of shear strength.

In turning to hillside slopes, it follows that the resistance against shear along any surface depends substantially on the pore pressures along the surface. Some understanding of the distribution of pore pressures in hillmasses is therefore important in any slope stability study.

The distinction between conditions above and below the main ground water table in a hillmass has already been outlined in chapter 3. It is clearly naive, however, to think in terms of only one underground water system. In times of prolonged rainstorms, temporary ground water systems, perched above the real water table, may occur. This is particularly the case when relatively permeable soil overlies a more impermeable bedrock. In such situations, at least during very prolonged rainstorms, percolating water is impeded at the soil-rock interface and saturated throughflow may occur. Pore pressures within the soil mantle will thus become positive, shear strength is temporarily decreased, and a landslide may occur. A number of research workers have emphasized the importance of this hydraulic discontinuity between the soil mantle and the underlying rock (Ruxton, 1958; Olmstead and Hely, 1962; Hewlett and Hibbert, 1967), but it is only recently (Carson and Petley, 1970) that it has been used to explain landslides.

In unconsolidated materials, with little hydraulic discontinuity between the parent material and the overlying residual mantle, it is unlikely that saturated throughflow occurs. Moreover, if the hillslopes are steep, it is also unlikely that the true ground water table ever rises to the ground surface. Pore pressures are thus unlikely to exceed zero at any time, and slopes should be stable at higher angles. Sandy slopes, for instance, may dry out completely in the surface layers and the pore pressure become zero. Soils with an appreciable amount of silt-clay particles may, on the other hand, be able to retain some moisture permanently in the pores; such soils should possess extra strength from this *capillary cohesion* or *capillary suction* and may be able to stand at quite high angles even though the intrinsic shear strength of the soil is low. Schumm (1956*b*) has suggested this in the case of certain badland slopes in South Dakota.

Capillary suction is not only important as a component of the strength that a soil mantle can muster against landslides, but also in the resistance that a soil surface offers against the stresses of rainsplash and overland flow. Sandy soils tend to be more easily erodible than clay soils. This is partly because they possess less genuine cohesion, as will be shown below, and partly because the large pore spaces in sandy soils prevent the development of much capillary suction. Irrespective of soil type, it is probably true that soil erosion is greatest when the soil is in its driest state, since, at that time, it usually possesses least capillary suction. This point will be discussed more fully in chapter 8.

Cohesion in intact rock

Suppose that a number of direct shear tests are performed on a sample of

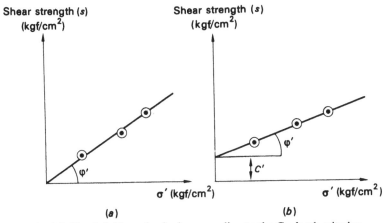

Fig. 4.9. The definition of cohesion according to the Coulomb criterion.

sand particles. The effective normal stress is made to differ between the tests by placing different loads on the plate on top of the sample. The plot of shear strength (s), in units of stress, against the effective normal stress (σ') is, at least for a very small range in σ', a straight line passing through the origin of the two axes. When the axes are scaled in the same units (fig. 4.9a) the angle which the line makes with the σ'-axis is the angle of internal friction. The fact that the line passes through the origin indicates that, when there is no effective normal stress acting on the sample, the shear strength is zero.

Similar tests on an intact rock specimen (fig. 4.9b) would also produce a straight line plot, but this time the line would pass the s-axis

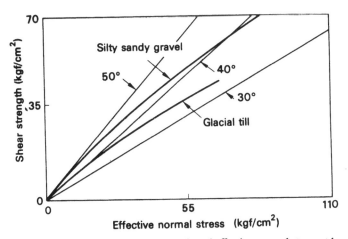

Fig. 4.10. The relationship between shear strength and effective normal stress at large stress values (after Bishop, 1966).

77

above the origin. This indicates that the sample would possess some shear strength even when $\sigma' = 0$. This strength is called *cohesion*. The cohesion in rocks is due to the cementing and fusion of the mineral particles. Actually, it is interesting to observe (fig. 4.10) that, as fragmented rocky material is subjected to increasingly large normal stresses, the shear strength characteristics of the material change from those of a cohesionless mass towards those of an intact rock specimen. As shown in the diagram, the ϕ' value gradually decreases and, correspondingly, a cohesion intercept develops. The reason for this is quite simple. Under low normal stresses, the mineral particles in the shear zone experience little difficulty in moving up and over each other (fig. 4.11a), and interlocking strength supplements plane friction between particles. At very high normal stresses (fig. 4.11b), this is not possible; upward move-

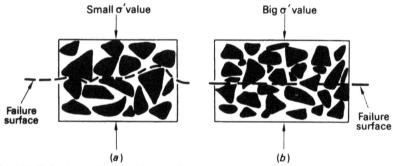

Fig. 4.11. Behaviour of interlocking particles during shear at (*a*) small and (*b*) high normal stress values.

ment is constricted by the large normal load and particles themselves tend to shear. Since each small particle behaves as an intact rock mass, with a definite cohesive strength, the particulate mass as a whole develops an apparent cohesion. At the same time the loss of interlocking strength results in a decrease in ϕ' towards the value of the plane friction coefficient. In this way the pattern shown in fig. 4.10 develops.

The shear strength characteristics of some intact rock specimens are provided in table 4.2. The ϕ value clearly reflects the size of the individual rock constituents; as will be evident later this is also true of unconsolidated debris, sands having high ϕ values and clays lower ones. The amount of cohesion, on the other hand, is determined more by the history of consolidation of the rock. It should be emphasized, however, that the shear strength of many rocks depends critically on the orientation of the shear surface relative to the direction of bedding or cleavage. Deklotz, Brown and Stemler (1966) have shown, for instance, that the shear strength of blocks of schistose gneiss varied by up to 50 per cent depending on the orientation of the sample. Similar observations were made by Price (1958) during tests on Snowdon sandstone: in tests with the plane of failure at an angle of about 65° to the bedding, the values of ϕ and c were, respectively, 40° and 400 kgf/cm² (5,700 p.s.i.); in tests with the shear surface at an angle of 17° the same parameters decreased to 35° and 343 kgf/cm² (4,900 p.s.i.).

In attempting to estimate the shear strength of a large rock mass there is a further problem that the strength of a small intact rock specimen may not be representative of the whole mass. This is not simply due to heterogeneity in the composition of the mass, but, in addition, there is the reduction of the strength of the rock mass as a whole by the presence of well-developed joints and bedding planes. These areas are likely to represent the weakest part of a rock mass. Accordingly there is a great deal of research (Jaeger, 1959; Ripley and Lee, 1961; Krsmanovic and Langor, 1964; Mencl, 1965; Patton, 1966; de Freitas and Knill, 1967;

TABLE 4.2. *Shear strength of intact rock samples*

Material	c kgf/cm²	ϕ_m°	Test	Source
Chalk	9	21	T 0–70	Meigh and Early, 1957
Chislet siltstone	210	29	T 0–350	Price, 1958
Pennant sandstone	350	44	T 0–350	Price, 1958
Sandstone	42–420	48–50	T 210–2100	Hendron, 1968
Limestone	35–350	37–58	T 210–2100	Hendron, 1968
Granite	97–406	51–58	T 700–2800	Hendron, 1968
Basalt–sandstone contact	10	59	D 0–20	Ruiz and Camargo, 1966

Test: direct shear (D) or triaxial (T); and range of normal stress or cell pressure in kgf/cm².

Byerlee, 1968; Ladanyi and Archambault, 1969) being directed to the shear strength of discontinuities in rock masses.

The science of rock mechanics is a relatively new one and many established ideas may have to be changed in the future. The interested reader may find the texts by Coates (1967), Farmer (1968), Stagg and Zienkiewicz (1968) and, particularly, Jaeger and Cook (1969) useful as introductions to the subject of rock strength. An excellent bibliography is provided in the appropriate chapter of Yatsu's (1966) recent book. Other sources of reference are the journal *Rock Mechanics and Engineering Geology* and the various proceedings of the congresses of the International Society of Rock Mechanics.

Cohesion in clays

Clays, unlike sandy soils, also possess cohesion, although in much

79

smaller amounts than intact specimens of solid rock. The existence of cohesion in clay soils is due to physico-chemical forces acting between the particles (Lambe, 1960), and, as might be expected, increases as the area of contact between the mineral particles increases. This is, primarily, the reason why cohesion is restricted to clay-size particles; the specific surface of soils containing larger particles is usually too small to produce any physico-chemical force. In addition to the fraction of clay-*size*

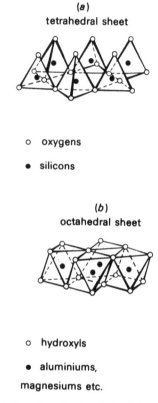

(a)

tetrahedral sheet

o   **oxygens**

•   **silicons**

(b)

octahedral sheet

o   **hydroxyls**

•   **aluminiums,**

magnesiums etc.

Fig. 4.12. The basic units of clay minerals (from 'Clay Mineralogy' by R. E. Grim, © 1953, McGraw-Hill Book Co. Used with permission of McGraw-Hill Book Co).

particles in a soil, the *type* of mineral is also very important. For this reason it is useful at this stage to provide some discussion of the more important aspects of clay mineralogy in the context of cohesive strength. The interested reader is referred to Grim (1953) for a more comprehensive treatment of clay mineralogy.

The term clay minerals usually refers to hydrated silicates of aluminium, iron and magnesium. Most of these are composed of two basic sheets. One is a tetrahedral sheet of silica (fig. 4.12a) with an oxygen atom at the apices of the tetrahedrons and a silicon atom enclosed within; the other is an octahedral lattice (fig. 4.12*b*) of oxygen atoms or hydroxyls (OH compounds) with an atom of aluminum, magnesium or some other element enclosed within. The three main types of clay mineral are compounds of these two main lattice structures. The *kaolinite* group comprises alternations of the silica and octahedral sheets; in the *illite* and *montmorillonite* groups the octahedral layers are individually sandwiched between two silica sheets.

The true nature of the physico-chemical forces that are responsible for cohesion among clay particles is still not fully understood. It was suggested by Lambe (1960) that van der Waals' forces are probably the main source of strength. These are forces that arise between two

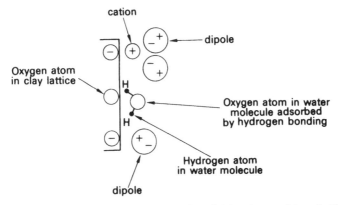

Fig. 4.13. Adsorption of water by a clay mineral (after Low and Lovell, 1959).

electrically neutral molecules which have their positive and negative charges distributed unsymmetrically. These forces are particularly important in orienting water molecules near to the surface of clay particles; the resultant thin layer of *adsorbed* water is believed to play a major role in developing cohesive strength in clays. Three mechanisms of water adsorption by a clay particle are shown in fig. 4.13. The simplest is the straightforward attraction of the positive pole of a water molecule to the negatively charged surface of the particle. Slightly more complicated is the attraction of a cation, such as calcium, magnesium, hydrogen or sodium, to the clay particle and, in turn, the attraction of dipolar water molecules to the cation as hydrated water. Lastly, a simple case of hydrogen bonding, in which a hydrogen atom in the water molecule is shared between oxygen atoms in the clay and in the water molecule itself, is also shown. According to Skempton and Northey (1952), one of the most important contributions made by ion adsorption by clay minerals is the development of a structural pattern among the soil

81

particles. This will be referred to in chapter 7 in a discussion of Bjerrum's (1954) ideas on earthflows in Norwegian clays.

The amount of adsorbed water held in a soil depends upon the type of cations adsorbed by the clay, the mineralogy of the clay particles and the fraction of these particles that occur in the clay-size range. The combined effect of these influences is often shown in empirical parameters known as the *Atterberg Limits* of a soil. The two most common limits are the Plastic Limit and the Liquid Limit and together they determine the range of moisture contents through which the soil behaves plastically. At moisture contents lower than the Plastic Limit the soil behaves like a brittle solid; at moisture contents higher than the Liquid Limit the same soil, in a fully disturbed condition, behaves more like a liquid. The difference between the Plastic Limit and the Liquid Limit is the *plasticity index* of the soil. As shown in fig. 4.14, for any given soil, there is a close, straight-

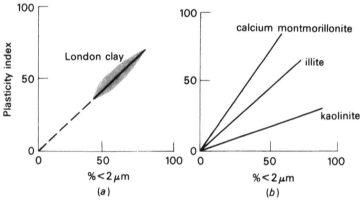

Fig. 4.14. The relationship between plasticity index and clay content: the activity ratio (based on data from Skempton, 1953).

line, relationship between Plasticity Index (I_p) and the clay-size fraction. The steepness of the line is a measure of the activity of the surface forces on the clay particles, and the tangent of the slope is called the *activity ratio* of the soil. The activity ratio varies considerably with the mineralogy and cation composition of the soil. Interesting and more detailed discussions of the relationship between clay mineralogy and the Atterberg Limits are provided by Grim (1962), and Yong and Warkentin (1966).

Along with the chemical make-up of a clay soil, the other main determinant of the amount of cohesive strength is the preconsolidation pressure to which the clay has been subjected. Strictly speaking, cohesion is only found in *over-consolidated* clays; it is not present in *normally-consolidated* clays. An over-consolidated clay is one which, at some time in the past, experienced pressures greater than those acting upon it today. A normally-consolidated clay is one that has not been subjected

to a pressure greater than that exerted upon it by the present overburden.

The significance of this distinction between the two types of clays is
brought out in fig. 4.15. This shows the change in moisture content and

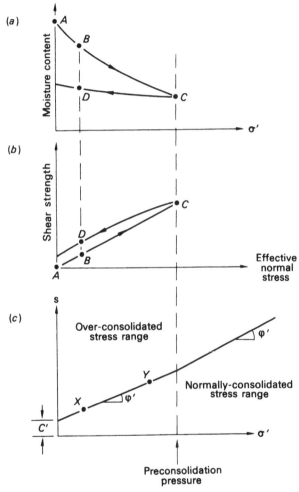

Fig. 4.15. The distinction between normally-consolidated and over-consolidated clay.

shear strength of a marine clay during part of its geologic history. The
change from point *A* to point *C* represents increasing load on the clay
during deposition. Associated with this build-up of load there is an
expulsion of some of the pore water, greater packing and, as a result,
greater strength. The change from *C* to *D* represents a reduction in the

83

effective normal pressure due to erosion of some of the surface clay after exposure. This reduction in pressure produces an increase in pore water content, an increase in void ratio and, as a consequence, a decrease in strength. Note that less pore water is taken up than, for the comparable effective normal pressure, was expelled previously. It follows that the decrease in shear strength in moving from C to D is not as great as the increase in strength in moving from B to C. Extension of the CD path to the shear strength axis shows that some strength still remains even when the effective normal pressure is zero; this is cohesion. Any tests that are made on samples from this clay, tested at effective normal stress values in the vicinity of D, will thus show a cohesion intercept. Now contrast this with a clay mass which is consolidated to the point C and undergoes no subsequent unloading. The shear strength envelope is given by the line AC in fig. 4.15b. Samples taken from this mass and tested at normal stresses greater than those at C would show a straight line passing through the origin indicating no cohesion.

In as much as almost all clays have, to some extent, been subjected to pressures in the past which were greater than the clay currently experiences, there are probably few clays which are truly normally-consolidated. Indeed, rather than speak of two different types of clay, it is probably better to envisage, for all clays, a particular preconsolidation pressure which divides the strength envelope of that clay into two ranges: an over-consolidated stress range and a normally-consolidated stress range. Cohesion exists in the former, but not in the latter stress range. Once again it is clear that the strength envelope is not a simple straight line, and the shear strength parameters depend on the stress range. Finally, it should be noted that the entire discussion above relates solely to the peak strength conditions. In any clay that is tested to its ultimate strength, the observations made by Skempton (1964) suggest that, not only does the angle of internal friction decrease from a peak value, but the cohesion disappears completely. This is a point anticipated long ago by Moseley (1855) and more recently by Taylor (1948).

The interpretation of the shear strength of clays in the over-consolidated stress range, by the classic Coulomb equation 4.1 creates a number of problems. The most serious is that it describes the cohesion as a unique property of a given clay; by definition the Coulomb c (or c') value is independent of the normal stress. This idea is rather difficult to accept. It seems very unlikely that, in fig. 4.15c, *all* the increase in strength between points X and Y is due to increased frictional strength. The magnitude of the interparticle electrical forces that constitute the phenomenon of cohesion would be expected to increase as the particles become closer together. Since the void ratio of the sample decreases as the normal stress increases, it seems probable that the cohesive force increases also. The increase in strength from X to Y is probably due to an increase in *both* cohesion and frictional resistance. This argument has been strongly put forward by Hvorslev (1937, 1960) who proposes that equation 4.1 be replaced by:

$$s = c_e + \sigma \cdot \tan \phi_e \qquad (4.8)$$

where ϕ_e is defined as the *true* angle of internal friction and c_e, *which varies with the void ratio*, is the *true* cohesion. If it could be assumed that the relationship between void ratio and normal stress were linear, this could be written:

$$s = \sigma\,(\bar{c} + \tan \phi_e) \qquad (4.9)$$

where $\bar{c}$ would be a measure of cohesion per unit normal stress; since the relationship between void ratio and normal pressure is not linear this is, in fact, a purely hypothetical point.

Notwithstanding the logic of the Hvorslev approach, it has been pointed out by Skempton and Bishop (1960) and others that the actual measurement of c_e and ϕ_e is a matter of some difficulty; moreover, any stability analysis based on these properties is likely to be rather laborious. Throughout this book shear strength is interpreted in terms of the empirical Coulomb parameters, c' and ϕ', sometimes termed the *apparent* strength parameters, rather than in terms of Hvorslev's components. Ideas on the real meaning of clay strength are still emerging, and a particularly good summary of some of the more recent contributions has been provided by Kenney (1967a).

Much of the previous discussion has been based on work by civil engineers; these workers are particularly interested in cohesion *at depth* in a soil mantle, in connection with slope stability. The cohesive strength of clay soils has also been thoroughly studied, from a different point of view, by agricultural engineers concerned with the resistance offered by the soil *surface* to the shear forces of surface water. In chapter 3 it was pointed out that the tractive force necessary to initiate particle movement by flowing water on slopes, assuming turbulent conditions, is a direct function of particle size; this does not mean, however, that, under the same forces, sandy soils are less erodible than silt–clay soils. In actuality, the converse is generally the case. Clay particles, unlike sand grains, do not occur as detached minerals on the ground surface; they cohere as aggregates to the main soil mass. In order that fluvial erosion can occur on a clay slope, the force of flowing water or raindrop impact must not only overcome frictional resistance among the particles, but also break down the cohesive bonds. It is not surprising therefore that agricultural engineers such as Middleton (1930), Musgrave (1947), Anderson (1951, 1954), and Woodburn and

Kozachyn (1956), have devoted so much effort to the study of cohesion at the soil surface. Some of this information has been summarized by Baver (1956); and a more recent study on the erodibility of soils has been provided by Bryan (1969).

The measurement of shear strength

This is not the place to enter into a detailed description of different types of shear strength tests. Manuals provided by Lambe (1951) and Akroyd (1957) do this adequately. Some understanding of the problems involved in the determination of shear strength is, however, necessary.

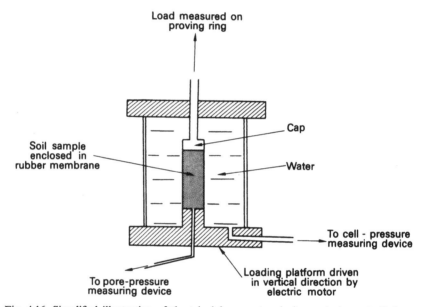

Fig. 4.16. Simplified illustration of the triaxial apparatus during a strain-controlled test.

The standard techniques for evaluating c and ϕ are the *direct-shear* and the *triaxial compression* tests. These are primarily laboratory techniques, although some engineers (e.g. Hutchinson and Rolfsen, 1962) have used the direct shear test in the field on *in situ* samples. The direct shear test has been briefly described above (fig. 4.5) and, here, attention will be focused mostly on the compression tests.

A simplified picture of the triaxial apparatus is given in fig. 4.16. A cylindrical sample is enclosed in a rubber membrane and subjected to a specified lateral water pressure. The actual test may be approached in one of two ways. In the *strain-controlled* method, the approach is rather similar to the testing procedure in most direct-shear tests. The base of the sample is connected to a moving platform, and the upper end to a proving ring. Compression of the sample, through the upward movement of the platform, increases the proving ring reading. Failure of the sample may take place through either shear displacement or bulging (fig. 4.17); at the time of failure the shear strength of the sample along the failure zone is computed from the lateral pressure and the stress in the proving ring. The alternative method is termed the *stress-controlled* approach: the platform under the base of the sample is fixed and the sample is subjected to an increasing vertical stress σ_1. Eventually the stress becomes so large that deformation of the sample, leading to failure, occurs; whereas in the strain-controlled test, the shear strength

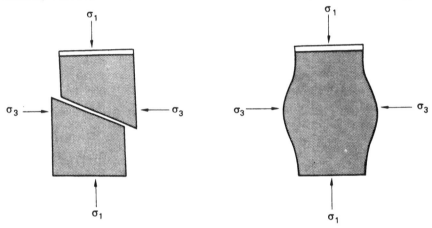

Fig. 4.17. Typical specimens after failure in a triaxial test.

is given by the maximum proving ring pressure, in stress-controlled tests, the shear strength is given directly by the stress acting at the time of failure.

One of the difficulties in the stress-controlled test is the actual definition of the instant of failure. This would be simple if the sample behaved as a perfect plastic, that is, if there was *no* deformation at all until a critical stress had been attained, and then sudden *rapid* deformation at this stress. In practice, most materials, especially clays, do not behave in this way. Initially there will be no deformation until a *yield stress* is reached, and then strain (deformation) will occur at a steady rate which increases with the σ_1 stress, until eventually the sample deforms at a very rapid rate, constituting failure. The term shear strength is usually reserved for this final stress. The yielding prior to failure is commonly referred to as *creep* and is discussed in chapter 10. Now, while the problem appears to be clearly defined in the stress-controlled test, it is more difficult to appreciate it in the strain-controlled

87

method of testing. It has been noted that, at stresses between the yield stress and the ultimate stress, the rate of shear depends on the magnitude of the σ_1 stress. Conversely, in the strain-controlled test the σ_1 stress (the measured strength) depends on the rate of strain. In the strain-controlled approach it is therefore impossible to define shear strength *except in relation to a particular rate of strain*, and, for this reason, the stress-controlled test is to be preferred. These, and other points, are discussed in more detail by Bishop and Henkel (1962).

TABLE 4.3. *Uniaxial (unconfined) compressive strength of intact rock specimens*

Rock	Strength kgf/cm²	p.s.i.	Source
Chalk (England)	40	575	Meigh and Early, 1957
Sandstone (Cliff House)	175	2,500	Schumm and Chorley, 1966
Sandstone (Gosford)	375	5,360	Weibols, Jaeger and Cook, 1968
Sandstone (Entrada)	392	5,600	Schumm and Chorley, 1966
Sandstone (Darley Dale)	404	5,780	Price, 1960
Sandstone (De Chelly)	505	7,200	Schumm and Chorley, 1966
Sandstone (Pennant)	1,575	22,500	Price, 1960
Shale	1,030	15,000	Bredthauer, 1957
Shale (Witwatersrand)	1,740	24,900	Cook *et al.*, 1966
Granite (Barre)	1,695	24,200	Robertson, 1955
Granite (Charcoal)	1,760	25,100	Cook, 1965
Granite (Westerly)	2,320	33,200	Brace, 1964
Limestone (Chico)	703	10,000	Bredthauer, 1957
Limestone (Solenhofen)	2,280	32,500	Wiebols, Jaeger and Cook, 1968
Limestone (Virginia)	3,360	48,000	Bredthauer, 1957
Gabbro	2,800	40,000	Hendron, 1968

A special version of the axial compression test, used for cohesive materials, especially rock samples, is the *uniaxial* or *unconfined compression* test, in which the cell pressure (σ_3) is zero. Typical values of the unconfined compressive strength for various rock samples are provided in table 4.3. The relationship between uniaxial compressive strength (q_u) and c and ϕ is given by:

$$q_u = 2c \cdot \tan(45 + \phi/2) \tag{4.10}$$

In the direct shear and triaxial apparatus, samples are usually tested in a wet state. Excess pore pressures will be produced by both consolidation under the normal load and the subsequent shearing process. In order to determine c' and ϕ', therefore, it is necessary to know the pore pressure in the shear zone at the time of failure. In the direct shear test there is no facility for measuring pore pressures: tests are therefore usually undertaken slowly. Sufficient time must be allowed for full consolidation of the sample, and shearing should be undertaken at a rate (Gibson and Henkel, 1954) that is sufficiently slow to permit dissipation of excess pore pressures. This method is termed the *consolidated-drained* test. Sometimes, rapid *undrained* tests are undertaken, but only with reference to *total stress analysis*, as against the use of the effective normal stress, as will be demonstrated in chapter 7. The triaxial apparatus contains provision for pore pressure measurements so that the results of both drained and undrained tests may be expressed in terms of effective stresses.

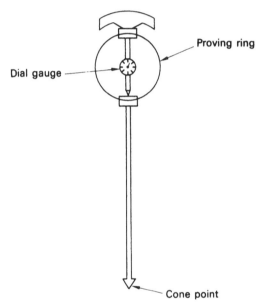

Fig. 4.18. The Vicksburg penetrometer.

Other techniques for determination of shear strength do exist and have been discussed by Skempton and Bishop (1950) and, more recently, by contributors to the *Proceedings of the Research Conference on the Shear Strength of Cohesive Soils* (American Society of Civil Engineers, 1960). A popular instrument among geomorphologists, described by Evans (1950) and used by Chorley (1959, 1964), is the Vicksburg penetrometer. This comprises a long rod tipped with a cone at one end (fig. 4.18) and bounded by a proving ring and handle at the other. The cone is manually driven into the soil and the amount of resistance offered against penetration is reflected in the compression of the proving ring as indicated by a dial gauge. In technical terms, the resistance to penetration that is measured is not shear strength, but *bearing capacity*, although this is strongly correlated with shear strength, and equations do exist (Terzaghi, 1943, chapter 8) linking the two. One problem in using the penetrometer is that there is no method of measuring pore pressures that are generated by the penetration process, so that the strength must be considered as *un*drained, and difficult to interpret in terms of effective stresses. It is also difficult to separate the strength into c and ϕ components using this test. The test should, in fact, be used only when nothing more than a rapid, crude indication of soil strength is needed.

The topic of shear strength is a large one and an entire text could be devoted to the subject. The reader who requires more information should consult those references already listed and, in addition, the relevant chapters in the texts by Terzaghi (1943), Terzaghi and Peck (1948), Taylor (1948), Scott (1963), Wu (1966), Lambe and Whitman (1969) and Lee (1969).

THE CHANGE IN SHEAR STRENGTH WITH WEATHERING

This text departs from the pattern of most of the standard works in geomorphology in its treatment of weathering. The two components of weathering, the transfer of solutes away from a rock mass and the concomitant weakening of the residual material, are discussed separately. The first constitutes a distinct process of debris removal and, as such, it is discussed in chapter 9. The second aspect, the breakdown in rock strength, is discussed here. In this treatment, emphasis is directed, firstly, to the manner in which the shear strength of natural materials is changed over time by weathering, and, secondly, to the rate at which this process of breakdown takes place. There is thus no separate discussion of the individual weathering processes, although some understanding of these is implicit in the two aims. In the context of this chapter it is the mode and rate of rock disintegration, rather than the actual weathering processes, that is important. Outlines of weathering, by process, have been provided by Reiche (1950), Keller (1957, 1963) and Chorley (1969) and others.

The mode of rock disintegration

The pattern of breakdown of a rock mass is linked closely to the general climatic setting, but far more to the character of the rock material itself. The climate may dictate the relative importance of chemical *vis-à-vis* physical processes (Sparks, 1960), but the rock type is probably more important in determining the character of the weathered debris at any stage in the destruction of the material. The jointing character of a rock complex is an especially important feature.

A rock mass with well-developed joints or fractures, spaced not too closely nor too far apart, will nearly always weather initially into a mantle (*talus*) of loose rock material. These fragments are simply smaller scale replicas of the original mineral assemblage; they are controlled in size by the spacing of the fractures in the unweathered rock. Once these latent fractures have been sufficiently enlarged to produce talus, the individual rock fragments are, in turn, broken down further, sometimes by splitting, but more commonly by the decay of weak minerals and the cementing matrix. Ultimately the loose rock is converted completely into a *colluvial* mantle of soil-size particles. Some of these soil grains are of the same overall composition as the original parent material; many, however, are now composed of a single mineral derived directly or in altered form from the bedrock.

Now, although the composition of the initial colluvium, and its subsequent alteration, depends to a large extent on the climatic environment (Jenny, 1941; Mohr and Van Baren, 1954), the early stages of this weathering sequence are probably valid for *jointed rocks* in almost all climates. The actual processes involved may differ among different environments, but the same basic pattern of breakdown still holds. In humid tropical areas spheroidal weathering probably dominates: water percolates through the rock mass along joint planes, removes soluble material and, eventually, produces a mantle of corestones in a matrix of residual soil. Berry and Ruxton (1959), Lumb (1965) and others have noted this in the tropical weathering of granites, but it is generally applicable to most well-jointed rocks in this climate. Talus production in Arctic areas (Rapp, 1960a) is more commonly through frost action thrusting blocks apart along fracture zones, and the resultant fracture blocks are angular rather than rounded, but the history of disintegration is substantially the same.

The weathering sequence of *non-jointed* rocks, or rock masses with very widely spaced joints, is radically different from the situation described above. In a sandstone, for instance, the major process of weathering is simply the removal of the cementing material in solution, leaving behind a mantle of sandy particles as the residual soil. The change from rock to colluvium is thus achieved, in this case, without intermediate stages of rock breakdown. Again, the subsequent weathering of the colluvial material will depend not only on the mineralogy, but also on the climate. Irrespective of this point, however, it is clear that the early stages of rock breakdown in non-jointed rocks are very different from jointed masses.

The breakdown of a rock mass into a soil mantle has attracted less attention among geologists than the more complex alteration of existing colluvial material through chemical processes. In the context of slope development, however, it is the initial phase of the weathering sequence, rather than this final alteration, that is most important. Most hillslopes in areas of consolidated material are mantled, not by pure colluvium, but by debris which, usually, still possesses coarse material. Moreover, in view of the frequency with which jointed rock masses break down through this sequence of talus, *taluvium* (mixed talus and colluvium) into pure colluvium, it is surprising that more attention has not been paid to the concurrent changes in shear strength during weathering.

In the breakdown of a mass of solid rock into an accumulation of talus, a term used in this book synonymously with *scree*, there are two main changes in shear strength. The most marked effect is the loss of cohesion; this is accompanied, usually, by an increase in the internal

TABLE 4.4. *Shear strength of talus (gravel fraction > 0·9)* (data from Marsal, 1967 and (*a*) Carson, 1967*b*)

Material	Median particle size cm	c' kgf/cm²	ϕ'_m	Void ratio (*e*)	σ' range kgf/cm²
Carboniferous limestone (*a*)	2	0	53	0·83	0–0·2
Basalt	1	0	45	0·30	0–10
Granitic-gneiss	5·5	0	36	0·32	0–10
Granitic-gneiss	11	0	31	0·62	0–10
Basalt	1	8	30	0·30	20–50
Granitic-gneiss	5·5	6	22	0·32	30–50
Granitic-gneiss	11	7	18	0·62	30–50

friction. Indeed this change must closely follow the curve given in fig. 4.10 as the effective normal stress is gradually decreased from very high values. Talus accumulations are thus characterized by minimal cohesion and high ϕ values. Peak ϕ values may be very high, as shown in table 4.4, although, for reasons cited earlier, this depends on the density of packing and on the range in σ' values. Values of ϕ_r, assuming them to be close to the angle of repose of such material, are much lower and, for a wide variety of materials, are close to 35°. The shortage of test data on the shear strength of coarse rocky material, even among civil engineers, is due primarily to the size of testing apparatus needed to test this material.

The change in shear strength during the weathering sequence from talus through to colluvium is extremely complex and poorly under-

TABLE 4.5. *Shear strength of taluvial material*

Material	Gravel fraction	ϕ'_m	e	Test	Source
Alluvium	0·7	41–4	0·3	T 3·5–24	Lowe, 1964
River deposit	0·2–0·6	45–8	0·4–0·55	T 0–1·4	Holtz and Gibbs, 1955
Quarry rubble	0·6–0·8	39–44	0·3	T 0–1·4	Holtz and Gibbs, 1955
Silty, sandy gravel		45	0·15	T 0–7	Hall and Gordon, 1963
Glacial till		37		T 0–7	Insley and Hillis, 1965
Slaty-greywacke	0·85	43	0·45–0·7	D	Schultze, 1957
Furnas Dam gravel		43	0·15–0·4		Queiroz, 1964
Exmoor slate	0·57	44	0·5	D 0–0·2	Carson and Petley, 1970
Shale Grit	0·4–0·6	43	0·4–0·5	D 0–0·2	Carson and Petley, 1970
Fluvial, glacial, talus, etc.	0·6–0·8	37		T 0–4	Pellegrino, 1965

e: void ratio
Test: direct shear (D) or triaxial (T); and range in normal stress or cell pressure in kgf/cm².

TABLE 4.6. *Shear strength of residual soils*

Parent material	c' kgf/cm²	ϕ'_m	ϕ'_r	$\% < 2\ \mu m$	e	Test	Source
Sand: Leighton Buzzard	0	33–43		0	0·5–0·7	T	Kirkpatrick, 1965
Molsand: Belgium	0	35–50				T	De Beer, 1965
Slate: Exmoor	0	42	42	0	0·6	D 0·3	Carson and Petley, 1970
Sandstone: Pennines	0	36	36	0	0·7	D 1·5	Carson and Petley, 1970
Granite-gneiss: Brazil	0	42		5		D	Vargas and Pichler, 1957
Granite: Hong Kong	0	36–7		5	0·4–0·6	T	Lumb, 1962
Sandstone: Brazil	0·1–0·5	30–3		6–22		T	Vargas, Silva and Tubio, 1965
Silt	0	32–6		10		D/T	Schultze and Horn, 1965
Mudstone: Shropshire	0·1	25	19	36		D 2·1	Henkel and Skempton, 1954
Gneiss: Brazil	0·2–0·4	26–9		6–42		T	Vargas, Silva and Tubio, 1965
Basalt: Brazil	0·4–0·5	29–30		10–47		T	Vargas, Silva and Tubio, 1965
Clay: London	0·15	20	16	55		T 2·8	Skempton, 1964
Shale: Bearpaw			6	50			Ringheim, 1964

e: void ratio
Test: direct shear (D) or triaxial (T); with maximum normal stress or cell pressure in kgf/cm².

stood at the moment. There is, however, enough evidence to suggest ϕ values for taluvial material in a loose state of packing, as on hillside slopes, commonly approximate to 43–45° for a wide variety of different materials. The consistency of this value (table 4.5) has extremely interesting implications for slope development in jointed rocks (Carson, 1969*b*); this will be developed in chapter 7 dealing with instability in soil masses. The amount of cohesion developed by taluvial material is more variable, depending on the clay content of the colluvial matrix, but in most materials tested by engineers it has been minimal.

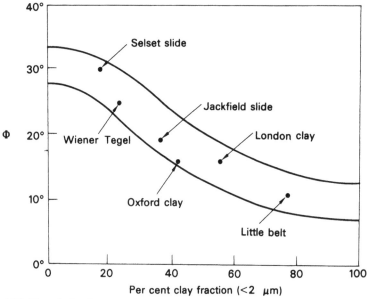

Fig. 4.19. The relation between ϕ and the clay content of soils (after Skempton, 1964).

The shear strength characteristics of pure colluvial material are much more variable. Table 4.6 and fig. 4.19 emphasize the importance of grain size on shear strength: sandy soils have minimal cohesion and large ϕ' values, whereas clay soils tend to gain cohesive strength and lose some of the internal friction. The importance of grain size may, however, be misleading. Data shown previously (table 4.1) indicate that, at least in terms of residual strength, mineralogy is much more important than particle size. The apparent importance of grain size in controlling shear strength in colluvial material may well, therefore, simply reflect the mutual association of both items with mineralogy in natural soils.

The change in shear strength through tables 4.4 to 4.6 probably summarizes the *general* change in the strength of material derived from the weathering of a well-jointed rock. The maximum internal friction occurs at an intermediate stage of weathering with a mantle of taluvial material. Unfortunately there is little known about the *exact way* in which *c′* and *ϕ′* change with time; the impression that is gained from tables 4.4 to 4.6 is that there are three distinct stages of weathering. Some evidence from engineers indeed suggests that the change in shear strength during this weathering sequence is a discontinuous one and not a gradual change. Vucetic (1958), for instance, has undertaken detailed tests to examine the response of the shear strength of clayey-schist material to changes in the *size* of the largest rock fragment. During tests with maximum particle size at 30, 15 and 8 mm, shear strength remained unchanged and *ϕ′* averaged 39° each time. Subsequent tests with maximum rock size at 4 mm showed a sudden drop in strength with *ϕ′* much lower at 25°. Tests by Holtz (1960), in which the *fraction* of gravel in soil-gravel mixtures was gradually varied, probably simulate the weathering sequence of coarse material more accurately. The tests, unfortunately, only examined the change from taluvial to colluvial material, rather than the complete range from 100–0 per cent gravel. None the less his findings are very interesting. A sudden decrease in *ϕ′* occurred at about 20 per cent gravel with little change either above or below this value. More tests are needed to simulate the change in shear strength through the weathering sequence, but, in combination, the tests by Vucetic and Holtz do suggest that the change in strength from talus, through taluvium, to colluvium, is discontinuous rather than gradual.

Once a rock mass has been reduced to an assemblage of soil-size particles, weathering may still produce further alteration of the debris, and changes in strength. This, admittedly, is probably unlikely with sandstones; once quartz particles have been released by removal of cement, little further change probably takes place. In many other rocks, however, mineralogical changes continue long past the stage when the mantle first becomes colluvium. Lumb (1962), for instance, has shown that the shear strength of residual soil derived from the tropical weathering of granite varies greatly with the extent of weathering. The effect of weathering on the mineralogical composition of soils has interested geologists and soil scientists for some time; it is all the more unfortunate, therefore, that the inter-relationship of this with changes in shear strength has been largely ignored.

The effect of weathering on shear strength cannot really be divorced from transport processes which, through selective removal of debris, also alter the character of the residual products. In order to emphasize this point, consider again the weathering sequence on well-jointed rocks that has been described above. This may be valid for humid areas, but it has long been argued by King (1951, 1953) that it is inapplicable in semi-arid areas. In this type of environment the fine soil particles, derived from the weathering of loose rock debris, are washed away almost as soon as they are produced. Hillsides, as a consequence, stay mantled by veneers of talus, and the colluvial products accumulate elsewhere. Under these circumstances there is little chance for taluvial type material to be produced on these slopes. In humid areas, in contrast, the more abundant vegetation cover would aid retention of the soil-size particles and allow a gradual alteration, at least in composition, from talus through taluvium into soil *in situ. Sub*surface transport of solutes and clay particles by moving water in humid areas may, however, produce an analogous separation of coarse and fine material in the *soil*, rather than the slope, profile. The character of the soil mantle, at least on gentle slopes, is rarely uniform with depth; leaching tends to produce a coarse sandy horizon overlying a more cohesive illuviated layer. The corresponding changes in shear strength within the mantle may be expected to alter the long-term stability of such slopes.

The rate of rock disintegration

Our knowledge of the rates at which rocks disintegrate under the action of weathering is more meagre than our understanding of the general pattern of such breakdown. As Kellogg (1941) emphasized, the production of a centimetre of soil (soil-size particles) may take anywhere between ten minutes and ten million years depending upon the type of rock and the bio-climatic environment.

Some indication of the rate of weathering under different settings may be obtained from the thickness of the soil mantle assuming that this is primarily a reflection of different weathering rates. At a worldwide scale, for instance, soils are generally thinnest in the polar areas and thickest in the humid tropics. These observations are usually taken to support the view that weathering proceeds at a much more rapid rate in moist, warm climates. The same argument might also be applied at a local scale. Shale areas usually possess thicker soil mantles than highly metamorphosed rocks, and this is not surprising, since, mechanically, weak shales should offer less resistance to the processes of weathering than metamorphics.

This type of argument is, however, a very dangerous one. The thickness of a soil mantle depends not only on the weathering rate, but also on the duration of weathering. As an example Pigott (1962) argued that the thicker soil cover on the southern part of the Pennine limestone block, as against the northern part, was essentially due to the contrasting glacial history of the two areas. The northern part was subjected to stripping of surface waste during the last ice advance and, as a consequence, weathering of a new mantle started only about 20,000 years ago. The southern part of the limestone is thought to have been south of the limit of the last ice advance, so that weathering of the limestone began at a much earlier date and, thus, the soil mantle is much thicker.

In addition, the approach above overlooks the feedback between weathering and processes of debris transport. The thickness of the soil mantle depends not only on the amount of weathering but also the rate of debris transport. This is especially important on slopes, but is also applicable on level terrain. Most limestone areas, for instance, possess

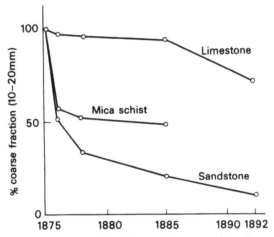

Fig. 4.20. Illustration of the rate of breakdown of large particles due to weathering for different rocks (after Jenny, 1941).

only a thin veneer of waste, but this is probably more a reflection of the strength of subsurface processes of transport than the weakness of the disintegration process. A great deal of limestone is so pure that almost all of it may be removed as solutes by subsurface water and very little may be left as residual material. This is a further reason therefore why an assessment of weathering rates through the thickness of the mantle of soil waste is a dubious procedure.

There is clearly a great need for direct measurement of the rates of weathering of rock under different conditions. An interesting summary of some early attempts to achieve this is given by Jenny (1941) in his discussion of time as a soil-forming factor. A pioneer attempt was that of Geikie (1880) who undertook a survey of dated tombstones in Edinburgh churchyards and discovered marked variance in the sharpness of inscriptions of comparable age according to the nature of the rock. One of the most interesting of experiments in weathering is that by Hilger

(1897) who exposed rock particles of different types, with diameters in the range 10–20 mm, to atmospheric forces for a period of seventeen years. The results of this are summarized in fig. 4.20 and illustrate very well the resistance of the limestone in comparison with the sandstone. It is unfortunate that these experiments began with a mixture of loose rock debris, rather than a block of solid rock, since the rate at which solid material breaks up into loose rocky material is itself an interesting feature. The conversion of consolidated rock into loose debris depends particularly on the joint-spacing of the rock mass, whereas the subsequent disintegration of the individual rock fragments into soil-size particles depends far more on the porosity of the individual fragments and the strength of the cementing material. The ability of rocks to withstand weathering will inevitably vary therefore according to the particular phase of the weathering sequence.

Much more work is needed on the determination of weathering rates of different rocks under different conditions. Ollier (1969) has presented a comprehensive summary of present-day thinking on the topic of weathering, but it is clear that much more experimental work is still needed.

PART TWO: PROCESS: THE INTERACTION OF FORCE AND RESISTANCE

CHAPTER 5

PROCESS: INTRODUCTION

TYPES OF GEOMORPHIC PROCESS

A geomorphic process is a distinctive mechanism or group of mechanisms for the transportation of debris. Basically all processes are alike in that material begins to move only when the forces involved become greater than the resistances; and how often this occurs is the frequency of the process. Once movement has begun, the mode of interaction of force and resistance may differ greatly from one process to another, and it is these differences that are commonly used to classify debris movement processes (Sharpe, 1938). A basic distinction is between movements in which neighbouring particles remain close together as they move, so that the debris moves in a coherent whole, or *mass movement*; and movements in which particles move as individuals, with little or no relation to their neighbours. Movements of the latter type can be called *particle movements* or *movement in solution*, according to the size and nature of the moving material.

There are three main types of mass movement: a slide, a flow and a heave. In a pure slide, resistance to movement drops sharply once an initial failure has occurred along a well-defined thin surface. The mass above the slide surface moves as a block with no internal shear (fig. 5.1a). Debris moves down the hillside until conditions (e.g. reduced slope angle) change to slow down and stop the moving mass: a brick, sliding down an incline is an example of an ideal slide. In a pure flow there is no sharply defined failure surface, but instead shear is distributed throughout the moving mass. At the base shear is usually at a maximum, but the velocity is very low and all the movement occurs as differential movement within the body of the flowing mass (fig. 5.1b). Movement may stop on reduced gradients, but may continue until the driving forces are reduced, for example at the end of a storm. Movement of a fluid corresponds closely to a pure flow. The third type of movement

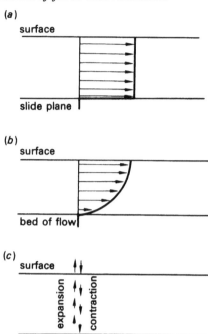

Fig. 5.1. Velocity profiles for ideal mass movement types. (*a*) Pure slide. (*b*) Pure flow. (*c*) Pure heave.

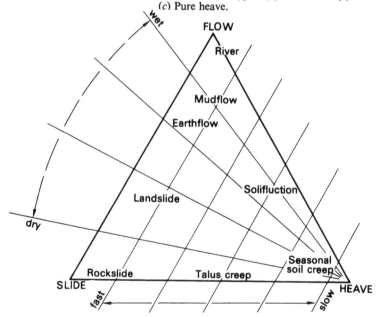

Fig. 5.2. Classification of mass movement processes.

is a pure heave, in which the soil expands perpendicular to the surface and subsequently contracts (fig. 5.1*c*). This motion is not a lateral transport in itself, but provides the basic mechanism for some processes, especially seasonal soil creep; and acts as a trigger to more rapid movements, providing the last small contribution which finally overcomes the soil resistance. In heaves, the forces acting are large, but they are exerted at right angles to the surface, and their energy is expended in movements of rather small amplitude, so that downslope movement in which a heave is the major component can never be rapid.

Actual mass movement processes are rarely, if ever, a pure slide, pure flow or pure heave, but all are a combination of the three, perhaps with some basal sliding, some internal shear, and some response to the triggering action of heaves. A triangular diagram thus provides a convenient basis for classifying the types of movement involved in the different processes (fig. 5.2). This diagram can also be used to indicate relative rates of movement during transport, and to show the moisture content of the movement. Flows tend to be moist, and slides dry, while heaves may be at any moisture content, so that a series of lines radiating from the 'heave' corner show relative moisture contents (fig. 5.2). Movements which are mainly a slide or a flow tend to be rapid, whereas movements which are mainly heave tend to be slow, so that relative rate isolines can also be drawn, parallel to the 'flow-slide' axis.

Particle movements cannot be rigidly separated from *mass* movements, because turbulent flows have a random particle movement superimposed on the overall mass movement; and any particle movement which carries large amounts of material at high concentrations has many of the properties of a mass movement. Nevertheless, some processes seem to be mainly particle movements, especially movement by rainsplash, by surface wash and by subsurface wash. In all of these processes, particles are entrained one at a time and have relatively low interactions with each other (particularly at low concentrations). Random diffusion movements in the soil, which may be responsible for some soil creep, also have most of the characteristics of particle movement.

Movement in solution can be distinguished from particle movement by either including or excluding colloidal suspension. On theoretical grounds, it may be preferable to treat the movements of colloids as a separate class of movement, but it will be seen in chapter 9 that most measurement methods combine colloids in measurements of 'dissolved load', and so the same convention is used in this book; '*solution*' being used to refer to both true ionic solution and also to movement in colloidal suspension.

MAGNITUDE AND FREQUENCY

The safety factor for a slope process is the ratio of resistance to force, and whenever this ratio falls below 1·0, a movement will begin. The frequency of movement is therefore closely related to the frequency distribution of force and resistance. At a point both are varying through time as the microclimate and especially the soil moisture changes, and over longer periods as a result of weathering and climatic change; likewise both force and resistance are varying throughout the area, with factors of gradient, flow concentration and soil properties. Because force and resistance often change together, for example in response to a change in the weather or the soil, it is difficult to study the frequency distribution of either separately, and instead the distribution of the ratio between them, that is the safety factor, will be considered. We have already said that movement is not the usual state for hillslope

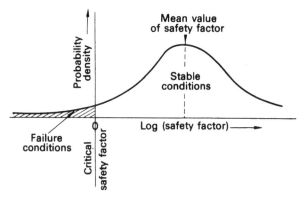

Fig. 5.3. Diagrammatic frequency distribution for the safety factor of a hillslope. (Safety factor ≈ resistance/force.)

debris: hillslope safety factors are therefore usually well above the critical value of 1·0, and the resulting distribution curve shows failure conditions (safety factor < 1·0) lying at one tail of the curve.

Fig. 5.3 illustrates the distribution of the safety factor as approximately log–normal, corresponding to independent log–normal distributions for force and resistance, but the exact form of the 'tail' region depends critically on the *exact* form of the two distributions and their correlation. In some cases, especially for landslides, this random variation in the safety factor is superimposed on a steady decline through time, until it reaches 1·0 and a movement occurs (Terzaghi, 1950). The random variation in the safety factor may therefore be quite small, even for major slides. For heaves, which begin every time it rains or freezes, failure is common, so that the *toe* area (shaded in fig. 5.3) of the frequency distribution is large. At the other extreme, overland flow produces very wide random variations in the applied force, so that it requires rather extreme storm conditions to produce appreciable movement, and the failure zone is a very small area of the frequency distribution. This varying balance between force and resistance determines how often movement will begin on a hillside.

The total geomorphic effect of a process is decided partly by how often it occurs, and partly by how fast it operates, having once started. These two aspects of process can be combined in a second frequency distribution; that of rates of debris transport. Fig. 5.4 illustrates this distribution and shows the proportion, $p(S)$, of the time for which debris is moving at rate of sediment transport, S. The total duration of movement is given by the area under this frequency curve, and will usually amount to only a small proportion of the total time in each year. The total amount of debris moved at a particular sediment transport rate is equal to that rate multiplied by the duration at that rate (equal to $S \cdot p(S)$), to give the broken line curve shown in fig. 5.4. Provided that the original

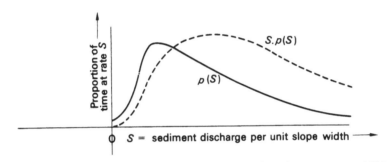

Fig. 5.4. Diagrammatic frequency distribution for the rate of debris transport on a hillslope.

frequency curve gives a finite mean rate of debris movement (which must always be the case), and is smooth and continuous, then the $S \cdot p(S)$ curve of total transport at a given rate has the properties: (1) It has a right-hand tail which tends to zero for large values of sediment transport rate, S; (2) If the frequency curve of $p(S)$ has only one peak, and a single point of inflection to the right of the peak, then the $S \cdot p(S)$ curve also has a single peak, which is to the right of the frequency curve peak. That is to say that there is a dominant rate of debris transport, or size of geomorphic event, which does more work than any other (Wolman and Miller, 1960); and this event is larger than the most frequent event. (If the frequency curve of $p(S)$ has several peaks, then the first peak of the $S \cdot p(S)$ curve is to the right of the first peak of the $p(S)$ curve.) For river channels, the dominant flow has been associated with the bank-full channel, and similar specific associations between form and process may exist for hillslope processes.

The concept that the spectrum of events of many different magnitudes can be considered to have the same geomorphic effect as a series of events, all of the dominant size, is an attractive one. The dominant size is larger than the most frequent event, but is still of moderate size, such as may be expected with a recurrence interval of about ten times that of the most frequent event. Two problems which may limit the usefulness of this concept are first, that the curves for $p(S)$ and $S \cdot p(S)$ may be multimodal, leading to a number of dominant event sizes; and second that larger events may have effects which are different in kind as well as in degree. The curves in fig. 5.4 may be multimodal even for a single process, and if it is considered to represent the sum effect of all processes together acting at a point, then it appears highly probable that there are a number of peaks on the total transport curve, each corresponding to a dominant event size for a particular process. Major landslides are an example of a process which may be very rare and may contribute only a small amount of material to the total debris transport, but which will mould the landscape to a depth and in a way which must always dominate a description of its site. In a similar way a major rainstorm can cause channel extension which will greatly increase the drainage density of an area and so modify the whole landscape to an extent which its contribution to the total debris transport belies. Although total debris transport is the most important single measure of landscape development, it is not the only one which is relevant. Instead the rational scale of sediment discharge might be replaced by an ordinal scale of effects, associated with progressively larger events. Movement of material in solution and fluvial silt transport would be at one end of the scale and channel head extension and major bedrock erosion at the other. Although there is still a dominant event *size*, it is not at all certain that the event associated with it is the most important in shaping the landscape. With these reservations however, the concept of dominant event size is a valuable one, and one which is widely used in the chapters on process below.

TRANSPORT-LIMITED AND WEATHERING-LIMITED PROCESSES

Processes have most commonly been classified by the way in which the debris moves, and their names reflect this classification. For landscape development, however, a more important distinction is based on the immediate source of the moving debris. Material is loosened from the hillslope bedrock by weathering and then moved downslope by transport processes. If transport processes are potentially more rapid than weathering, only a thin soil cover is able to develop because debris is removed as fast as it weathers loose. Movement is then said to be '*weathering-limited*'. If, instead, weathering rates are potentially more rapid than transport processes, a soil cover develops and the movement is said to be *'transport-limited'*. This dichotomy, first described by Gilbert (1877), is basic to an understanding of hillslope development. Where the removal is *transport-limited* then hillslope development depends on the transporting capacity of the relevant processes, and the rate of weathering is reduced to an equilibrium value which is less than its potential maximum by an increase in soil thickness. Where the removal is *weathering-limited*, hillslope development depends on the variations in weathering rate, and the rate of transport is reduced to the rate at which fresh material weathers. These quite different sets of controls lead to quite different sequences of slope development, and there is a general associa-

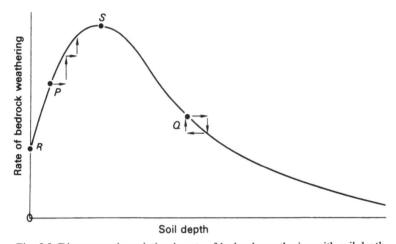

Fig. 5.5. Diagrammatic variation in rate of bedrock weathering with soil depth.

tion between *weathering-limited* slopes, relatively thin soils, the existence of important threshold slope angles, and parallel retreat at these angles; and between *transport-limited* slopes, a well-developed soil cover and convexo-concave slopes which become progressively less steep without significant threshold slope angles. The extent of these associations and the reasons for them will be discussed below in the context of individual processes, and again in part three.

Weathering of bedrock at the base of a soil depends mainly on the circulation of water. In *very* thin soils, water runs off rapidly because there is insufficient pore space in the soil to accommodate it, so that weathering rates are low: in very thick soils water circulates so slowly that rates of weathering are again low. Soil weathering is therefore at a maximum at intermediate soil thicknesses (fig. 5.5), a concept first proposed by Gilbert (1877). (This argument is considered again in chapter 9, in the context of equilibrium soils.) The shape of the soil thickness–

rate of weathering curve is important, because the behaviour of soils to the left and right of the peak is quite different. If a soil represented by point Q in fig. 5.5 has a local increase in soil depth, due for example to animal burrowing activity, then this soil increase will cause a reduction in the rate of weathering at that point. If weathering is in balance with transport processes, then the reduction in the rate of weathering will lead to a thinning of the soil which will tend to obliterate the original increase in soil depth. This is a stable situation, in which small local irregularities tend to be eliminated. In contrast, at point P, to the left of the peak of the weathering curve, a small increase in soil depth will lead to a local increase in the weathering rate, which will tend to further increase the soil thickness, and the situation is unstable, only reaching equilibrium at the peak point, S. Similarly a small local decrease in soil depth will lead to instability in the opposite direction, reaching equilibrium at point R, under conditions of zero soil. In this way the whole left-hand side of the weathering curve, between points R and S, is unstable, and soils tend to polarize either towards no soil, at point R; or else to a soil of at least a minimum depth, corresponding to the peak point S.

The importance of this unstable range of soil thicknesses for hillslope development processes is that there is minimal overlap between weathering-limited and transport-limited removal. At zero soil thickness, removal is entirely *weathering-limited*; and at finite soil thicknesses, removal is mainly *transport-limited*. Surface removal would be entirely transport-limited, but most processes transport some material below the surface, and may thus have a weathering-limited element in them. For example a steep slope may develop through slides which remove the entire thickness of soil, and which occur whenever soil weathers to a critical thickness (Wentworth, 1943). This removal is weathering-limited, despite the presence of a soil.

The arrangement of the following chapters on individual processes is based on the extent to which each is weathering-limited; beginning with processes which are entirely weathering-limited, like rock-falls, and ending with seasonal soil creep which can only occur in a continuous soil cover and is almost entirely transport-limited. Soil mass movements and wash processes are intermediate, the former depending strongly on the properties of the soil but in the extreme case like that described by Wentworth, becoming mainly weathering-limited. Wash processes are mainly transport-limited, but surface wash is in some ways analogous to a weathering-limited removal on a micro-scale, because each rainstorm can readily detach and transport a thin layer of debris loosened from the soil mainly by drying and cracking, and further erosion is very much more difficult.

THE CONTINUITY EQUATION IN PROCESS–RESPONSE MODELS

One of the purposes of process studies is to show how each process influences hillslope form. To do this, the process rate is expressed in terms of topographic factors, especially slope gradient and distance from the divide, and the resulting expression is included in a continuity equation, which for hillslopes takes the form: (Debris transport in) – (debris transport out over a unit length of slope profile) – (increase of soil thickness due to expansion during weathering of bedrock) = (decrease of elevation of land surface). Or in differential equation form:

$$\frac{\partial S}{\partial x} - (\mu - 1). W = -\frac{\partial y}{\partial t}, \qquad (5.1)$$

with the notation of fig. 5·6 and in addition μ = volume of mineral soil

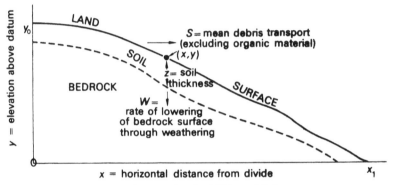

Fig. 5.6. The continuity equation for hillslope debris transport.

produced from the weathering of unit volume of bedrock (Kirkby, 1971). This equation describes the simple case of a ridge on which it is assumed that lines of steepest slope are parallel and contours are straight lines. To be more generally applicable, the mean rate of debris discharge, S, must be treated as a vector in the direction of movement, and the first term in the equation replaced by the divergence of the vector S. If x and the transport rate are measured along lines of greatest slope, then this first term becomes $(\partial S/\partial x - S/\rho)$, where ρ is the contour radius of curvature, considered positive if the contour lines are concave downslope.

A second continuity equation can be used to describe the change in soil thickness. This states that, (Increase of soil depth) = (Increase of elevation of land surface) + (reduction of elevation of bedrock), or in differential terms:

$$\frac{\partial z}{\partial t} = \frac{\partial y}{\partial t} + W = \mu. W - \frac{\partial S}{\partial x}. \qquad (5.2)$$

107

Two special cases of these equations provide a workable basis for process–response models for the two cases of transport-limited and of weathering-limited removal. Under weathering-limited conditions, the soil cannot grow thicker so that the rate of increase, $\partial z/\partial t$ is zero. The second continuity equation can then be written: $\partial y/\partial t = -W$, and this equation gives the rate of hillslope lowering in terms of the varying rate of bedrock weathering. In the transport-limited case, the mean rate of transport, S, can be replaced by the capacity of the process, denoted by C, which is defined as the mean rate of operation of the process under conditions where very much more material is available than is ever transported by that process. Eliminating the weathering rate, W, the continuity equation becomes:

$$\frac{\partial C}{\partial x} = -\mu \frac{\partial y}{\partial t} + (\mu - 1) \cdot \frac{\partial z}{\partial t}. \tag{5.3}$$

In this form, the last term can often be ignored, because both of its components tend to be small. Except for calcareous rocks and tropical conditions, the bulking factor during weathering, μ, tends to be close to 1·0 (Twenhofel, 1939); and soil thickness remains at a few centimetres or metres while the land elevation is reduced by tens or hundreds of metres, so that the rate of change of soil thickness is small compared to the rate of surface lowering, $-\partial y/\partial t$. In this approximate form, the equation simply relates the rate of surface lowering to differences in rates of debris transport which control the slope development.

Process–response models based on the continuity equation will show the relationship of process to form for any hillslope, if we also know the initial hillslope profile and the conditions of removal at the base of the slope. Because we need to know these things before using the continuity equation, we shall generally use the simplest case of a straight initial profile and of unimpeded removal at a fixed basal point, to allow us to compare the effects of the different processes as they are discussed in the chapters which follow.

Exact solutions to equation 5.3 can only be obtained in a few cases, but we can obtain an *approximate* solution which is most relevant to cases where there is no threshold slope angle. If C, the (capacity) rate of debris transport can be expressed in the form:

$$C = f(x) \cdot \left(-\frac{\partial y}{\partial x} \right)^n; \tag{5.4}$$

that is as a function of the distance from the divide multiplied by a power function of the tangent slope; then it can be shown that there exists an asymptotic solution towards which the real slope form *tends*, and to which it approximates closely by the time the initial relief has

108

been reduced to half. This solution is called a 'characteristic form' (Kirkby, 1971), because it depends on the *process* and not on the *initial form*.

A reasonable *approximation* to the characteristic form can be obtained for the transport-limited case ($S = C$) and for $\mu = 1 \cdot 0$. It is:

$$y_0 - y \propto \int_0^x \left\{ \frac{x}{f(x)} \right\}^{1/n} \cdot dx, \tag{5.5}$$

where y_0 is the elevation of the divide, which alone remains a function of time, often an inverse exponential function. If $f(x)$ in equation 5.4 is

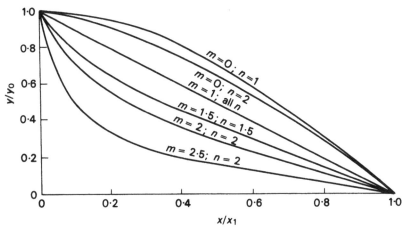

Fig. 5.7. Approximate 'characteristic form' slope profiles; obtained as solutions to the continuity equation 5.3 in the *transport-limited* case, when all trace of the initial form has been removed; and process laws are of the form $S = C\infty x^m$. (slope)^n. The curves correspond to different values of m, n. For the slope profile at $m = 1$, see Appendix B.

proportional to a power of distance, x^m, then equation 5.5 can be further simplified to:

$$y = y_0 \left\{ 1 - \left(\frac{x}{x_1} \right)^{\left[\frac{(1-m)}{n} + 1 \right]} \right\}. \tag{5.6}$$

Equations 5.5 and 5.6 are derived in Appendix B.

Characteristic forms can be obtained more exactly, and for less restrictive conditions, but equations 5.4 to 5.6 form a basis for relating process to form, to which we will refer in the chapters on process. Without pre-judging the identification of particular processes with particular versions of equation 5.4 for the variation in rate over the hillslope, fig. 5.7 shows the approximate characteristic forms obtained for a variety of values of the exponents m ($f(x) \propto x^m$) and n.

The topics considered here – of process definition, of frequency distributions and of the response in hillslope form to different processes, form a background to the rest of part two, in which each process is examined in some detail. They have been considered here because each is relevant to a number of processes. The order of the chapters follows a sequence from mainly weathering-limited processes to mainly transport-limited processes, not only because there is a gradation of hillslope forms corresponding to the gradation in process types, but also because the sequence corresponds roughly to the sequence of dominant processes as a profile develops through time. After rapid uplift, slopes are initially steep and instability processes, first in rock masses (chapter 6) and later in soil masses (chapter 7) dominate the slope profile development. Only after the slopes are too stable to allow slides, do the transport-limited processes (chapters 8–10) become important.

The division into weathering-limited and transport-limited processes also corresponds to the division into processes which do (chapters 6, 7) and do not (chapters 8–10) depend on threshold conditions. For the first group there is a rapid adjustment of slopes to a threshold angle or series of angles; and for the second group a slow evolution of form towards a characteristic form. It follows that for *threshold-dependent* processes the key element in understanding slope form is to know what controls the threshold angle, and this is what chapters 6 and 7 are ultimately about; whereas for *threshold-independent* processes, the key element in understanding slope form is to know how the characteristic form is produced, and this is what chapters 8, 9 and 10 are concerned with. Among each group of processes the sequence of chapters has no special significance, local rock and soil conditions determining which process is most important at a particular time and place. Within each chapter, the nature of the process is discussed, and then the techniques which may be used to measure its mechanism and rates of action, evidence of its rates and their frequency and spatial distributions, and models of the process mechanism: lastly process–response models link the process to hillslope development.

The way in which the process studies are presented below clearly reflects our position on the role and aims of process studies. It is not enough simply to measure the rate of a process at a point, or even to obtain an average rate. We must seek to understand how the process works, and hence how its rate of action varies in time and space. The time-distribution is related to the problem of magnitude and frequency and so to the extrapolation from necessarily short measurement periods. The space distribution is important both on the macro-scale to understand the effects of climate and lithology; and on the meso-scale to understand the variation of transport and erosion rates within a slope profile or a valley and thus connect the processes to the landforms.

CHAPTER 6

INSTABILITY PROCESSES IN ROCK MASSES

MECHANICS OF THE PROCESSES

The underlying process in all aspects of hillslope development is the weathering of the surface of a consolidated rock mass into relatively loose debris. In many instances this process of rock destruction is, itself, insufficient to produce movement of the material from the hillside, and the role of weathering is simply to cloak the rock mass with a veneer of residual material. Actual movement of this debris is then initiated by subsequent mass failure, soil erosion and slow soil creep. In other circumstances, however, the mere detachment of a mineral grain or rock fragment from the main body of a rock mass is, together with the ever-present force of gravity, sufficient to produce rapid movement of the debris away from the hillside. In this case there is no accumulation of a cloak of weathered material, except possibly at the base of the hillside, and the evolution of the hillslope depends on the rate of weathering at each point on the slope. The development of the hillside is thus controlled or limited by the pattern and rate of weathering upon it. In devoting this chapter to a discussion of these *weathering-limited* processes, we are, by implication therefore, restricting ourselves to bare *rock* slopes, as against *debris-mantled* hillsides or slopes in *weakly-consolidated* material. Instability processes in the two latter situations are discussed in the succeeding chapter.

The mechanics of these various processes have in common one important feature: weathering reduces the strength of the material at the surface of the hillside so much that the force of gravity removes it virtually instantaneously. In this chapter a number of such processes are grouped together. At first glance, they may appear to possess little in common, but, in fact, they are merely members of one spectrum of process. An obvious example of the *weathering-limited* process is the instantaneous stripping of a sheet of rock along a joint-plane, through the exfoliation process, common in rocks where the main joint pattern is parallel to the surface. A similar mechanism is the slide of a rock wedge in a rock mass where the rock is so highly fractured that it acts virtually as a densely-packed mass of angular debris. A third process is the gradual removal of the surface layer of a rock face through the fall of isolated fragments of rock, and this again is common in rocks with a dense fracture pattern. A final process, exclusive to rocks with no real joint pattern, such as some sandstones, is the detachment and release of discrete mineral grains from the cliff face. These processes are essentially very similar and the major criterion in their separation is the scale of the rock particles involved in the process. The four mechanisms are depicted in fig. 6.1 and are considered separately below under the headings of (*a*) slab-failure, (*b*) rock avalanches, (*c*) rockfall and (*d*) granular disintegration.

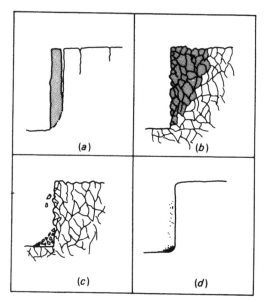

Fig. 6.1. Types of instability processes in rock masses: (*a*) slab-failure, (*b*) rock avalanches, (*c*) rockfall and (*d*) granular disintegration.

Slab-failure

This process is important in rocks where the macro-jointing pattern is more prominent than jointing at the micro-scale. Such situations are not common, but where they do occur the process and resultant landforms are extremely spectacular.

The simplest example of this process is probably the slumping of bank-material into a stream channel when deepening of the channel has resulted in the development of cracks in the alluvium parallel to the stream bank. It is fairly easy to visualize this as a result of the removal of lateral pressure on the newly-exposed bank face following cutting down by the stream. Nevertheless, a full understanding of the process can only be achieved through reference to the theory of earth pressure.

Envisage a vertical rock slope bounded by a frictionless wall. Any element of the rock (fig. 6.2) is subject to both vertical and horizontal stresses. The vertical pressure is given simply by the weight of soil above a unit area and is equal to

$$\sigma_z = \gamma . z \qquad (6.1)$$

where γ is the bulk unit weight of the material and z is the depth below the surface. The horizontal pressure is given by

$$\sigma_x = K . \sigma_z \qquad (6.2)$$

where K is called the coefficient of earth pressure. Any movement of the retaining wall slightly away from the rock mass will reduce the horizontal earth pressure and, assuming that the earth behaves in a per-

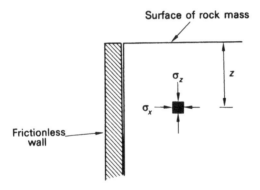

Fig. 6.2. Stresses at a point in a rock mass: two-dimensional case.

fectly elastic manner, the change in the horizontal pressure will be given by

$$\Delta\sigma_x = E . \Delta\varepsilon_x \qquad (6.3)$$

where $\Delta\varepsilon_x$ is the strain in the x direction and E is the modulus of elasticity. Continued movement away of the wall will not produce the same response indefinitely. The horizontal pressure in the rock and, thus, the pressure exerted by the material on the wall, would soon reach a minimum value.

Various approaches have been used to compute the distribution of horizontal earth pressure in a solid mass *at this stage* using different assumptions in each case. The theory of Rankine (1857) indicated that, in the condition of limiting stability,

$$\sigma_x = \gamma . z . \tan^2 (45 - \phi/2) - 2 . c . \tan(45 - \phi/2) \qquad (6.4)$$

113

where c is the cohesion and ϕ the angle of shearing resistance of the material, using these terms as defined in chapter 4. The pressure pattern in the rock mass is depicted in fig. 6.3.

It is not proposed to discuss the ideas of earth pressure in any more detail here and reference should be made to the standard texts in soil mechanics, such as Terzaghi (1943), Terzaghi and Peck (1948), Scott (1963) and Wu (1966), by the interested reader. It is sufficient to state

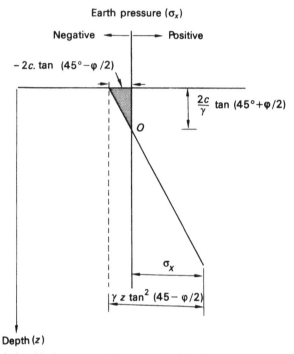

Fig. 6.3. The relationship between earth pressure and depth. (Stippled area represents tension zone; at point O the earth pressure is zero.)

that the above treatment corresponds to the *active state of stress*; the situation is slightly different in the *passive* state where a *maximum* pressure on the retaining wall is built up by a forward movement of the wall against the rock mass.

The major feature of interest in fig. 6.3 is the development of a tension zone in the uppermost layer of the mass. The thickness of this layer is given by Terzaghi (1943, p. 37) as equal to

$$z_0 = \frac{2c}{\gamma} . \tan(45 + \phi/2) \tag{6.5}$$

using the symbols as before. In nature, soil and rock masses may experience great difficulty in withstanding these tensile forces which are tending to split the rock. Usually cracks develop as a consequence, and so alleviate the force. In the cases where the material has zero tensile strength, a crack will develop to a depth

$$z = z_0 \qquad (6.6)$$

and this is not uncommon in certain clay masses. Stronger rock masses do not produce such deep *tension cracks*.

The above situation in which a retaining wall is slowly moved away from a rock mass may seem to possess little relevance to the development of natural slopes, but it is, in fact, closely analogous to the situation where a cutting is made through a rock mass by stream erosion. The elements of rock in the banks of the stream channel or valley were previously subjected to definite horizontal pressures and the removal of material by stream undercutting results in the relaxation of these pressures in a similar way to the movement away of a retaining wall. Cracks will develop in the material at the side of the bank or the valley, and the stability of this material depends closely upon the depth of the cut relative to the depth to which tension cracks will develop.

In material with little ability to withstand tensile stresses, such as stream alluvium, a small cut will result, in time, in the development of a tension crack deep enough to produce instability. Stronger rocks may, in contrast, allow deep gorges or canyons to develop without any instability of this type occurring.

An interesting recent discussion of this mode of slab-failure in loess has been given by Lohnes and Handy (1968) working in western Iowa and Tennessee. The stability analysis employed by these workers is as follows. In fig. 6.4, the downslope component of the weight of the block $[\frac{1}{2}\gamma.(H_c'+z).x.\sin(45+\phi/2)]$ is resisted by shear strength along the potential failure plane. This strength, as a force, comprises both friction $[\frac{1}{2}\gamma.(H_c'+z).x.\cos(45+\phi/2)\tan\phi]$ and cohesion $[c.x/\cos(45+\phi/2)]$. At limiting equilibrium, the shear force and shear strength are equal, so that we obtain (appendix A 1)

$$H_c' = \frac{4.c}{\gamma}.\frac{1}{[\cos\phi - 2.\cos^2(45+\phi/2).\tan\phi]} - z \qquad (6.7)$$

where H_c' is the maximum height of the cut before slab-failure and z is the depth of the tension crack. Note that this expression is identical to the more familiar equation presented by Terzaghi (1943, p. 153) and quoted, for instance, by Wu (1966, p. 279)

$$H_c' = \frac{4.c}{\gamma}.\tan(45+\phi/2) - z \qquad (6.8)$$

115

which is easily derived from the case where β (in fig. 6.4) is zero. (The identity of equations 6.7 and 6.8 is proved in appendix A 2.) In fact equations 6.7 and 6.8 are merely special cases (vertical cliff) of the relationship between critical height and properties of the material (c, ϕ and γ) and slope angle. The more general equation, in its various forms, is derived in appendix A 3; in the present discussion it is convenient to treat just the case of the vertical cliff. Note that, assuming $z = z_0$ (the worst case from the viewpoint of the stability of the slope), equation 6.8 simplifies further to

$$H'_c = \frac{2c}{\gamma} \cdot \tan(45 + \phi/2) \tag{6.9}$$

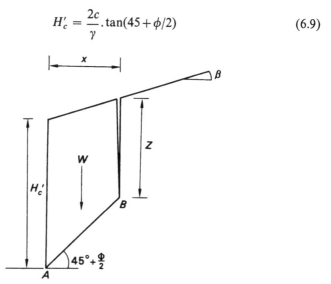

Fig. 6.4. Stability analyses of a slope subject to slab-failure.
AB represents the potential sliding surface; its length is $x/(\cos 45 + \phi/2)$; weight of block is $W = \frac{1}{2} \cdot x \cdot (H'_c + z) \cdot \gamma$; downslope component of weight of block $= W \cdot \sin (45 + \phi/2)$; component of weight normal to $AB = W \cdot \cos (45 + \phi/2)$.

It should be noted, lastly, that the assumption that the potential failure plane must pass through the slope base at an angle $(45 + \phi/2)$ to the horizontal stems directly from the active Rankine pressure state.

When a cut is made, by man or by a stream, the slope will be stable provided that the depth does not exceed H'_c. In the Iowa loess the critical height before slab-spalling occurs was found, through field survey, to be about 4·5 m. Employing values of $c = 0·09$ kgf/cm^2, $\phi = 25°$ and $\gamma = 1240$ kgf/m^3, Lohnes and Handy predicted this value using reasoning outlined above. Agreement between theory and practice in this case however is only achieved by using equation 6.8 and assuming z equal to zero, despite reported tension cracks as deep as 2·3 m. The implication, in the case of this particular study, is that tension cracks develop at the

116

instant that slab-failure is about to occur, by which time the cliff has attained a height approximately twice that possible if the tension crack had developed earlier.

Tension cracks are common at the top of many clay slopes and are often associated with large, deep-seated circular landslips. The actual direct contribution made by the presence of the cracks is, however, mostly very small, and a discussion of these slides would be out of place in this chapter. Instability is discussed in this chapter only in those instances where it is closely linked to joint or crack development and may thus be attributed to a large extent to weathering processes.

Slab-spalling in the manner of the loess of western Iowa is also very common in massive sandstones in arid areas. A particularly well-documented instance of this process is given by Schumm and Chorley (1964) in the case of the fall of Threatening Rock, a huge monolith of Cliff House Sandstone in the Colorado Plateau area of New Mexico. This famous rock slide took place in the area of the Chaco Canyon National Monument. The upper walls of the canyon are cut in Cliff House Sandstone, the upper member of the Mesaverde group in this area, and it overlies the mostly shale Menefee formation. Along the north side of the canyon, huge joint-blocks of the sandstone occur in various stages of detachment from the cliff. This slab erosion was thought by Bryan (1954) to be primarily due to rainwater passing down through the porous sandstone and seeping out along the junction with the shales. It was this process that was thought to account primarily for the fall of Threatening Rock. Although slab-failure is thus, to some extent, due to particular geologic conditions in the area, it is, nevertheless, also dependent upon the more widespread mechanism of the development of vertical joints behind the cliff face. Once a block has slumped down, the relaxation of pressure on the newly-exposed face results in the gradual development of a new joint system, and the process continues in this way.

The fall of Threatening Rock is especially interesting since movement of it has been fairly well documented by National Park Service personnel and it affords some insight into the processes involved. The block was, prior to its fall, about 50 m long, 30 m high and about 10 m thick. It was completely detached from the cliff with a vertical split which was about 4 m wide at the top and 1 m at the base. The monolith rested directly on shale. The wider gap at the top is partly due to overturning and partly the product of differential weathering. At the base there was an overhang of about 2 m resulting from undercutting of the shales. A plot of rock movement (widening of the gap) against time is given in fig. 6.5 and it illustrates both the exponential nature of the trend and the acceleration of movement in the winter months. This latter feature suggests that frost action is possibly important, along with melting of winter snow accumulation in the gap between slab and canyon wall, which would wet the underlying shale and encourage undermining of the rock.

Slab-failure is common in many areas. It is widespread in sandstone areas in dry climates and it is also very noticeable in chalk cliffs undercut by the sea along the south coast of England. The two prerequisites are some mechanism of undercutting and a rock mass susceptible to the

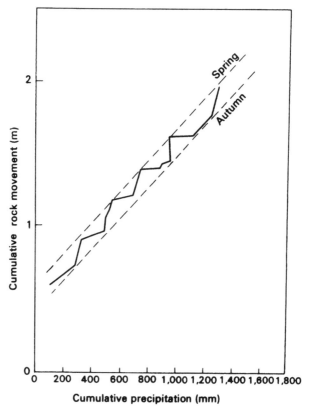

Fig. 6.5. The movement of Threatening Rock prior to its fall (after Schumm and Chorley, 1964).

development of large joints through lateral pressure release. It is unfortunate that so little is known about the mechanics of these processes, although engineers are now beginning to turn their attention to the problem.

A recent paper by Bjerrum and Jorstad (1968) affords a particularly lucid insight into the problem. These workers summarize almost two decades of research by the Norwegian Geotechnical Institute in matters relating to rock slides in Norway. Most of the rock slides in that country occur on the slopes of western coastal valleys oversteepened by glacial action. These authors conclude that more than one factor must be considered in the gradual time-dependent processes which culminate in a rock slide. A considerable number of the Norwegian rock slides occur where rock slabs have been detached by major joints from the main mountain mass. They emphasize that these joints are completely independent of structure and geological boundaries, and usually run parallel to the surface of the slope. These *'valley joints'* are interpreted as products of the stress changes which accompanied erosion of the valleys, and may, again, be thought of as pressure-release joints. A mountain wall usually possesses several sets of joints at various distances from the surface and it is suggested that these different sets correspond to different stages in the erosion of the valley. Although the individual joints rarely extend over the full height of the slope, potential sliding surfaces coincide with these joints as far as they extend.

Two other features relate to this joint system. Water naturally tends to accumulate in these open joints and affects the stability of the mass directly in at least two ways. Water pressure will reduce the effective stress along the sliding surface (as outlined in chapter 4) and thus reduce the friction between the sliding body and the base. In some cases also, water in the joints will exert an outward pressure on the sliding body, encouraging overturning. This direct effect of water pressure is supplemented by the creation of rock fatigue through variation in the water pressure in the joints. The fact that the rock mass does not behave completely elastically means that continual change in the water pressure in the joints is likely to lead to extension of the joint; this will be augmented by the wedging of crushed rock in the gaps, thus preventing the joints from returning to their original position after opening slightly under higher water pressure.

This slow breakdown of rock strength and extension of the joints by water pressure fluctuations is probably the most important item in understanding the time scale of these slides. Some steep slopes in western Norway have been stable and unchanged since the retreat of the last ice masses. Joints are slowly extending into the mountain mass, however, and when the intact area along the potential sliding surface becomes so small that the shear stresses due to gravity, water pressure and internal stresses exceed the sum of the shear strength of the intact rock and frictional resistance along the sliding surface, a slide will occur.

Bjerrum and Jorstad (1968) point out that, since the distance between valley joints increases with depth, successively longer periods of time will be necessary to initiate subsequent slides, although, with our

119

meagre present knowledge of these phenomena, it is impossible to predict the circumstances which will result in complete stability of a rock cliff with respect to slab-failure.

The classic illustration of the slab-movement process is, of course, the exfoliation of igneous domes, especially granite, in semi-arid areas, and the standard textbooks of geomorphology, such as Strahler's *Physical Geography* (1966, p. 315), contain excellent photographs of the spectacular landforms which ensue from this process. Although it is possible that exfoliation may occur on a small scale through thermal effects, the large-scale peeling of slabs in onion-like layers, so typical of the massive igneous domes of Yosemite and other areas, is, once again, more accurately explained in terms of pressure release jointing parallel to the surface of the main body of rock.

The previous treatment of slab-failure has focused attention on the time-dependent aspects of the problem associated with changes in stress and strength which result from the weathering process. It is clearly impossible to discuss all the relevant variables which enter into the issue and the interested reader is referred to a concise and lucid exposition by Morgenstern (1967) of some of the items which affect the shape of the slide surface, the strength of a rock mass and the development of water pressures, and thus the general stability of rock masses. Our current knowledge of this whole field is exceptionally meagre, and there is a great need for deeper understanding of the nature of rock strength and the mechanics of large-scale instability in rock masses.

Rock avalanches

Although in some cases, such as the sandstones and loess noted in the previous section, the major joint-pattern is a widely-spaced one associated primarily with pressure-release, most rocks possess a fairly random micro-jointing pattern superimposed on this larger network. Slopes cut in this type of rock mass are subject to two types of weathering-limited processes. These are rock avalanches and rockfalls. The latter type is the subject of the next section and the former is discussed here.

Terzaghi (1962*a*) uses the term rock slide rather than rock avalanche, but here the term rock avalanche is employed to distinguish it from slab-failure because both may be considered as rock slides. Terzaghi's paper is an outstanding pioneer work in this sphere of rock mechanics; not only are various modes of analysis outlined and relevant problems indicated, but also there is a useful set of definitions. Most of this section on rock avalanches is taken from Terzaghi's paper.

In chapter 4 on resistance it was shown that intact blocks of rock may possess very great unconfined compressive strengths, and that the bulk of this strength is cohesion implanted by various forces associated with the crystalline nature of the rock. Terzaghi points out that the strength of *intact* rock is, however, at least in the case of well-jointed rock masses, usually irrelevant to the problem of slope stability. This may be appreciated from the following points. The critical height of an unweathered rock is approximately given by the formula:

$$H'_c = q_u/\gamma \qquad (6.10)$$

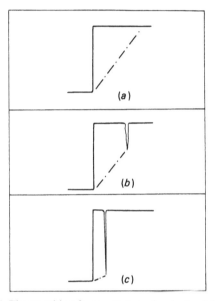

Fig. 6.6. The transition from a rock avalanche to slab-failure.

where q_u is the unconfined compressive strength of the rock and γ is its unit weight. This formula is the one presented by Terzaghi in his paper. Since it differs from the ones derived in the previous section, an explanation is in order here. Instead of considering the critical height of a completely intact rock mass (fig. 6.6a), the formula relates to the critical height of the outermost column of the rock mass (fig. 6.6c), and the assumed failure plane is taken to pass through the bottom block of this column. The unconfined compressive strength, as a stress, is simply the weight of the rock per unit area of the base ($H'_c \cdot \gamma$), and, from this, we derive the previous equation 6.10. In fact, of course, equation 6.10 is identical to equation 6.9, since the numerator of the right-hand side of that equation (6.9) is a common expression for the unconfined compressive strength of intact rock. Terzaghi, in employing equation 6.10,

121

is considering the worst case possible, with tension cracks at maximum depth.

If we take $q_u = 350$ kgf/cm^2 (5,000 p.s.i.) as a minimum value for a strong rock, together with $\gamma = 2720$ kgf/m^3 (170 lb/ft^3), and insert these values into the formula above, we obtain

$$H'_c = 1{,}300 \text{ m.}$$

This figure, it should be appreciated, relates to the weakest rock under the worst condition; intact hard rocks such as granite should be able to stand vertically at much greater heights. The fact that extremely few vertical slopes exist at even this height in nature must indicate that the critical height of unweathered rock is reduced substantially by mechanical defects such as joints and faults.

Joints divide a rock mass into individual blocks which almost fit

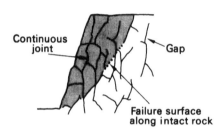

Fig. 6.7. The concept of effective joint area (based on Terzaghi, 1962*a*).

together; across the joints the cohesive bond, so prominent in intact rock, is equal to zero. Terzaghi distinguishes between continuous joints and gaps (fig. 6.7), and defines the term *effective joint area* as the total area of a section through a rock mass which is not in intact rock. A shear failure along such a section is resisted partly by friction along its joints (an up-to-date discussion of the strength characteristics along joints is given by Ladanyi and Archambault, 1969) and cohesion in the intact rock between them. The term effective cohesion is defined as

$$c_i = \frac{c(A - A_g)}{A} \tag{6.11}$$

where c is the cohesion of the intact rock, A is the total area of the section through the rock and A_g is the total area of joints and gaps within the section. The maximum height of a rock mass thus depends, at least in theory, on the effective cohesion and the pattern of jointing of the mass. In practice, Terzaghi argues, the surface zone of a rock mass is so fractured that generally c_i is, in fact negligible, so that the strength is entirely internal friction, and the pattern of jointing is all-important.

As a river cuts down into a rock mass, the valley-side walls get bigger and, consequently, the shear stresses on any potential failure plane passing through the base gradually increase also, since more weight is acting upon the incipient surface of sliding. The initial strength of the rock mass is predominantly due to cohesion in the portions of intact rock in the mass. The shear stresses are thus concentrated here, rather than on gaps and joints, and as they increase, so, locally, the intact rock splits in places due to this stress concentration. As the rock becomes more fractured so the shear stresses are concentrated on fewer pieces of intact material and the stresses on these remaining parts of the rock are inevitably greater still. As a result more splitting takes place and the process accelerates through a type of chain reaction. Ultimately there will arrive a time when the surface material is really simply a densely-packed cohesionless aggregate of angular blocks fitting together. The steepest slope cut in the rock mass then depends entirely on the nature of the joint pattern (whether it is random or rectangular, for instance) and the orientation of the pattern relative to the slope. The steepest slope angle may be anywhere between the vertical and about 30° depending upon these considerations. As the valley continues to cut down, the slope must recede at this angle through a process of ravelling, whereby the blocks located next to the slope drop out of the slope, individually or several at a time.

This sequence described by Terzaghi is an important part of the geomorphological evolution of a landscape and it is refreshing to come across a description based on the underlying principles of mechanics in contrast to the speculative, non-quantitative and confused thinking of early geomorphologists attempting to deal with this issue.

Although the rock mass in the state above closely resembles, except for scale, a mass of angular sand grains densely packed together, the immense scale of this rocky material prevents any attempt to subject it to standard testing procedures in order to determine the ϕ value. A number of possible solutions are, however, proposed by Terzaghi in his paper.

It is suggested, first of all, that the macrostructure of a rock with a random joint pattern is simply a large-scale model of the microstructure of an intact rock specimen, since crystalline rocks such as granite or marble consist essentially of irregularly shaped particles interlocking with one another, just as the blocks between the joints of a large rock mass. In the intact rock there are, of course, cohesion bonds, but the angle of shearing resistance depends purely on the size and shape of the interlocking particles, that is, the jointing pattern. The tests by Ros and Eichinger (1928) on marble indicated that with a normal stress on the failure plane of about 100 kgf/cm^2 (1,420 p.s.i.), the ϕ value would be about 40°. It has already been pointed out, however, that the ϕ value usually increases as the normal stress decreases and, since the average pressure on a potential failure plane near the surface of a rock mass would, even in the highest slopes, be much less than in these tests, this estimate for the ϕ value is probably on the low side in the context of a large rock mass.

Terzaghi argues that another model of the fractured rock mass is the shear strength of crushed rock aggregate. In chapter 4, it was noted that, at least in a very loose state, the ϕ values of rockfill material are commonly about 43–45°. In a more densely-packed state, the internal friction coefficient is likely to be still higher, and Silvestri (1961) reports ϕ values at about 65°. Since the porosity of a jointed rock mass (porosity is defined as the ratio of the void space to the total volume of the rock mass) scarcely exceeds a few per cent, Terzaghi believes that the ϕ value of an average rock mass with a random pattern of jointing is probably about 5° higher than Silvestri's figure for rockfill and suggests a figure of about 70°. This expectation appears to be confirmed by the observations made by MacDonald (1913) who examined a large number of natural rock slopes in canyons in the U.S.A. and Central America. The typical angle of stable rock slopes in unweathered granite, quartzite and similar rocks was about 71° which is to be expected if this is also the ϕ value of the rock mass and the cohesion has been entirely dissipated. This work by MacDonald represents the only piece of research known to the authors where a quantitative explanation may be advanced for the steepness of natural rock slopes, although, in fact, MacDonald's work was not so much to explain natural slopes on the basis of the shear strength parameters of the material, as to infer the latter from the former.

Terzaghi concludes that 'the critical slope angle for slopes underlain by hard massive rocks with a random joint pattern is about 70°', on the assumption that the walls of the joints are not acted upon by seepage pressures. This assumption is not valid for soil and weathered rock on more gentle slopes, as will be shown in the next chapter, but, in general, it is more applicable to rock slopes. Notwithstanding the simplicity of Terzaghi's ideas, it should be emphasized that they are still rather speculative, and it is probable that a range of ϕ values from 45° to 75° is more appropriate for fractured rock systems, rather than a single value of about 70°.

In addition to Terzaghi's paper, mention should also be made of a more recent article by Brawner (1966) which gives a useful and concise summary of the more important aspects of slope stability in open-pit mines in this type of rock. At present it is the work of engineers, as exemplified by these two papers, that has contributed most to our understanding of the role of these various weathering-limited processes in the evolution of hillslopes. The process of rock avalanches and the stability of highly-fractured rock masses is a central issue in the development of the landscape in the early stages of stream dissection. Some indication of the importance of this type of stability analysis will be given in part three, especially in connection with the geometry of badland landscapes.

Rockfall

Many rock masses with a fairly close jointing pattern as described in the previous section may be stable with respect to rock avalanches, but still subject to the process of rockfall, that is, the fall of small blocks and fragments of rock from a cliff face, sometimes separately and sometimes in large numbers, but always derived from a *very narrow* superficial part of the rockwall. This is the essential difference between this and rock avalanches. In the process discussed in the previous section, the major mechanism responsible for fracturing of the rock mass and dissipation of this cohesion of the intact rock was concentration of the gravity stress at local points within the mass. This mechanism is applicable to a fairly wide zone of the rock mass. The process of fracture initiation and enlargement, in rockfall, is, in contrast, primarily due to atmospheric weathering forces confined to the layer of rock immediately behind the cliff face. This is an important distinction, since rock avalanches may be expected to involve a thick wedge of material, whereas rockfall is probably a process of parallel stripping of material from the rock face.

The mechanics of the weathering process are fairly well understood and appear to be much simpler than the instability of large slabs, at a larger scale, previously described. It is useful to be aware of two different roles played by weathering in this process. One is the actual *creation* of small fractures at the surface of a rock mass, and the other is the *enlargement* of these cracks and complete physical separation of the small block.

The small-scale cracking of intact rock by temperature changes has long been accepted as one of the basic processes of physical weathering, although, at different times, doubt has been expressed as to the efficacy of the process. The usual argument in favour of the mechanism is that, since rock is a poor conductor of heat, the effects of daytime heating are confined only to the surface. As a consequence the outer layers of a rock mass tend to expand more than the interior, stresses build up within the rock and, if these exceed the yield strength of the rock, splitting should occur. This theory is not supported by the laboratory experiments of Blackwelder (1925) or Griggs (1936) who failed to achieve splitting of rocks, purely through thermal effects, despite subjecting specimens to conditions which, at least on a short time scale, were probably more severe than those experienced in nature. The relevance of these experiments to natural processes is still, however, disputed, and the issue is by no means resolved.

The role of weathering through initiating small fractures is, however, probably not a fundamental one, since most rocks possess small cracks for a wide variety of other reasons. The more important role is in the enlargement of these fractures and in this process, the action of water-freezing is all-important. Water seeps into the fractures and, whenever it

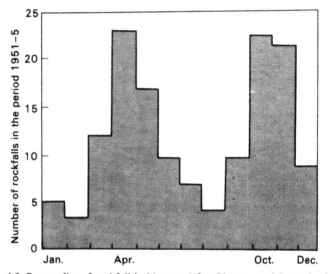

Fig. 6.8. Seasonality of rockfall in Norway (after Bjerrum and Jorstad, 1968).

freezes and ice lenses form, the fractures are extended and widened into open splits. Eventually the shear strength of the incipient rock fragment is reduced below the level of the gravitational and other shear stresses acting upon it and, unless there is some confining support, it will fall, assuming the slope is steep enough, in a manner similar to the failure of the larger slabs.

Other agencies may be important in this process, such as the extension of plant roots into rock fractures wedging the rock apart, but in areas where there is daily or seasonal freezing and thawing of water, this must rank as the major mechanism. Some evidence for this viewpoint is provided by Bjerrum and Jorstad (1968) in the article referred to previously. In fig. 6.8 the occurrence of rock falls in Norway is plotted according to the time of year; there are noticeable peaks at both spring and autumn. In these two seasons there is continual change of temperature about the freezing point and frost action is at a peak. Similar results are reported by Rapp (1960*b*) also working in Scandinavia, and are summarized by Leopold *et al.* (1964, p. 89).

In a well-known work on rockfalls in the Alps, Heim (1932) suggested that it was useful to distinguish between two phases of the process: the *preparation* of material for rockfall by various weathering processes and the *release* of debris by momentary shock such as earthquakes, heavy rains, snowblock falls and frost-bursting. This view is, however, possibly misleading. Earthquakes may occasionally be instrumental in rock release but cannot be considered as a continual, universal mechanism, while there is little evidence to indicate that heavy rains are particularly important in this specific mode of slope instability. Snow

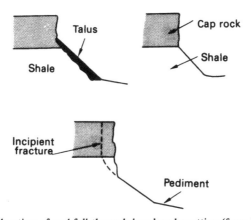

Fig. 6.9. Acceleration of rockfall through basal undercutting (from Koons, 1955).

avalanches are, moreover, as noted by Rapp (1960*b*), more effective in disturbing existing debris from rockfall on talus slopes than in stripping debris from rock faces. Frost-bursting seems to be an important mechanism in the process, but there appears to be little point in separating this aspect from the work of frost in preparing material for rockfall. And, finally, many rockfalls seem to occur when interstitial ice has thawed and the adhesion of the rock-ice assemblage has been lost. It is perhaps more realistic to visualize the release of rock debris simply as that instant in time when the various mechanisms responsible for the breakdown of a rockwall into small fragments reduce the shear strength of the surface material below the level at which it is capable of countering the stresses imposed by gravity and, as a result, a rockfall materializes.

In some areas the process of rockfall is considerably accelerated by undercutting at the base, as in the case of slab failure, and, indeed, it is sometimes difficult to adequately distinguish between the two processes.

127

One example of this is provided by Koons (1955) working in New Mexico. Often in this area the major scarps comprise strong, although highly fractured, rocks underlain by weaker shales. Instances of this are basalt overlying the Pierre Shale, and the sandstone of the Mesa Verde formation capping the Mancos Shale. The existence of escarpment profiles such as those depicted in fig. 6.9 led Koons to argue that the intermittent process of scarp retreat depended very much on undercutting of the shale at the base of the scarp. A rockfall would litter the talus slope with material and in the subsequent period, gullying in the shale would not only strip away the talus, but also cut back into the base of the scarp undermining the cap rock. Incipient fractures develop and enlarge in the cap rock, in the area behind the cliff face, until the strength of the rock is diminished so much that rock collapse once again takes place. Since the rock collapses as a mass of small fragments rather than one distinct slab, the process is described by Koons as rockfall, although it is clearly very similar to the slab-failure of Threatening Rock and the process of rock avalanching discussed previously.

Since the rockfall process is fairly familiar and discussed, at least qualitatively, in most of the standard texts, it is not proposed to treat it any further here. Some measurements of actual rockfall rates in nature are given in the next section, and models of slope development utilizing this process are discussed in the concluding part of the chapter. It remains to be emphasized that rockfall is probably the most universal process discussed in this chapter, since granular disintegration demands rather rare conditions, and, although many rocks are susceptible to rock avalanches and slab failure, it is not uncommon for rockfall to occur also on these rock masses and dwarf, in importance, the larger scale processes.

Granular disintegration

Some rocks do not possess a small-scale fracture pattern and so cannot break down into the small rock fragments involved in rockfall. These rocks are usually weak mechanically. This arises from the fact that intense pressures responsible for really strong rocks usually also impose some network of fractures on the rock. Such rocks are the weak sandstones common in the south-western U.S.A., and other similar materials where the strength of the rock is primarily due to the presence of a cement. Weathering of such rocks usually takes the form of slow release of individual grains from the surface through the destruction of the cement bond.

There are, however, many different mechanisms involved in this process of grain release. An obvious one in sedimentary rocks is solution of the cement detaching the grain from the main body of the rock mass and leaving it more susceptible to the gravity force. This is, thus, very similar to some of the processes involved in slab failure and rockfall, albeit at a much smaller scale. The change of scale is, however, very important, and complicates the situation. As noted in chapter 4, the presence of moisture at the contact between two rock pieces imparts capillary cohesion to the material provided that the curvature of the water films is sufficiently sharp. This curvature increases as the size of the rock pieces decreases so that it is negligible with the slabs and fragments involved in the three previous processes, but exists in the case of grain release. The movement of water into the pores of a sedimentary rock, through the solution of the cement, may result in detachment of the grain and yet, at the same time, still bind the grain to the main rock surface. It is usually only when the surface of the rock mass dries out that grains fall from the rock slope, since then capillary cohesion is reduced or lost, and the gravity force prevails. The release of grains as the surface of a material gradually dries out is easily observed when a cutting is made into moist silty-sand. At first the cut may be perfectly stable, but, slowly, as capillary cohesion is lost at the surface, grains begin to pour down and accumulate at the slope base. The same process may be seen at a larger scale in semi-arid areas in the first few days after a rainstorm.

It is sometimes suggested that thermal effects may produce granular disintegration through differential absorption of heat by the different minerals of a rock. This is particularly the case in igneous and metamorphic rocks which are often composed of minerals with radically different specific heats. The validity of this notion is often questioned, but, irrespective of the *absolute* efficacy of granular disintegration due to thermal effects, there is little doubt that in igneous and metamorphic rocks, with well-developed fracture systems, this process is, relatively speaking, unimportant in relation to other weathering-limited processes.

A more effective mechanism of granular decay is apparently frost action. This may seem a little surprising since it is usual to think of frost action in connection with fractured rocks, and granular disintegration with temperature changes. Some experiments by Schumm and Chorley (1966) nevertheless reveal the potency of frost action on rocks which crumble rather than shatter into smaller fragments. Samples of rock were taken by these workers from cliff-forming sandstones in the Colorado Plateaux and were exposed to natural weather conditions, supplemented by artificial rainfall, at Denver, Colorado during the period 1962–4, in an attempt to determine the rate of weathering and its association with weather conditions. Specimens of relatively fresh rock were dried, brushed to remove loose grains and weighed before exposure to the atmosphere. This procedure was repeated eight times during the two-year period to determine the amount of loose granular material produced in each period. The amount of weathering is plotted against cumulative precipitation in fig. 6.10 for three different sets. Sets 1 and 2 comprise similar rocks and are separated only because set 2 was added at a later date than set 1. The changes in the two sets approximately parallel one another. The Entrada Sandstone specimens are separated on account of their much greater susceptibility to decay. The

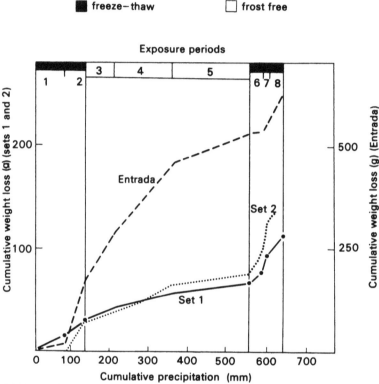

Fig. 6.10. The influence of frost action on the weathering of sandstones (after Schumm and Chorley, 1966).

changes in steepness of these plots represent acceleration or deceleration of weathering rates relative to the amount of precipitation, and the plots show a number of interesting points. The two main sets both show peak weathering rates during the two periods of freeze–thaw activity. There were forty-eight freeze–thaw cycles in exposure periods 1 and 2, and over fifty in the last three exposure periods. The Entrada set also shows the same acceleration of weathering in the freeze–thaw periods, although in this case the extent of weakening appears to be so great that considerable weight loss was still experienced after the first freeze–thaw, although at a decelerating rate. It should be pointed out that a small amount of fracturing did occur in the experiment, but, overall, this was dwarfed by the process of granular decay.

In the same experiment it was noticed that other processes were effective in releasing the surface grains on these specimens in addition to frost action. A large amount of granulation of the Entrada Sandstone occurred after a long cold spell simply through the impact of falling drops of water. Evidence of the importance of rainbeat on natural sandstone slopes exists in many parts of the Colorado Plateau province where vertical lobes of clay, silt and fine sand adhered to the cliff face as the water transporting the sediment was absorbed by the main block of porous sandstone.

It is, of course, possible to include other types of processes in this discussion of weathering-limited modes of debris removal from hillslopes. Classic rock slides and rock avalanches are, for instance, often associated with slides along bedding planes and an excellent discussion of this mode of instability is included in the paper by Terzaghi (1962*a*) already referred to. In addition, reference must be made to the extremely comprehensive studies of the complex Vaiont slide by Müller (1964, 1968) and others. The emphasis in this chapter has, however, been on processes applicable to rock masses of a fair degree of homogeneity, and those modes of instability which depend upon rather rare geological settings have been left aside.

These four processes of weathering-limited debris removal may be found in most parts of the world where mountain slopes are steep enough, although they tend to be most spectacular in arid and cold environments. In humid temperate areas and humid tropical areas, the relative ease with which vegetation colonizes even rock slopes makes the processes less apparent. The effect of a vegetation cloak is twofold. In the first place it may reduce the potency of some of the processes of rock detachment, as in the case of grain release due to the impact of raindrops, and secondly, and more important, it allows the weathered mantle to withstand the shear stresses due to gravity for a much longer period of time due to the strength afforded by penetrating root systems. The subsequent removal of the debris then hinges much less on the rate of weathering and belongs to the discussion of rapid soil mass movements, soil wash and soil creep.

RATES OF INSTABILITY PROCESSES IN ROCK MASSES:
TECHNIQUES OF MEASUREMENT AND CURRENT EVIDENCE

A discussion of the rate of operation of weathering-limited transport processes immediately highlights the central feature in any consideration of rates of geomorphic processes: the distinction between processes which involve the transport of only a small amount of material in unit time but which operate, at least on a geologic time scale, almost continuously; and those processes which are extremely sporadic, but, which, when they do operate, move vast amounts of material in a short space of time. Since few processes operate continuously, although some may occur for at least the majority of the time, this resolves itself, as noted in chapter 5, into an issue of magnitude and frequency.

An extremely instructive essay dealing with the magnitude and frequency of forces in geomorphic processes has been provided by Wolman and Miller (1960) where they argue that in the case of many processes, most of the work of evacuating material from the landscape is achieved by events of moderate magnitude operating fairly often rather than the more continual smaller-scale events or the rare catastrophes. This conclusion was linked specifically to fluvial processes such as the scour of alluvial channels, and may well be valid, in general, for these processes.

The critical factor in determining the relative importance of events of different magnitude and frequency, as transport processes, is the magnitude of the *threshold* stress, or force, necessary to initiate the process, relative to the actual range of the stress encountered in nature. It is probably true that, in general terms, the stress necessary to initiate sediment transport or scour in stream channels is fairly low relative to the range of stresses which exist, and, as a consequence, small-scale events may be very important.

An examination of the processes operative on hillslopes indicates that many of these demand threshold stresses which are so large that they occur only rarely, and, in these instances, events of a catastrophic nature may be all-important. There is also a further point, rather overlooked by Wolman and Miller, that stress is not the only dynamic element in this context. The *resistance* of the landscape to withstand these stresses is often a function of time also. The progressive weakening of a rock mass, through the extension and enlargement of joints within it, is one instance, previously discussed, of this phenomenon. Nevertheless it is clearly invalid to argue that the most important events responsible for the removal of material from hillsides are always catastrophic ones. This depends upon the *type* of process which prevails on individual hillslopes, and this is nowhere better illustrated than in the case of weathering-limited removal.

Slab-failure

The rate of denudation attributable to the slab-movement process

132

clearly varies considerably according to the climatic setting, but, above all, to the nature of the rock mass, and the general geologic environment. This is borne out in the survey of rock slides by Bjerrum and Jorstad (1968) in Norway. They noted that in some areas a large number of rock slides occurred in the period immediately after the retreat of the ice, while, in other areas, slopes have remained stable for about 10,000 years since. In many parts of the world, this process is not operating at all and, in others, it is so slow that it is dwarfed by other processes. There are, however, some areas where slab-failure does appear to be a fairly rapid contemporary process. Unfortunately measurements of this process are difficult to make and naturally rather rare. A lone example is afforded by Schumm and Chorley (1964) in the Chaco Canyon area of New Mexico using data provided by the National Park Service personnel working there.

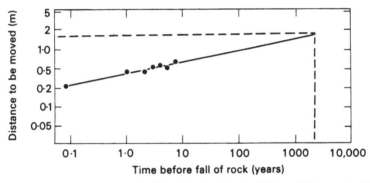

Fig. 6.11. Computation of the length of time involved in the failure of Threatening Rock (from Schumm and Chorley, 1964).

Alarmed at the apparent movement of Threatening Rock, the custodians of the Chaco National Monument installed, in 1935, two steel bars in the sandstone, one on the cliff and the other on the slab. At monthly intervals during almost the whole period of the time between installation and the collapse of the slab in January 1941, measurements were made of the gap between the slab and the cliff face, using the steel bars at the top of the rock. The gap between the stakes continually widened during this period and, even more important, apart from seasonal fluctuations, the rate of widening of the gap increased at a systematic rate throughout the period. The plot of distance to be moved by the slab against time before its fall, on logarithmic paper (fig. 6.11) is, in fact, approximately linear. In the month before Threatening Rock toppled, it moved 25 cm, and altogether a total of 55 cm in the 5 years prior to its fall. Assuming that the line linking movement against time in fig. 6.11 may be extended up to 180 cm (72 in) to be moved, the width of the gap prior to the fall, this would give the length of time before the fall in which the monolith of sandstone was gradually moving away from the cliff face. This would suggest that movement of the slab away from the canyon wall began about 2,500 years before its actual fall. On the assumption that the crack in the sandstone formed immediately subsequent to the previous fall which exposed the outer face of Threatening Rock, and assuming that movement began immediately, it is possible to infer that the process of slab-failure is responsible for a retreat of the rock wall at a rate of ten metres (the thickness of the slab) in 2,500 years, or about one metre in 250 years, which is extremely rapid by geologic standards.

The assumptions made are obviously rather dubious. In addition there is the implicit assumption that any climatic change has had little influence on the rate of movement. Nevertheless the approach is a novel one, and it provides us with an order of magnitude of the rate of cliff retreat. At the same time it illustrates some of the problems involved in assessing the rates of operation of intermittent processes.

The rates of operation of geomorphic processes often depend, to some extent, on the nature of the landform produced by the process. This is the familiar notion of 'feedback' between form and process. In the case of slab-failure there is the obvious question of whether slab retreat of a cliff proceeds, on a geologic timescale, at a fairly constant rate, assuming that climatic conditions do not change, or whether the nature of the process *itself* involves a gradual decline in the rate of retreat until the cliffs become stable. It may be recalled that Bjerrum and Jorstad (1968) suggested that the frequency of rock slides on a given cliff, at least in Norway, probably decreases over time so that eventually the slope will become stable. The landscapes of the south-western U.S.A., on the other hand, possess cliff faces which have apparently retreated many kilometres under the process of slab-failure. An important feature here is probably the presence of shale underneath the sandstone cliffs. As a consequence there is continual undercutting of the massive sandstone and it is possibly this process, rather than the development of vertical joints on its own, which is responsible for continual retreat of these cliffs.

A further point is that in order for the process of slab-failure to produce continual retreat of the cliff, there must be some mechanism by which the fallen slabs are removed from the base of the slope. In situations where this condition is not satisfied, material will accumulate in front of the steep slope at higher and higher levels until the face is completely cloaked by debris. This point is treated more fully in the last section of this chapter, and raised again in chapter 13 on the development of slopes in semi-arid areas.

134

Rock avalanches

In the discussion of slab-failure the concept of *feedback* between the rate of operation of a process and the nature of the resulting landform was introduced. This is very relevant in the case of rock avalanches and differentiates it from the other weathering-limited processes. The mechanism of rock avalanches, as described in the earlier part of the chapter, is a slide of a wedge of the rock mass arising from the breakdown of the cohesive forces in the mass. The effect of this is to replace an initially very steep slope by a slope of gentler inclination; the new slope is then stable with respect to this type of process, although it may still be subject to rockfall. It is true, admittedly, that the first slide may produce a new slope which, although gentler than the initial slope, is still greater than the equilibrium slope, but eventually another slide will take place. Accepting the assumption that the effect of the rock avalanche is to replace the existing hillslope by one of gentler inclination must mean, however, that *eventually* the slope will become completely stable with respect to this process. In this situation then the notion of a particular rate of operation for the process is rather irrelevant.

There is, however, still a time factor involved here. This is the rate at which the shear strength of the rock mass is dissipated by fracturing of the rock, since a vertical cut into an unweathered rock mass does not produce an immediate rock avalanche. This is a very complex issue. It depends on the initial degree of fracturing of the rock mass and the rate and depth of the cut into it. Undoubtedly it varies from rock type to rock type, but unfortunately there is little quantitative information available on this. A search through the literature of the stability of open-pit mines may help in this respect, although since these mines are limited to relatively shallow depths, it is probable that much of the rock mass involved is already weathered to a certain degree. Nevertheless, there is little doubt that a comprehensive search of existing literature in rock mechanics generally would assist the geomorphologist interested in the development of steep valley-side slopes.

One situation does occur, of course, where rock avalanches may be thought of as a continuing process, and this is where the rock slope is continually undercut by a stream, or some other agent, at its base. Since in this situation, however, the rate of rock avalanching is to a large extent controlled by the rate of undercutting of the slope, there is little point in delving further into this.

Rockfall

The process of rockfall is, as with slab-failure, a very sporadic process, and the main problem in determining long-term average rates is, again, the issue of frequency. It is, at least relatively, a simple matter to compute the amount of material involved in a rockfall when it occurs, but it is much more difficult to establish a reliable frequency from the records of a few years.

In many instances it is clear that rockfalls occur from some part of a rock face every year, whereas in other situations rockfalls may occur sporadically at intervals of years. In addition, even in those cases, where rockfall is an annual event, the size of the rockfall usually varies markedly from year to year. It is possible, for instance, that rockfalls associated with particularly severe weather conditions which occur only, on average, once in twenty or thirty years may move far more material than rockfalls in a 'typical' year. As a consequence, any records in a short period of time which fail to include a rockfall of this frequency will underestimate the process rate; similarly, a short record which includes it, will overestimate it. One of the greatest problems in attempting to assess the rates of hillslope processes which act intermittently, and with variable magnitude, is thus to obtain a sufficiently long record so that it is not biased by the absence or presence of infrequent events. The problem, however, is probably not as serious in the case of rockfall as with slab-failure, and a record of ten to fifteen years should, allowance being made for long-term climatic change, give a reasonable estimate of contemporary process rates.

Various methods have been employed to calculate the amounts of material involved in rockfalls. One of the longest records was obtained by Guillien (1960) from 1935–60 working on rockfall from a sandstone face in France, but, unfortunately, the methods were purely qualitative and based on visual inspection. Jackli (1957) working in the Swiss Alps employed periodic air photography to locate rockfalls, while Starkel (1959) attempted to derive estimates of rockfall rates in the Polish Carpathians by mapping at a scale of 1:10,000 over fairly small areas. The greatest contribution to quantitative measurement of rockfall rates has undoubtedly been from Rapp (1960*a*, 1960*b*) working in Spitsbergen and northern Scandinavia. In his Spitsbergen study, Rapp also used air photographs and, in addition, made calculations from the amount of debris in the medial moraines of mountain glaciers and the material accumulating in talus cones. In the Kärkevagge study in Lapland (1960*b*), installation of wire nets over talus slopes was used to estimate the volume of material in fresh rockfalls. Since most of the rockfalls occurred in the spring season in Kärkevagge at a time when the bases of many cliff walls were still covered with snow, a fairly accurate estimate of new rockfalls may also be made by determining the amount of debris on the snow surface.

In the period from 1952 to 1960, using these techniques, Rapp concluded that rockfall in Kärkevagge was responsible for the loss of fifty cubic metres of debris from the rockwalls of the studied area in the average year. Since the area of vertical rockwall in the particular valley studied amounted to 900,000 square metres, the average amount of rockwall retreat annually approximates to 0·06 mm. It was pointed out by Rapp that this figure might be an overestimate, but it was claimed that the minimum estimate would be no less than 0·04 mm in a year. In some areas of active rockfall, the rockwall must have been undergoing retreat at about 0·15 mm per year on the basis of Rapp's evidence. Since this figure might have been derived in a period of abnormal rock-

TABLE 6.1. *Rates of rockwall retreat in different environments*

Location	Rate of rockwall retreat in mm per year			Rocks	Author
	(min.)	(max.)	(mean)		
Mt Templet (Spitsbergen): last 10,000 years	0·34	0·50		{ limestone sandstone	Rapp (1960*a*)
Mt Langtunafjell (Spitsbergen): recent time	0·05	0·50		{ limestone sandstone	Rapp (1960*a*)
Kärkevagge (Lappland): 1952–60	0·04	0·15		{ mica-schist garnet–mica- schist	Rapp (1960*b*)
Austrian Alps: last 10,000 years			0·7– 1·0	{ gneiss schist serpentine	Poser (1954)
Brazil: recent 12-year period		20	2	{ granite gneiss	Freise (1933)
S. Africa: last 10–100 million years			1·5	granite	King (1956)

fall activity, it was fortunate that Rapp was able to compare his figures for the eight-year period from 1952, with records of rockfall from natural slopes onto the Kärkevagge–Narvik railroad where observations date back fifty years. These records suggested that the 1952–60 period was not exceptional, and that, although the frequency of rockfalls in this period was a little above average, the figure of 0·06 mm per annum is probably representative of the current rate of rockwall retreat. Accepting this figure as a long-term value, it implies that in this area, rockwalls have retreated about 1–2 m since glacial times. All these figures relate to an area in mica schists and garnet–mica schist. Some measurements were also made on an amphibolite face and substantially higher rates, in the order of 0·3–0·6 mm per year, were obtained there.

In view of the apparent frequency of rockfalls in the Kärkevagge area, and the severity of the climate in comparison with other parts of the world, these estimates of about 0·1 mm of rockwall retreat per year, or one metre in 10,000 years, indicate the slowness of even the more rapid of geological processes on a human timescale. In table 6.1 other figures for rockwall retreat, as summarized by Rapp (1960*b*), are given for arctic and alpine environments. It will be noticed that they are about one order of magnitude smaller than the figures reported by Freise (1933) and King (1956) in hotter areas, although it is doubtful if the processes involved are directly comparable.

Granular disintegration

The process of grain-release, described earlier in this chapter, is encountered much more rarely than the rockfall process, although there appears to be little doubt that many sandstone cliffs in desert areas are retreating in this way. There are as a consequence extremely few observations on the rate of this process.

The ease with which some sandstones, especially the Entrada Sandstone in the south-west of the U.S.A., disintegrate through granular disintegration suggests that the process may take place quite rapidly under favourable circumstances. This is supported by the casual observations made in the Colorado Plateaus by Schumm and Chorley (1966), although unfortunately they were unable to acquire sufficient data to permit a definite estimate of the rate of scarp retreat.

One indirect way of estimating scarp retreat suggested by the authors of this paper is to measure the width of major valleys in relation to the time that has elapsed since the incision of the valley. As an example, stream incision in the Grand Canyon area is supposed to have been initiated by uplift in early Miocene time, about twenty million years ago. The average width of the canyon at the rim is now about 12 km, although it varies from less than 8 to over 25 km in places. Using these figures, the average rate of rock wall retreat appears to be about 6 m in the last 10,000 years, a rate which far exceeds the value for retreat under rockfall reported by Rapp (1960*b*) in northern Europe.

Similar approaches were employed by King (1953) in South Africa, and Pugh (1955) in western Africa, in assessing rates of scarp retreat. As an example, the Natal Drakensberg, which is now 240 km inland, is supposed to have originated along a coastal flexure forming the margin of south-eastern Africa during Jurassic times. Accepting this view implies an average rate of scarp retreat of about 17 m in 10,000 years, a figure which exceeds even the estimated rate in the Grand Canyon. Other instances suggest that this figure is a reasonable average for the major continental scarps in the drier parts of the African continent.

These long-term estimates of the rate of scarp retreat have, unfortunately, rarely been accompanied by detailed observations on the nature of contemporary processes and the rates at which they operate. Indeed, since scarp retreat in the Drakensberg seems to be mainly by gullying and rill action, and wall retreat along the rim of the Grand Canyon is effected by rockfall at least as much as granular disintegration, it may be misleading to include these observations here, although they do serve a comparative purpose.

In concluding this subsection on rates of weathering-limited processes of debris transport, it is perhaps useful to emphasize the existence of two radically different approaches. One method is, through detailed observation, instrumentation and measurement, to attempt an estimate of current rates of activity. The other method seeks an indication of the long-term average rate through the amount of material which has accumulated or been removed since a distinct geological date. Ideally the two approaches should be employed together.

MODELS OF SLOPE DEVELOPMENT UNDER INSTABILITY PROCESSES IN ROCK MASSES

It has been argued elsewhere (Carson, 1969*b*) that there are two necessary and sufficient requirements for the construction of a model aimed at predicting, or, conversely, understanding, the change in the geometry of a hillslope profile over time, under a specific set of processes. The first is an understanding of the behaviour of the endpoints of the slope profile, and, in particular, the conditions which exist at the slope base. The second is an appreciation of the way in which the rate of ground-lowering between the two endpoints varies with position along the slope and with time.

In relation to the first point, it is useful to separate three different situations. One is the state of *impeded removal* of debris (Savigear, 1956), in which material moving off a hillside is allowed to accumulate on a platform at the base of the slope; this situation arises when the processes responsible for the removal of material from the base are proceeding only slowly relative to the rate of debris transport on the slope above. A second case is *unimpeded removal* of debris, in which the material moving off a hillside is cleared from the base of the slope by other processes at a rate comparable to the rate at which it is being shed from the slope. Lastly, there is a more extreme case in which stream channel processes are not only competent enough to completely remove any material brought down from the side slopes, but are capable, also, of lowering the channel bed and thus the base of the side slope.

The case of unimpeded debris removal is the simplest of the three, and, in this situation, it is necessary to know only the mechanics of the processes operating on the hillside to be able to predict the change in slope geometry over time. In the case of impeded debris removal, it is important to understand the way in which material accumulates at the base, since this process in effect adds a new component to the hillside profile. In addition, the rate of debris removal at the base *relative to the rate at which material is shed from the main slope* is likely to be a very important element controlling the overall profile. In the situation where there is concomitant stream downcutting and side-slope erosion taking place, the endpoint of the slope is again a moving one: continual stream lowering extends the slope profile. In this case there is an additional complication in the intimate dependence of the side-slope processes on the undercutting action of the slope base, as illustrated previously in the discussion of slab-failure and rock avalanches.

Within the limits of the two endpoints of the slope profile, changes in profile geometry over time are determined by the variation in $\partial y/\partial t$ (using the notation established in chapter 5) with position along the slope profile. An understanding of the basic mechanics of slope processes does allow us to make sensible assumptions about these values, and thus to construct plausible process–response models, but until there are empirical data on the rates of operations of these processes along the whole length of a slope profile, these assumptions will remain questionable.

The development of models of hillslope development under weathering-limited processes is perhaps best illustrated from the work of Fisher (1866) and Lehmann (1933), subsequently summarized and expanded by Wood (1942), Bakker and le Heux (1946, 1947, 1950, 1952), Bakker and Strahler (1956) and Scheidegger (1961a, 1961b). These models refer specifically to rockfall rather than the whole range of weathering-limited processes, but they adequately demonstrate the general techniques and problems involved in this type of work. The case of weathering-limited processes in more general terms will be dealt with subsequent to this discussion.

The most accessible summary of these ideas is to be found in the paper by Wood (1942). At the same time it should be emphasized that this is also the least quantitative of the various treatments. In this article, Wood considers the case of a straight rockwall retreating under the process of rockfall with a platform at the base of the slope upon which the talus may accumulate. On the assumption that the rockface is homogeneous, it is argued that there is no reason why the rate of retreat should vary over the rockface, and, as a result, the initial rock slope retreats parallel to itself. Talus accumulates at the base of the slope at the angle of repose of the material, and, with the retreat of the initial slope, the talus slope is extended upslope until in fact it completely obliterates the initial slope. This sequence is depicted in fig. 6.12, and it is clear that the major aspects of the slope development pattern follow immediately from the two major assumptions made. The problem, however, is to determine the form of the bedrock slope underneath the veneer of talus.

As the scree builds up at the base of the slope, it must protect the lowermost part of the initial slope from further weathering and rockfall, so that as the rockwall above the scree retreats, a part of the initial slope is left protruding into the scree. This process must continue at all stages of the sequence, since each rockfall adds more to the scree and protects a further portion of the upper slope. As Wood points out, if the scree

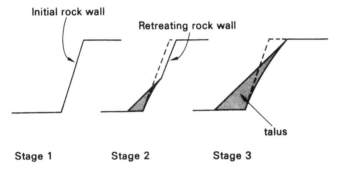

Fig. 6.12. Theoretical cliff and talus sequence assuming no removal of talus from the base.

increased in height at a constant rate relative to the rate of retreat of the rockwall, then the buried bedrock slope would be uniform. This, however, cannot be the case. At late stages in this sequence the rockwall is smaller than before and must therefore contribute smaller amounts in each rockfall. In addition at later stages the area of talus upon which new rockfall material accumulates is much greater than in the earlier stages. As a consequence, a continually smaller increase in the height of the scree is associated with the same amount of retreat of the rockwall as the sequence progresses, and it is easy to visualize that the buried rock slope must therefore be convex in profile.

In this article, Wood briefly mentions other considerations which affect the steepness of the convex buried slope. He points out that since the volume of talus must appreciably exceed the amount of material lost from the rockwall (due to the larger voids ratio of talus *vis-à-vis* un-weathered rock), assuming there is no means by which talus is moved away from the base, the talus slope will grow more rapidly than en-visaged in the previous case and the subsurface slope will thus be steeper. On the other hand, in those cases where the initial slope is less than the

141

vertical, there will be a smaller total amount of material lost from the rockwall and thus the buried convexity will not be as steep. It may be pointed out here that the talus slope is termed by Wood the *constant* slope and it corresponds with the *steilwand* of Walther Penck (1924) and Meyerhoff's (1940) *gravity* slope.

After describing the creation of the constant slope, Wood then discusses the subsequent evolution of the slope profile under other processes of debris transport. This section of the paper is out of context here and a criticism of the relevance of the overall ideas of Wood, and subsequently King (1951, 1953, 1957), will be deferred to the chapters of part three. It is now appropriate to consider one of the more mathematical treatments of the rockfall process. The first comprehensive attack on the problem mathematically was by Lehmann (1933) and his ideas are summarized briefly in Scheidegger's (1961*b*) *Theoretical Geomorphology*.

The assumptions made are essentially the same as those of Wood, although expressed more mathematically. The initial rock slope is inclined at an angle β to the horizontal. The exposed initial slope retreats parallel to itself and debris accumulates at the base at the angle of repose (α) of the debris. The volume of debris which accumulates at the base is, as was pointed out before, not necessarily the same as the volume of rock moved from the initial slope, and this is expressed in the equation

$$v_d = c \cdot v_r \qquad (6.12)$$

where v_r is the amount of material moved off the initial slope by rockfall in a given period of time, v_d is the amount of talus deposited at the base of the slope in the same time, and c is a coefficient (different from that used by Lehmann). In the case where there is no removal of material from the base of the slope, the value of c may be expected to be greater than 1, since the increase in voids ratio for the scree, relative to the intact rock, will demand v_d greater than v_r. On the other hand, in those cases where rapid removal of talus is taking place at the base of the slope, the value of c may lie between 0 and 1.

As before the problem is to determine the shape of the bedrock slope buried under the talus. In this section the mathematics of the Lehmann model are outlined in a fair amount of detail.

In fig. 6.13, let the rock slope underneath the talus veneer be given by *AB*. In this sketch the slope is drawn as convex upwards, but the shape of the profile in the diagram is irrelevant to the solution of the problem; since the exact geometry of this profile is determined algebraically, and not geometrically, the way in which it is drawn in the figure has no influence on the ultimate solution. The problem is to compute expressions for the volume of material lost from the cliff face, and for the volume of rock accumulating at the base. Insertion of these expressions into equation 6.12 will provide a further relationship describing the geometry of the *AB* slope profile.

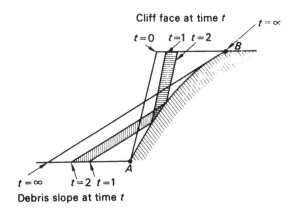

Fig. 6.13. Illustration of the process of cliff retreat according to Lehmann (1933).
≡ material removed from rockwall in time interval t_1 to t_2.
/// debris accumulated at base in time interval t_1 to t_2.

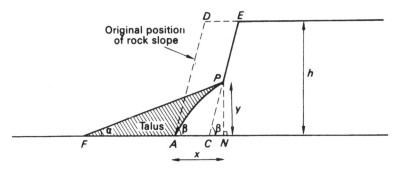

Fig. 6.14. Computation of the volume of talus accumulation in the Lehmann model.

Fig. 6.14 represents the extent of cliff retreat, from its initial position *AD*, after a time *t* has elapsed. The point at the top of the rock core, *P*, is denoted by its coordinates $x \, (= AN)$ and $y \, (= PN)$. Applying equation 6.12 to the total amounts removed up to time *t*:

Volume of scree = c.volume of rock removed.

or $$AFP = c.ADEP.$$

Adding the area *APN* to both sides:

$$AFP + APN = c.ADEP + APN,$$

or $$FPN = c.(ADEC + PCN) + (1-c).APN.$$

143

Each component in this equation can now be evaluated, giving:

$$\tfrac{1}{2}y^2 . \cot \alpha = c\{h(x - y . \cot \beta) + \tfrac{1}{2}y^2 . \cot \beta\} + (1 - c) . \int_0^x y . dx. \qquad (6.13)$$

Differentiating with respect to x; and transferring terms in dy/dx to the left hand side:

$$\{y(\cot \alpha - c . \cot \beta) + h . c . \cot \beta\} \frac{dy}{dx} = c . h + (1 - c) . y \qquad (6.14)$$

Rearrangement of this equation gives:

$$dx = \left(\frac{\cot \alpha - c . \cot \beta}{1 - c}\right) . dy - \frac{c . h}{1 - c} . \frac{(\cot \alpha - \cot \beta)}{c . h + (1 - c) . y} . dy, \qquad (6.15)$$

which can be integrated in general to give:

$$x = \left(\frac{\cot \alpha - c . \cot \beta}{1 - c}\right) . y$$

$$+ \frac{c . h}{(1 - c)^2} . (\cot \alpha - \cot \beta) . \log_e \left\{\frac{c . h}{c . h + (1 - c)y}\right\} \qquad (6.16)$$

This equation (6.16) is the expression for the bedrock surface under the talus slope when x and y are the horizontal and vertical dimensions respectively for any point on this slope profile relative to the origin at the base of the slope. The equation is quite general and specifies no particular values of c and β. Now let us examine two particular limiting cases.

The case with $c = 1$ and $\beta = 90°$ is interesting historically since this is the particular set of circumstances investigated by Fisher (1866) over a century ago. The condition $c = 1$ implies that the volume of loose debris which accumulates at the slope base is exactly the same as the volume of intact rock fallen from the initial slope. This situation might occur where the cliff face was the result of faulting, or stream erosion accompanied by a sudden movement of the stream away from the slope. The important point is that there is no mechanism by which talus is moved away from the slope. The $c = 1$ assumption is still, even then, difficult to maintain because of the inevitable increase in voids ratio from intact rock to talus in most cases. Inserting $c = 1$ and $\cot \beta = \cot 90° = 0$ in equation 6.13 gives

$$\tfrac{1}{2}y^2 . \cot \alpha = h . x \qquad (6.17)$$

This, expressed in the form $y^2 = 2h . x . \tan \alpha$, is the parabola derived by Fisher.

The case where $c = 0$ is, however, geomorphically more interesting since it corresponds to the situation where all the debris is lost and none accumulates at the foot of the slope. This is the common situation where a stream at the base of the slope is capable of moving away all the talus which would accumulate at the base, although the stream itself does not undercut the slope.

Inserting this value of c into equation 6.14, we obtain:

$$y \cdot \cot \alpha \cdot \frac{dy}{dx} = y$$

or
$$dx = \cot \alpha \cdot dy. \qquad (6.18)$$

On integration, we derive

$$x = y \cdot \cot \alpha + A$$

and, with $A = 0$,

$$x = \cot \alpha \cdot y \qquad (6.19)$$

which is the equation of a straight line of gradient α to the horizontal. The bedrock surface produced by the retreat of the initial rock slope is thus, in this case, a straight hillside with a veneer of talus, standing at the angle of repose of the rocky material. This is an extremely interesting result because, as will be shown later, there are many slopes in the land-scape which are completely straight and mantled only by the thinnest veneer of debris and standing at the angle of repose of this material.

This type of work initiated by Fisher and Lehmann has been developed extensively by Bakker and le Heux, and other Dutch workers, and the ideas have been applied to peaked as well as flat-topped hills and non-parallel as well as parallel retreat of the mountain slope. The most interesting observation by these workers is contained in the paper by Bakker and le Heux (1952) where they demonstrate that, under conditions with $c = 0$, the emergence of a straight rock hillside at the angle of repose of the talus is an inevitable development *irrespective of the type of mountain wall recession*. In the words of the authors, 'in every type of approximately continuous recession by the free play of weathering removal on steep mountain walls of well nigh homogeneous fairly resistant rock over a terrace, a denudation slope with a rectilinear rock-profile and a slope angle equalling the screes-angle of the rock concerned will be formed, provided this screes-angle is not subject to noticeable changes during the recession-process and no, or hardly any, scree is deposited at the foot of the initial wall'.

Notwithstanding a definite elegance to this mathematical treatment of slope development under rockfall, there is no doubt that the assumptions are rarely satisfied. One of the assumptions in the derivation of the 'remarkable new geomorphological law' of Bakker and le Heux described above is that 'the screes-angle is not subject to noticeable change during the recession process' and there are many objections to this. Current laboratory experiments by Kirkby suggest that only under special circumstances will rockfall produce a straight talus slope at the angle of repose; usually a concave talus slope profile results. In high-latitude areas, snow avalanches produce appreciable downslope movement of the talus while it is accumulating, and in other areas similar processes may achieve the same effect.

No mention has been made of the case where stream downcutting and rockfall take place simultaneously. According to many of the early geomorphologists interested in slope development, this situation exists

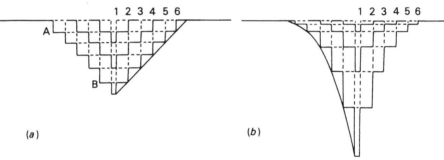

Fig. 6.15. Slope development under a steady rate of rockfall and (*a*) a steady rate of stream-lowering and (*b*) an accelerating rate of stream downcutting.

commonly in the early stages of landscape development. All the talus produced in this situation is removed by the stream, and the interest centres on the profile of the slope subjected to rockfall. If it is assumed that stream downcutting occurs at a steady rate, and that weathering is uniform over the rockwall, then it is easy to appreciate that a straight hillside will develop. It is straight (fig. 6.15*a*) because a higher point on the slope (*A*) has retreated further from the channel than a lower point (*B*) in proportion to the amount of time that elapsed during downcutting from the level at *A* to the level at *B*. It is also easily appreciated that the angle of this slope depends on the rate of channel lowering relative to the rate of rockfall, and may be expected to vary with rock type. As was pointed out in chapter 2, Walther Penck and others argued that in the early stages of downcutting the rate of channel lowering is likely to accelerate, at first, and, as a result, a convex hilltop (fig. 6.15*b*) will develop; in fact, this seems unlikely. The entire model outlined in fig. 6.15 is not only speculative, but probably irrelevant. In the *early* stages of landscape development on jointed rock masses, rockfall is unlikely to be important, because other weathering-limited processes, particularly rock avalanches, are likely to dominate. The rockfall process will probably only assume importance when slopes have become stable with respect to rock avalanches, and since this will occur only when stream downcutting is very slow, the model of concomitant stream-lowering and rockfall on the side slopes is probably a purely hypothetical situation.

In conclusion it is perhaps necessary to add that comparative models have not been developed for slab-failure and granular disintegration, although this is not serious since these two processes are, relatively speaking, rather rare. In addition, *in terms of the development of the slope profile*, it is perhaps unimportant to separate granular disintegration and slab-failure from rockfall; the important distinction is between those processes (rock avalanches) *where debris removal affects a whole mass of material, not just the surface skin*, and those processes (rockfall, slab-failure, granular disintegration) *where removal of debris is confined to successively exposed surface layers*. A model invoking rock avalanches is fairly simple: according to the mechanics outlined by Terzaghi (1962*a*), once avalanching has reduced a slope to its stable angle, continual undercutting merely lowers the slope at an unchanged angle. Once undercutting ceases, there is no further avalanching, and other *weathering-limited* processes then take over. In the case of the superficial *weathering-limited* processes, the development of the slope profile over time depends not so much on the individual type of process (rockfall, granular release or slab-failure), but on the rate of weathering and associated process *relative to* the rate of debris removal at the base of the slope. Talus slopes rarely develop in front of sandstone cliffs undergoing granular disintegration, for instance, but the form of these profiles may be little different from slopes where rockfall is taking place but where talus is rapidly broken down at the base and quickly removed. The problem of slope profile development is now clearly a *multi*-process one, and some comment upon this will be made in chapter 13.

CHAPTER 7

INSTABILITY IN SOIL MASSES

The processes of instability, or mass failure, in soils have been strangely neglected by geomorphologists until very recently. It is true that Sharpe (1938) long ago produced a classification of mass-movements, and that his ideas have been presented in almost every textbook in geomorphology since then; but the usefulness of Sharpe's ideas is not immediately clear. Little attention was paid to the mechanism of these processes, and even less to the role they played in shaping the sides of valleys and hills. Mass movements have, for too long a period of time, attracted attention merely because they constitute landforms in themselves, rather than because they contribute in a special way to the denudation of the landscape. Even in the recent text by Leopold, Wolman and Miller (1964), the ideas presented on instability processes in soil masses have not changed substantially since Sharpe's work. Our understanding of the nature of instability in soil masses has only been advanced by the efforts of soils engineers; the works by Ward (1945), Skempton (1948) and Terzaghi (1950), in particular, mark the first real progress.

THE TIME SCALES OF INSTABILITY

Soils engineers became interested in problems of mass failure when it was necessary to modify natural slopes, or to create artificial ones as in embankments, cuttings and other excavations. In creating new slopes, which were usually bigger or steeper than the existing natural slopes, many landslides were accidentally induced by increasing the shear stresses in the soil mass. The mechanism of this is shown in fig. 7.1. On the bigger slope, the shear stress on any potential failure plane is much larger, relative to the shear strength along it, than in the case of the smaller slope. The reason for this is quite simple. The shear stress increases directly with the weight of the block of soil above the plane; the increase in the strength of the soil, as the weight of the block above increases, is much less, because only the frictional part (ϕ') is related to the effective normal stress, the cohesion element (c') being constant with depth. This has already been discussed briefly in chapter 4. Artificial cuttings by engineers thus created slopes where sometimes the shear stress was increased so much that it exceeded the strength along a critical surface in the soil mass, and a failure occurred. Engineers naturally became very concerned about the stability of man-made slopes, and stability analyses were made to make sure that the new slope would be stable when it was made. Such stability analyses are referred to as *short-term* analyses; implicitly it was usually assumed that if a cutting was stable at the end of the construction period, it would not fail afterwards.

Subsequent experience demonstrated, however, that many new slopes which were stable at the end of construction nevertheless became unstable at some time in the future. Since many of these slides took place without any sudden external cause, it could only be assumed that, as the shear *stresses* had changed little since the end of construction, a gradual reduction of the shear *strength* had taken place, hence the failure. In this way, engineers became concerned with the *long-term* stability of man-

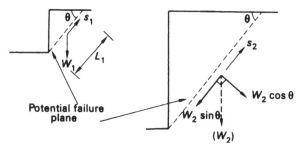

Fig. 7.1. Shear force and shear strength on a plane failure surface. Shear strength $(S) =$ $c'.L + W.\cos\theta.\tan\phi'$; shear force $(T) = W.\sin\theta$.

made slopes as well as the *end-of-construction* case. One of the effects of this was the trend towards the use of effective stress analysis, rather than the total stress mode of stability analysis which had been used for short-term failures. Although the effective stress principle has been emphasized in chapter 4, it is still perhaps worthwhile here to distinguish between the short-term (total stress) stability analysis and the long-term (effective stress) method. This is illustrated by reference to fig. 7.2. The treatment is adapted from Bishop and Bjerrum (1960).

A cut is made rapidly in the clay mass, as in the diagram; the effects of this are summarized in fig. 7.3. The first result is clearly a drop in the average overburden pressure and in the total normal stress at point *A*, as shown in fig. 7.3*a*. In addition, there is also a gradual increase in the shear stress at *A* during the construction period, as shown in fig. 7.3*b*. Another effect of the unloading is that there is a tendency for the clay mass to expand, due to the release of much of the confining pressure. This has an important effect on the pore pressures in the clay mass. If it could be assumed that water could enter the clay mass at the same rate

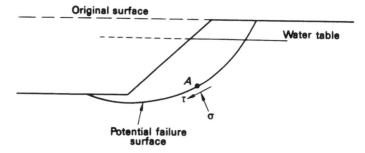

Fig. 7.2. A cutting in unconsolidated material.

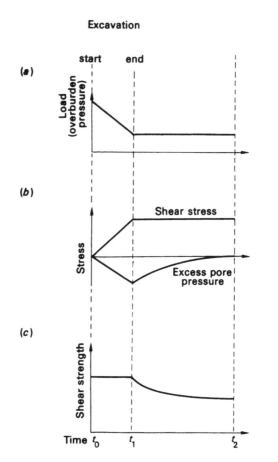

Fig. 7.3. Changes in shear stress and shear strength over time in a cutting in unconsolidated material (adapted from Bishop and Bjerrum, 1960).

as unloading took place, the pore pressures would not change appreciably. Most clays are very impermeable, however, and water would not be able to enter at this rate; as a consequence, pore pressures must fall, creating excess negative hydrostatic pressures (chapter 4). Although this may be difficult to envisage immediately, it is really just the opposite of the case where the clay mass is consolidating. The consolidation of a clay, due to the imposition of an extra load, is simply the result of expulsion of porewater from the soil mass; if the application of the load is very rapid and the permeability of the clay mass is very low, expulsion of the porewater will take place too slowly to maintain the original pore pressures, and these pressures will inevitably rise. In the limiting case where there is no expulsion of water, there is, of course, no consolidation, and all the extra load is taken up by the build-up of excess positive hydrostatic pore pressures. To return to the stability problem of fig. 7.2, there is thus a decrease in pore pressure during the construction phase. In the extreme case where there has been no drainage into the clay during excavation, the change in pore pressure is the same as the change in the overburden pressure; the shear strength at point A would therefore be unchanged.

This is the important point in short-term stability analyses: although the total normal stress at A has changed during construction, there will have been, in the limiting case, no change in the shear strength of the clay mass. In other words, the shear strength is independent of the *total* stress during the construction phase, and, thus, in effect, behaves as though $\phi = 0$. Indeed this type of analysis is known as the $\phi = 0$ *method*; moreover, since it uses only the *un*drained *cohesive* strength of the soil, and no pore pressure data, it is extremely simple to use.

Although the total stress method is valid for the short-term stability case, it breaks down during the period after construction. There is then time for drainage to take place and, gradually, the negative excess hydrostatic porewater pressures will become dissipated, and, as shown in fig. 7.3c, the pore pressure will slowly rise until it reaches the hydrostatic condition which existed before construction. This means that the shear strength at A must decrease because there is, during this time, no change in the total overburden pressure to counteract changes in the pore pressure. In order to obtain the correct strength in the long-term state, therefore, it is essential that the pore pressure is known, so that the *effective* normal stress may be used in conjunction with the *drained* shear strength parameters c' and ϕ', rather than the *un*drained strength ($s = c$) previously. The appropriateness of the $c'\phi'$ *stability analysis*, as against the $\phi = 0$ method, in the long-term stability of clay slopes has been demonstrated empirically in a number of papers (e.g. Hutchinson, 1961) in the last two decades. The article by Bjerrum and Kjaernsli

(1957), dealing with natural clay slopes in Norway, in particular, offers a neat summary of a fairly large collection of case studies.

This concern with porewater pressure changes is not, however, the only development associated with the interest in the long-term stability of man-made slopes. The development of a cutting such as the one in fig. 7.2 may also have a definite effect on the value of the shear strength parameters themselves. The important point here is that, in producing such a cutting, the shear stresses within the soil mass are disturbed, and a progressive reduction in strength upslope from the base may take place. This is considered briefly in the next two sections on shallow slides and deep-seated slips.

The importance of the distinction between short-term and long-term stability to natural, as distinct from man-made, slopes may not be immediately apparent. The undercutting of a slope by Man is, however, in many ways analogous to stream incision in the landscape. As streams cut down, they lengthen the side walls of valleys, and eventually instability occurs. Most stream undercutting occurs in times of large floods and is often accompanied by landslides; these are short-term mass movements. Sometimes, the effect of undercutting produces no immediate response and instability is delayed until some time after the flood; this is the long-term case. Some instances of mass failure occur, however, on slopes which have not been undercut for a very long period of time, measurable in thousands rather than tens of years, and which clearly do not fall into the *engineers'* category of long-term. On a time scale much greater than the engineers' long-term phase, weathering and related processes produce changes in the shear strength of soil masses which may result in instability. In addition to the short-term and long-term failures discussed above, we need also to add those that belong to the geologic time scale. Indeed it would be possible to organize this chapter on the basis of these three time scales, but, as in the case of *weathering-limited* processes in chapter 6, we prefer to discuss the various processes in terms of the mode of instability that occurs. In the following section, we shall discuss in turn: (*a*) shallow slides (*b*) deep-seated slips (*c*) earthflows and (*d*) micro-scale instability features.

THE MECHANICS OF INSTABILITY PROCESSES

Shallow slides

Shallow slides have, in general, received less attention in soil mechanics textbooks than the more spectacular deep-seated slips discussed next, and yet shallow slides are probably the most common form of instability on natural slopes. Most slopes in temperate areas are debris-mantled slopes, that is, the main hillmass is strong rock, but is mantled

152

by the products of weathering. Such mantles are usually thin (1–2 m) (although in tropical areas they may be 30–40 m thick) so that deep-seated slips are clearly precluded, and any form of instability must be of the shallow type. Shallow slides are not, however, confined to residual soils, but have been documented and analysed on thick clay deposits deep enough to produce rotational slips. One of the earliest analyses of a shallow type of failure was by Henkel and Skempton (1954) on a slide in a residual soil mantle overlying shales of the Coal Measures series near Jackfield, Shropshire. The slide was about 200 m long but only 5 m beneath the surface, occurring on a fairly straight valley-side slope inclined at about $10\frac{1}{2}°$.

Stability analyses of shallow slides in which the failure surface is

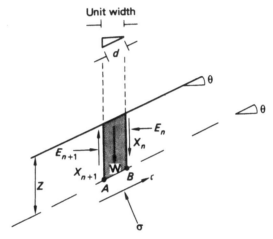

Fig. 7.4. Stability analysis of an infinite slope subject to shallow slides.
$E_n = E_{n+1}$; $X_n = X_{n+1}$; vertical stress $= W/d = \gamma.z.\cos\theta$; shear stress $= \tau = \gamma.z.\cos\theta.\sin\theta$; normal stress $= \sigma = \gamma.z.\cos^2\theta$; $\gamma =$ unit weight of soil; $d = 1/\cos\theta$.

approximately planar and parallel to the surface are fairly simple, and illustrated in fig. 7.4. It should be emphasized that the end and side effects are ignored, and the analysis is considered in terms of an infinite layer of clay on an inclined plane; the justification for this treatment is the great length of the slope relative to the depth. With this method we need consider only the shear stress and strength at the base of one column of soil. The reason for this is as follows. The failure plane is assumed to be at a constant depth beneath the surface along the slope; the soil mantle is assumed to be uniform along the slope; and it is further assumed that end and side effects are negligible. It follows that the stability of a single column of soil, of unit lateral dimensions, should be a reasonably accurate indicator of the stability of the slope as a whole.

153

The stability of the slope is measured by the ratio of the shear strength resisting a slide and the shear stress acting to induce movement. In this case, the *factor of safety* (with respect to shearing strength), F_s, is given by

$$F_s = S/T = s/\tau \tag{7.1}$$

where s is the shear strength along the base of the column AB, expressed as a stress (S as a force), and τ is the shear stress (T as a force) along AB. The shear stress along AB is given by

$$\tau = \gamma . z . \sin \theta . \cos \theta \tag{7.2}$$

where γ is the bulk unit weight of the soil, z is the depth to the failure plane, and θ is the angle of slope; this expression is simply the downslope

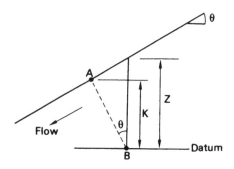

Fig. 7.5. The relation between pore pressure and depth in a soil mass with groundwater flow parallel to the ground surface (from Carson, 1969a). Pore pressure at $A = 0$, and at $B = u_1$; positional potential at $A = \gamma_w . k = \gamma_w . z . \cos^2\theta$, and at $B = 0$; total head at $B =$ total head at A, $\therefore u_1 = \gamma_w . z . \cos^2\theta$; $\gamma_w =$ unit weight of water.

component of the weight of the column of soil, as a stress. The shear strength, per unit area of the base of the column, is given by

$$s = c' + (\sigma - u) . \tan \phi', \tag{7.3}$$

assuming effective stress analysis, where σ is the total normal stress and u is the pore pressure along AB. This may be expanded to

$$s = c' + (\gamma . z . \cos^2 \theta - u) . \tan \phi', \tag{7·4}$$

substituting for the total normal stress.

In the Jackfield slide area during the winter the groundwater table is very close to the surface. Observations in wells in the area at the time of the slide indicated this, and confirmed that pore pressures were hydrostatic, that is, there were no artesian pressures involved in the slide. In the stability analysis it was therefore assumed that the water table was at the surface, and that ground water flowed freely downslope parallel to the ground surface. Under these conditions, assuming iso-

154

tropic permeability conditions, the relation between pore pressure and depth, as shown in fig. 7.5, is given by

$$u = z \cdot \gamma_w \cdot \cos^2 \theta \qquad (7.5)$$

where γ_w is the unit weight of water. Substituting this value for the pore pressure into equation 7.4, we have

$$s = c' + z \cdot \cos^2 \theta \, (\gamma - \gamma_w) \cdot \tan \phi'. \qquad (7.6)$$

Samples were taken from the failure plane to determine values for the remaining variables in this equation. These were approximately

$$c' = 0 \cdot 073 \text{ kgf/cm}^2$$
$$\phi' = 21°$$
$$\gamma = 2,100 \text{ kgf/m}^3$$

and substitution into equations 7.2 and 7.6 allows the calculation of τ and s, and, hence, the determination of the safety factor.

Using these values in fact produces an F_s value of 1·4 which is clearly meaningless, because, at the time of the slide, the shear stress should have *just* exceeded the strength of the material, and the safety factor should have been unity. One possible explanation is that the effective stress method of analysis is inappropriate here, but, as expected, the total stress analysis produces an answer which is even more in error. Use of effective stress analysis with a value of c' equal to zero produces an F_s factor very close to unity, suggesting that, at least in the case of the Jackfield slide, the clay mass behaved at the time of failure as though it were cohesionless. The authors hypothesized that the negligible cohesion intercept in the clay might have been due to the presence of many deep fissures in the soil mass which might have led to a progressive reduction of the strength of the clay.

Support for this idea was provided by later work by Skempton and DeLory (1957) in the London Clay. An extensive survey of natural slopes in the London Clay area by these authors indicated a distinct boundary between unstable and stable slopes at about $10°-10\frac{1}{2}°$; very few slopes above this angle were stable and no unstable slopes were noted below this angle. The unstable slopes in the area were mostly scarred by slides of the shallow variety. As in the case of the Jackfield slide the length of these slides was very much greater than the depth, so that it seemed reasonable to assume that the failure surface approximated to a plane, and the same type of stability analysis was applied. The weathered zone of the London Clay on these slopes has the following average properties:

$$w = 33 \quad LL = 80 \quad PL = 29$$
$$c' = 0 \cdot 12 \text{ kgf/cm}^2 \quad \phi' = 20° \quad \gamma = 1,900 \text{ kgf/m}^3$$

Substituting the values of $\gamma, z, \theta, c'.\phi'$ and u into equations 7.2 and 7.4 would provide us with an estimate of the F_s value of these slopes. An alternative approach is to assume that F_s equals unity and solve for θ, the slope angle; this would produce the theoretically maximum stable slope with respect to shallow slides, and enable a comparison with the empirically observed value of about 10°. Again the most acute problem is the assumption of the appropriate value of pore pressures. As at Jackfield, however, many slopes undoubtedly had water tables very close to the ground surface during the winter, and the pore pressure model of fig. 7.5, giving the previous equation 7.5, was used.

Earlier experience with the Jackfield slide had suggested that in strong fissured clays, slopes behave as though the cohesion value is zero. Since the London Clay is also a strong, fissured clay, it was assumed that a zero c' value was also more appropriate here. On the assumption that $F_s = 1$ and $c' = 0$, equations 7.2 and 7.6 yield:

$$\gamma.z.\sin\theta.\cos\theta = z.\cos^2\theta\,(\gamma - \gamma_w)\tan\phi', \tag{7.7}$$

simplifying to

$$\tan\theta = \left(\frac{\gamma - \gamma_w}{\gamma}\right).\tan\phi'. \tag{7.8}$$

With $\gamma = 1,900$ kgf/m^3 and $\gamma_w = 1,000$ kgf/m^3, this approximately yields:

$$\tan\theta = \tfrac{1}{2}\tan\phi'. \tag{7.9}$$

Insertion of $\phi' = 20°$ into equation 7.9 produces a θ value which is very close to 10°, agreeing remarkably well with field evidence.

Further evidence was thus provided to support the use of effective stress analysis, with $c' = 0$, for the long-term stability of natural slopes cut into stiff, fissured clays. It is important, however, to emphasize that these results appear to be valid only for fissured clays. Analyses of slides in non-fissured (intact) clays, such as at Lodalen, Norway (Sevaldson, 1956) and Selset, Yorkshire (Skempton and Brown, 1961), which will be discussed in the section on deep-seated slips, indicated that in these clays the whole cohesive element appears to be fully mobilized during failure.

Many of these points were summarized and drawn together by Skempton (1964) in an outstanding paper on the long-term stability of clay slopes. He emphasized, in particular, the importance of distinguishing between the peak and residual values (chapter 4) for the shear strength parameters of a clay. In the past, almost all stability analyses had used peak values, and little attention had been paid to the residual strength concept. Skempton emphasized that peak and residual strength values may be very different in clays, especially for the cohesion element, pointing out that invariably, in the residual state, the c' value was negligible. Without appreciable error, it appears that

$$s_r = \sigma' \cdot \tan \phi'_r \qquad (7.10)$$

where the subscript r emphasizes the residual condition. The immediate conclusion from this is that the stiff, fissured clays at Jackfield, and around London, behaved as if c' were zero simply because they existed in the residual and not the peak state. In assuming that $c' = 0$ in these two cases, Skempton and his co-authors had thus implicitly used the residual and not the peak value.

At this point, it is pertinent to pose a number of questions. What produces the loss of strength between the peak and residual states? Why is it that fissured clays appear to be pushed past the peak strength condition to the residual state, and why does this not take place with intact clays? The answers to these questions have still not been settled. One mechanism of the reduction in strength in moving to the residual state, suggested by Skempton (1964, p. 81), is an increase in water content; this arises from the fact that over-consolidated clays expand during the shear process, especially after passing the peak. The mechanism by which clays moved past the peak state was thought to be as follows. Assume that, at some point in a clay mass, the clay is forced past its peak; the strength at that point will drop and will throw additional stress on neighbouring points. As the strength at these points passes the peak, a chain reaction will develop leading to a type of progressive failure. The problem, of course, is to initiate the process. Skempton argued that fissures act as stress concentrators in clays in an analogous way to cracks in solid materials (Cottrell, 1963), and, in this way, in such clays, points within a clay mass may pass the state of peak strength even though the mass as a whole is stable. Other possibilities do exist, however, and the interested reader is referred to the original paper. More recently Bjerrum (1967) has offered an alternative mechanism for progressive failure which is not dependent upon the existence of fissures. Shortage of space unfortunately prevents discussion of Bjerrum's concept here, and, again, the reader is referred to the original article.

A further interesting outcome of the work on shallow slides in the London Clay is the suggestion that, under particular ground-water conditions, the limiting stable slope approximates, in tangent form, $\frac{1}{2} \tan \phi'$, assuming that the mantle behaves as though it is cohesionless. Support for this idea has more recently been provided by Carson (1967) working on gritstones, shales and limestones mantled by residual soils in parts of the southern Pennines and northern Exmoor. This work will be referred to again later in this chapter, but it is important to emphasize a very significant point here. Many hillslides occur in strong rocks which are mantled by residual soil mantles. Deep-seated slips are precluded by the shallowness of the mantle. The stability analysis used in the Jackfield slide is therefore appropriate. In addition, as pointed out in chapter 4, the existence of a loose permeable soil mantle overlying relatively impermeable parent rock affords ideal conditions for the creation of perched water systems with subsurface flow parallel to the ground slope, thus satisfying the pore pressure model given in equation 7.5. Under these conditions the state of limiting equilibrium is given by equation 7.8 and the threshold slope, in tangent form, will approximate to $\frac{1}{2} . \tan \phi'$. In addition, as will be pointed out in the last section of this chapter, since most slopes in the landscape probably occur near limiting equilibrium, slopes at about this angle are probably very common.

A word of warning is, however, necessary here. The applicability of this type of stability analysis to shallow slides looks very simple, but it rests on a number of key assumptions, and most important is the choice of the appropriate pore pressure model. In the analysis of actual landslides, observations of pore pressures, via wells and piezometers, are often available and these may be used in the determination of the pore pressure pattern. On the other hand where the problem is to determine the angle of the maximum stable slope on a geologic time scale, as in the London Clay study by Skempton and DeLory (1957), as against the stability analysis of an actual slide, the problem is more complex. In this case we need to know the *worst* pore pressure conditions which will occur, and, since they may occur very infrequently, short-term observations on pore pressure patterns may provide little assistance. The landslides associated with the 1952 Exmoor storm (Gifford, 1953; Dobbie and Wolfe, 1953), for instance, appear to have been brought about by weather conditions with a return period of greater than 150 years. In many cases it is very probable that the worst pore pressure conditions are given by the existence of the water table at the surface and subsurface seepage parallel to the ground slope (equation 7.5), and this should be especially common where a residual soil mantle of high permeability overlies strong impermeable rock. On the other hand, there are many known instances where this kind of model is not valid; an impermeable mantle over a more permeable bedrock might, for instance, lead to artesian pressures, as pointed out in chapter 4. In the case of thick soil masses, rather than shallow residual soils, ground water flow patterns may be extremely complicated, as indicated in an excellent theoretical treatment of this situation by Deere (1967). At the present time, the choice of the appropriate pore pressure model is probably the Achilles heel in analyses of the long-term stability of natural slopes, and much more work on the nature of subsurface water patterns is necessary.

Solifluction lobes are sometimes regarded as a special type of shallow landslide or earthflow, and it may be that this mode of instability depends upon the development of high pore pressures in permafrost areas during melt conditions. So far, however, the limited evidence available (e.g. Williams, 1966) indicates that excessive pore pressures are not common in solifluction lobes, and that it is better to regard these features as a special category of soil creep phenomena. The treatment of solifluction lobes is therefore deferred until chapter 10.

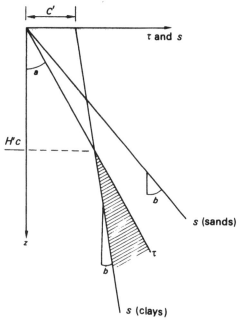

Fig. 7.6. Change in shear stress (τ) and shear strength (s) with depth on an infinite slope. $a = \tan^{-1} (\gamma . \sin\theta . \cos\theta)$; $b = \tan^{-1} ([\gamma - \gamma_w] . \cos^2\theta . \tan\phi')$ assuming equation 7.6. H_c' is critical depth for deep slips, z is depth below surface.

Deep-seated slips

Some of the mechanics underlying deep-seated failures in strong rocks have been discussed in connection with slab-failures and rock avalanches in chapter 6. It was pointed out there that stable vertical cuts may be made in rock, and that, provided the depth of the cut is less than a critical height (H_c'), the wall will not fail until weathering reduces the shear strength of the rock mass. It was also noted that the existence of vertical cliffs is directly due to the cohesive element of the rock mass, and that, once this disappears, the wall will become unstable irrespective of the depth of the cut. Similar mechanics may, with modification, be applied in the analysis of deep-seated failures in soil masses.

159

There are three important differences between deep-seated failures in rocks and in soil masses. In the first place, deep-seated instability does not occur in all soil masses, even if the mass is thick enough. Deep slips are confined to clayey soils and do not occur in sands. Secondly, although deep-seated failures in rocks often occur on plane surfaces, most deep slips in clay masses are rotational in type, and deep planar slides are very rare. And, lastly, the stability analysis of equations 6.7 to 6.10 made no reference at all to the pore pressures in the rock mass; in soil masses, as emphasized in the previous section, stability analyses of long-term failures cannot ignore pore pressures. These points are elaborated upon in turn below.

The reason why deep slips are confined to clays and do not occur in sandy soils is quite simple and may be understood by reference to fig. 7.6. This diagram is meant to refer to the type of situation depicted in fig. 7.4 on a slope of infinite length, and it shows the effect of increasing the depth of the failure surface. The shear stress clearly increases with depth in accordance with equation 7.2; the shear strength also increases with depth in a soil mass, but the rate of increase is much less in the case of a clay since the ϕ' part of the shear strength is much smaller than in

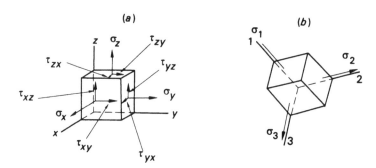

Fig. 7.7. (*a*) Stresses on an element referred to arbitrary axes of direction; (*b*) principal stresses on an element (after Wu, T.H., 'Soil Mechanics', 1966, p. 119, Allyn & Bacon, Inc.).

sands. At the surface of a clay mass there is a certain amount of strength due to cohesion; shallow slides will not take place in the situation depicted in fig. 7.6. The shear stress increases more rapidly with depth than the strength of the clay, and at greater depths is larger than the shear strength; this is the mechanism behind deep-seated slips in clay. In the case of the sandy soil the strength line never crosses the stress line: both radiate from the origin of the graph, since the shear strength at the surface of a sandy mass is theoretically zero. This arises from the absence of cohesion in sandy soils, and the fact that, at the surface of the mass, the normal stress which generates friction is also zero. The strength line for sands is therefore *either* entirely to the left (unstable at all depths) *or* entirely to the right (stable at all depths) of the shear stress line.

The explanation of curved rather than straight failure surfaces in deep-seated slips in clays, in contrast to the planar slide surfaces described in chapter 6 in strong rocks, is a little more complicated, and demands that the reader understands the concept of principal axes and principal planes within a soil mass. This is therefore very briefly summarized below.

In fig. 7.7 is shown the generalized case of stresses at a point within a soil mass. There exist shear and normal stresses on each of the six faces of the cube. It is not difficult to envisage that, for the particular set of stresses shown in fig. 7.7a, there must exist three mutually perpendicular planes *on which* there exist only *normal* stresses. These planes (fig. 7.7b) are called *principal planes*. The normal stresses acting on the principal planes are called *principal stresses*. A two-dimensional example of this concept has already been given in chapter 4 in the discussion of the triaxial test, illustrated in fig. 4.8a. The stresses in the direction *around* the sample and the *vertical* stress tending to consolidate the sample are both principal stresses; there is no shear stress on the surfaces perpendicular to these stresses. Shear stresses only appear along planes inclined at an angle to the vertical, such as the plane on which failure takes place. There are three principal stresses in a soil mass, and, ranked in order of magnitude, these are the major (σ_1), intermediate (σ_2) and minor (σ_3) principal stresses. In the case of the triaxial test just mentioned the σ_2 and σ_3 stresses are equal since the all-round pressure in the cell is constant.

The important point about the major principal plane is that when a soil mass fails through a build-up of the shear stress within it (as against a decline in soil strength), the slip surface is inclined at an angle $\alpha = 45° + (\phi/2)°$ to the major principal plane (Wu, 1966, pp. 210–11). It will thus be immediately apparent that in those cases where the principal plane does not change direction within the soil mass, the failure surface will be planar and inclined at an angle α to the major principal plane. This is precisely the case in the triaxial test (fig. 7.8a): the major principal plane is horizontal, and the slip surface is inclined at an angle of $45° + (\phi/2)°$ towards it. In the case where streams cut down into a strong rock mass and produce blocks of rock bounded by vertical sides (fig. 7.8b), as described in chapter 6, an analogous situation exists: the major principal axis is vertical and acts on a horizontal plane surface; the build-up of shear stresses within the mass, through further incision by the streams, will produce slipping on a plane surface at an angle α to the horizontal as indicated previously.

Process: the interaction of force and resistance

Now, in the case of valley-sides which are not vertical (fig. 7.8c), the direction of the principal axis is not constant within the soil mass, as shown in the diagram, and, thus, the direction of the major principal plane is continually changing also. It is therefore clear that the failure surface cannot be planar, but must be curved, and from the pattern of the major principal axis in fig. 7.8c it can be appreciated that the failure surface will approximate to the arc of a circle. (An elementary discussion of the pattern of failure surfaces in natural materials has been provided

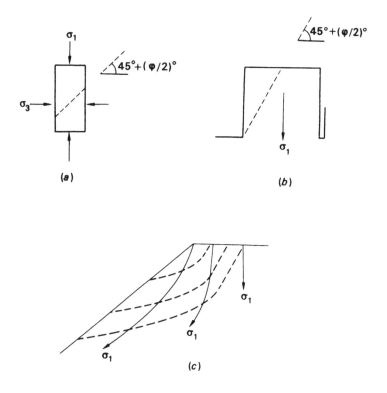

Fig. 7.8. The relationship between orientation of the major principal stress and potential slip surfaces in different situations (case (c) adapted from Sokolovski, 1956).

elsewhere (Carson, 1971b) and the reader who is not familiar with the basic ideas contained in the Mohr circle approach is referred to this treatment.) Observations in the field (e.g. Lohnes and Handy, 1968, p. 249) lend support to this reasoning: in the same material, high-angle slopes usually fail on a plane passing through the toe of the slope, whereas lower-angle slopes tend to fail along a circular arc or a logarithmic spiral. The reason that strong rocks seem to fail most commonly on plane surfaces, whereas deep-seated slips in clay masses are curved, appears to be simply that stream incision into strong rock masses is quite capable of producing vertical gorges (chapter 6), whereas, in clay masses, downcutting is *accompanied by* a rapid lowering of the angle of the side slopes produced. The implication is that denudational processes such as shallow landslides produce low-angle slopes in clay masses

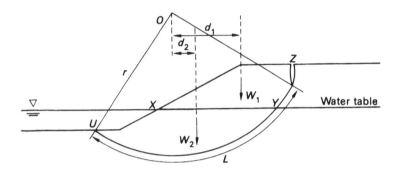

Fig. 7.9. Stability analysis of a deep-seated slip: the circular arc method.
Notation: W_1 is the weight of area XYZ; W_2 is the submerged weight of area XUY; acting through the centres of mass; s is the shear strength per unit area along the arc. Disturbing moment $= W_1 . d_1 + W_2 . d_2$; restoring moment $= s . l . r$; $F_s = (s . l . r)/ (W_1 . d_1 + W_2 . d_2)$.

during the early stages of stream downcutting, and, then, when the depth of the valley approaches the critical depth of the soil mass, deep-seated failures occur on a curved slip surface. Skempton (1953) describes such a sequence based on field observations; it is discussed more fully in the last section of this chapter.

Most of the pioneer work in the study of slope failures in clay masses was by Swedish engineers; they were the first people to emphasize the curved nature of deep-seated slip surfaces. The first published analysis of slope stability using a circular slip surface was by Fellenius (1927, 1936), and it is termed the Swedish circle method. Subsequently Taylor (1937, 1948), in particular, tested many slips using stability analyses based on circular arcs, or logarithmic spirals, and found close agreement between theory and reality.

One method of undertaking the stability analysis of a rotational slide is illustrated in fig. 7.9. The slip-surface shown in the diagram has been arbitrarily chosen to demonstrate the analysis. The failure is a *base failure*, using the term as suggested by Terzaghi (1943, p. 147) to denote

163

the situation in which the slip circle cuts the ground surface *in front of* the slope base. (Slips in which the failure surface passes directly through the slope base are termed *slope failures*, although, in this book, we also use this term in a general way to describe instability on slopes.) The tendency towards instability in the soil' mass, in fig. 7.9, is due to a moment about the centre of the circle which includes the failure arc. This moment is due to the weight of the wedge of soil above the arc tending to pull it downslope about the point O; this force is termed the *disturbing moment*. The unit weight of the clay mass can be determined in the field, and the disturbing moment can thus be calculated from the geometry of the diagram. In the case of fig. 7.9, the moment is subdivided into two parts because part of the wedge is beneath the water table and has a lower (submerged) unit weight. The disturbing moment is being resisted by the shear strength of the clay mass along the failure surface; the total moment due to the strength of the clay is called the *restoring moment*. The ratio of the restoring moment to the disturbing moment is, as before, the safety factor of the slope. The method described above is sometimes referred to as the *circular-arc method* and is popular in the $\phi = 0$ case. In the $c'\phi'$ analysis, where the shear strength clearly varies at different points along the arc, it is more common to subdivide the wedge into vertical slices, and variations of this method are discussed in connection with the Lodalen slide (Sevaldson, 1956) later.

One point that should be emphasized, before moving on to the importance of pore pressures in deep-seated slips, is the arbitrary nature of the choice of the appropriate slip circle. There is no *a priori* way of easily locating the most critical slip circle in any given slide, and, in practice, one must select a large number of arcs and examine the safety factor on each before commenting on the stability of the slope. In recognition of this problem, Fellenius (1927) and Taylor (1937) produced a system of stability charts which gives immediately the worst stability-case of the slope, in terms of the shear strength and the unit weight of the material, and the height and steepness of the slope, assuming a circular slip in uniform material. An example of these charts is given in fig. 7.10. The parameter on the vertical scale is the dimensionless stability number (sometimes termed stability factor) of a slope defined by

$$N_s = \gamma . H/c; \qquad (7.11)$$

the purpose of the chart is to provide the maximum stable slope in terms of the stability number and the ϕ value. When a cut is made in a clay mass to a depth H, the steepest stable slope can easily be determined through the three soil parameters and reference to this chart, without the analysis of many different slip circles. The relationship between critical height and slope, in relation to the soil mass properties, is slightly different in the case of rotational slips compared to the planar slide (Culmann wedge) analysis noted in chapter 6. Taylor (1948, p. 457) provides an interesting comparison of the theoretical stability numbers predicted by these two, and other, methods for different

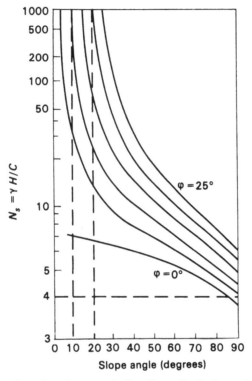

Fig. 7.10. Stability chart for deep-seated slips (from R. F. Scott, 'Principles of Soil Mechanics', 1963, Addison-Wesley, Reading, Mass.).

values of slope angle and the ϕ value. Taylor's charts are only valid, however, for short-term failures since they assume total stresses; it is only recently that Bishop and Morgenstern (1960) have produced comparable charts in terms of effective stresses using pore pressure values. This is discussed below, but, before passing on to it, one should note that the application of the high-speed computer to soil mechanics (Little and Price, 1958; Morgenstern and Price, 1965) has removed much of the problem of searching for the critical failure surface.

The stability charts prepared by Bishop and Morgenstern (1960) are more generalized than those presented by Taylor. The essential difference is that the stability analyses incorporate pore pressures through the parameter r_u, *the pore pressure ratio*, which is given by

$$r_u = u/(\gamma . z) \tag{7.12}$$

and is simply the ratio, at any point in the ground, of the pore pressure to the total weight of soil above. The authors emphasize that the pore pressure ratio may vary widely among different slopes, but they do point out that particular values of the ratio are probably very common. In the pore pressure model given by equation 7.5, it is easy to see that the r_u value corresponds to $(\gamma_w/\gamma) \cos^2 \theta$, which, in the case of a 45° slope, would be about 0·3, depending on the actual value of the bulk density. The special case of ground water at the surface, together with horizontal flow, corresponds to a pore pressure ratio of about 0·6. The stability charts presented by Bishop and Morgenstern are used to calculate two stability coefficients m and n; the safety factor, F_s, is then neatly given by the equation

$$F_s = m - n (r_u). \tag{7.13}$$

The stability analysis of a rotational slip is best illustrated with reference to an actual slide; the Lodalen slide of 1954 (Sevaldson, 1956), already referred to, is an especially well-described instance. The slide was analysed in terms of effective stresses by two different approaches: the conventional method of slices developed by May and Brahtz (1936) (a modification of the general circular-arc method (Wu, 1966, pp. 241–7)) illustrated in fig. 7.11; and Bishop's (1955) method of slices shown in fig. 7.12. The derivation of equations 7.14 and 7.15 (see fig. captions) is given in Appendix A4–A5. The conventional method, ignoring lateral earth pressures on each side of the slices, is the simpler of the two approaches, but may be as much as 15 per cent in error. The shear strength parameters, c' and ϕ', for the Lodalen stability analysis were determined from samples; the weight of the wedge was computed from the bulk unit weight of the clay and the geometry of the slip surface; and the distribution of pore pressures along the slip surface was determined from piezometers installed after the slide. These values were then substituted into equations 7.14 and 7.15. The analysis was applied to three sections along the slip and a weighted average F_s value computed in both cases; using the conventional method, the F_s value was 0·85, and, for the Bishop method, it was 1·05. The use of Bishop's modified method of slices, in terms of effective stresses, thus proved extremely successful, and subsequent investigations of other slides have emphasized its use in the analysis of long-term failures.

The results of the Lodalen slides are important in another respect: a successful stability analysis was achieved with the full field value of the

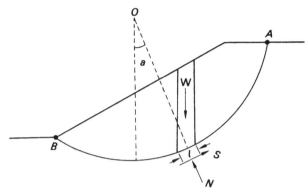

Fig. 7.11. Stability analysis of a deep-seated slip: the conventional method of slices. Notation: W = weight of slice; N = total normal force at base of slice; u = pore pressure at base of slice; O = centre of circle which includes arc AB; s = shear strength at base of slice.

$$F_s = \frac{\sum_B^A [c' \cdot l + (W \cdot \cos a - u \cdot l) \cdot \tan \phi']}{\sum_B^A W \cdot \sin a} \qquad (7.14)$$

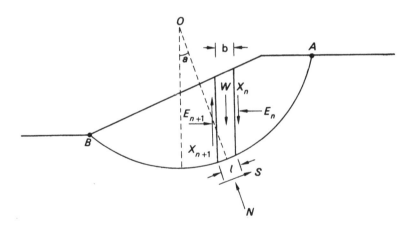

Fig. 7.12. Stability analysis of a deep-seated slip: the Bishop method (from Bishop, 1955). Notation: E_n, E_{n+1} – resultants of the total horizontal forces between the slices. X_n, X_{n+1} – resultants of the total vertical forces between the slices.

$$F_s = \frac{\sum_B^A \{[c' \cdot l + (W/\cos a - u \cdot l) \cdot \tan \phi']/[1 + (\tan a \cdot \tan \phi'/F_s)]\}}{\sum_B^A W \cdot \sin a} \qquad (7.15)$$

167

cohesion element used in computation. This is in contrast to the previously discussed shallow slides at Jackfield and in the London Clay and other deep-seated slips in the London Clay (Henkel, 1957) where the clay behaved as though $c' = 0$ at the time of failure. The Lodalen slide occurred in intact, non-fissured, over-consolidated clay, unlike the other slides, and it is now thought that intact clays do not lose the cohesive element of their shear strength in the way that fissured clays do. Support for this view has been provided by a number of slides subsequent to the Lodalen slip and reference, in particular, should be made to the article by Skempton and Brown (1961) describing the slide in boulder clay at Selset, Yorkshire. As was noted earlier, Skempton (1964) suggested that the essential difference between intact and fissured clays is that the former fail at peak strength values and the latter at residual strength values. He also hinted that the movement of fissured clays from the peak to the residual strength condition was associated with the stress concentration along the fissures.

One final point must be emphasized in connection with deep-seated rotational slips in clays. Throughout the discussion above, it has been assumed, in order to simplify the mechanics, that the clay mass is fairly uniform. In nature this is often not the case, and many spectacular slips occur in clays where strata of radically different permeability occur in contact with the clay. Some of the most impressive examples occur in the Weald of south-eastern England (Gossling, 1935; Gossling and Bull, 1948), where the Upper Greensand beds overlie Weald Clay. Similar well-defined slips occur along the south coast of England where chalk overlies the Gault Clay, and have been analysed by Hutchinson (1969) using effective stress analysis. It is a pity, however, that, in the standard textbooks of geology, attention is usually focused on these particularly unusual slips that depend upon special geologic conditions in the treatment of deep-seated mass-movements. The majority of earthslips in clays do not depend upon such unusual conditions and are explicable directly in terms of the laws of statics applied to the dissection of a clay mass by stream or marine erosion. Hopefully the previous pages will counter some of the traditional bias in the treatment of landslips as presented in most geomorphology texts.

Earthflows

Earthflows are rare occurrences and, for this reason, do not merit extensive treatment here. They are, however, extremely spectacular, and form very distinctive landform features in certain areas of the world, such as the St Lawrence Lowlands of eastern Canada and the coastal areas of southern and western Norway. An earthflow is defined here as any type of instability feature, whether initially a shallow or deep-seated slide, in which the majority of the movement takes place as a viscous liquid rather than the sliding downslope of a solid slab of earth.

The major part of this treatment of earthflows is taken from the excellent articles by Bjerrum (1954*a*, 1954*b*) dealing with the stability of natural slopes in quick clay in Norway. Since the vast majority of earth-flows occur in quick or extra-sensitive clays, the present discussion will begin with an explanation of the term *sensitivity* as applied to clays.

Although the shear strength of soil is usually measured on samples which are hopefully *un*disturbed, in some cases it is useful to have information on soils in the disturbed state. Samples which have been fully disturbed are termed *remoulded* samples, and, usually, show lower strength values. The decrease in strength between peak and residual state (chapter 4) is, in fact, a special case of the loss in strength due to remoulding during a shear test. The ratio of the undisturbed strength to the remoulded strength is termed the sensitivity (S_t) of the soil (Terzaghi, 1944). Most of the clays mentioned previously in this chapter have very low sensitivity values. The term sensitive clay is strictly applied to clays with S_t values in the range 4·0–8·0 (Skempton and Northey, 1952); this section is mostly concerned with highly sensitive clays ($S_t > 8$) and quick clays ($S_t > 16$). The type examples of these clays are the Leda Clays of eastern Canada ($S_t \simeq 25$) and the late glacial marine clays of Norway which are even more sensitive. In these extreme cases, the soil sample changes from a plastic solid to a viscous liquid, and, quantitatively, may lose more than 90 per cent of the undisturbed strength on remoulding.

A very important characteristic of extra-sensitive clays is that, unlike most soils, the natural water content of the clay exceeds the *Liquid Limit*. This is shown in table 7.1 with data for quick clays and, for comparison, the more usual plastic clays as well. It is perhaps useful to repeat here that the Liquid Limit is the moisture content of a clay at which, *in a remoulded condition*, it first attains the properties of a viscous fluid, rather than a solid. In the case of the extra-sensitive clays noted in table 7.1, the natural water content is in every case greater than the Liquid Limit and, thus, assuming that it is possible to explain the mechanism by which the clay becomes remoulded in nature, there is no further difficulty in accounting for the flow character of the failure in highly sensitive clays. Before considering the mechanisms by which remoulding takes place on a natural slope, it is interesting to consider the reasons why extra-sensitive clays possess natural moisture contents higher than the Liquid Limit value, in contrast to most other clays. There is clearly more than one explanation.

In the case of the Norwegian clays, the natural water content values

169

are not especially high, in comparison with other clays, and the peculiar feature is the low values of the Liquid Limit. Bjerrum (1954*a*, 1954*b*) has pointed out that low Atterberg limits are characteristic of late glacial marine clay deposits and are due to leaching of salts from the pore-water of these clays following uplift of the sediments above sea level. Salt concentrations in the porewater of the Ullensaker slide (table 7.1) ranged from 1·2 to 2·9 g/l in contrast to probable values in the range of 30 g/l at the time of deposition. Experiments indicated that addition of

TABLE 7.1. *Values for Plastic Limit, Liquid Limit and natural water content for different types of clay*

Clay	Plastic Limit	Liquid Limit	Natural water content	Source
Norwegian late-glacial				
marine clays				
Manglerud	20	27–31	35–42	Bjerrum, 1954*b*
Ullensaker	15–20	24–7	30–40	Bjerrum, 1954*b*
Bekkelaget	15–17	25–6	32–9	Eide and Bjerrum, 1954
Leda Clays, E. Canada				
St Thuribe	21	33	44	Peck, Ireland and Fry, 1951
Nicolet	23	55	65	Hurtabise and Rochette, 1956
Hawkesbury	25–8	53–72	61–90	Eden and Hamilton, 1957
Green Creek	27–30	55–75	70–5	Crawford and Eden, 1967
Breckenridge	25–8	55–70	75–80	Crawford and Eden, 1967
Low-sensitivity clays				
Lodalen, Norway	18	36	31	Sevaldson, 1956
Boulder clay, County Durham	14	28	14	Skempton and Brown, 1961
Boulder clay, Selset	11–15	23–7	10–15	Skempton and Brown, 1961
London Clay	20–8	59–75	23–31	Hutchinson, 1967

salt to the clay to a concentration of 35 g/l increases the Liquid Limit from 26 to 34 per cent. Observations in the field and in the laboratory, moreover, indicate that the process of leaching alters the natural water content of these clays very little. The decrease in the Liquid Limit value due to leaching of these salts, together with the unchanged natural water content of the clays, appears to be capable of increasing the sensitivity from values of about 5 to values in excess of 50.

The Leda Clays of eastern Canada are believed to have been deposited in brackish water during a late glacial stage, and, like the Norwegian clays, possess unusually low salt contents in the porewater. The distinguishing feature of the Leda Clays is not a low Liquid Limit value, however, but the very high natural water content. The special character of the Leda Clays has attracted a great deal of attention from the National Research Council of Canada, and the articles by Crawford (1961, 1968) provide valuable summaries of these studies. One of their findings is that, notwithstanding uniform salt concentrations of 1–2 g/l in the porewater of the clay (Crawford and Eden, 1965), sensitivity varies widely. Interesting work by Penner (1965) has revealed that sensitivity is only indirectly related to the salt concentration of the pore-water and the important direct control appears to be the electrokinetic potential of the clay–water system. Evidence is also gradually emerging to support the idea that the high voids ratio and associated high water contents of undisturbed Leda Clay are due to the 'card-house' fabric of the clay in contrast to the strongly stratified arrangement of most clays. X-ray diffraction methods used by Quigley and Thompson (1966), for instance, revealed little or no particle orientation in samples subjected to low consolidation pressures, and this has been supported by Penner's (1963a) thermal conductivity tests. The high natural water content of the Leda Clays thus appears to be a result of special environmental conditions, at the time of deposition, which produced a very open clay fabric.

A slightly different point has been emphasized by Yatsu (1966) in a study of earthflows in Canada, Scandinavia and Japan. Analysis of many clay specimens by X-ray diffraction techniques and by electron microscope showed that in every case the material involved in earthflows, as distinct from the more usual type of landslip, contained a large percentage of those clay minerals, such as montmorillonite, which swell appreciably when in contact with water, whereas the more usual mode of landslip seemed to be independent of the nature of the clay.

A great deal more could be written about the peculiarities of extra-sensitive clays (a comprehensive recent article has been provided by Mitchell and Houston, 1969), but it is more important here to consider other aspects about instability in such clays. There remains, in particular, the problem of explaining the change from the undisturbed state to the remoulded condition of these clays in nature. One possibility is that the structure of these quick clays is altered by disturbances such as earth tremors, although, in general, it is difficult to associate the occurrence of earthflows with any particular external cause. It seems more probable that any large failure in a mass of sensitive clay will *itself* produce sufficient disturbance to remould these clays and change the mode of failure from a slide to a flow. Eye-witness descriptions of these flows tend to confirm that they begin as slides.

A special feature of earthflows in sensitive clays is that the initial failure is usually quite small and then instability extends headwards at an extremely rapid rate. The typical scars of these flows are bowl-

171

shaped (fig. 7.13) and may be formed in a few minutes. The retrogressive backward development is thus another particular characteristic of these earthflows and presents another problem that is not easily explained. It seems very possible, however, that it is closely linked to the change from plastic failure to viscous flow in the initial slip. In most deep-seated rotational slips, such as those discussed earlier, the total movement of material is quite small and involves simply the rotation of a

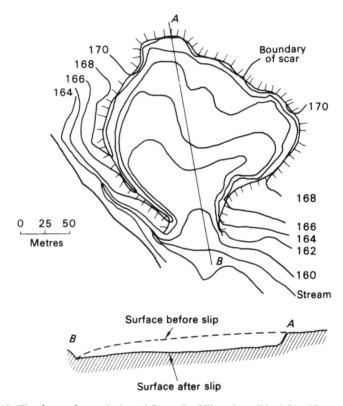

Fig. 7.13. The form of a typical earthflow: the Ullensaker slide (after Bjerrum, 1954*b*). Contours are in metres above sea level.

fairly intact block. This wedge still provides a great deal of counterweight on the clay behind the slip and thus lessens the chance of a subsequent failure. In the case of a slip in quick clay, however, the material involved in the initial slide is rapidly changed into a fluid and quickly moves away from the main soil mass. As a result, the soil mass behind is offered little, if any, support and a new failure is induced. The process is presumably repeated many times judging by the size of these flows, although it appears to take place almost simultaneously. Sometimes these earthflows are described as progressive slips because of this retrogressive character, but this is misleading since it is doubtful if any progressive failure of the type discussed earlier in this chapter is involved. The idea that most earthflows in quick clays consist of a rapid succession of rotational slips acting retrogressively was suggested by Bjerrum (1954*b*) and has been supported by Kenney (1967*c*) working on the slide at Selnes, Norway. It is thus becoming clear that although these earthflows with their vast extents relative to their depth have the appearance of shallow movements, they are really a succession of deep-seated slips and this perhaps explains why they are capable of developing on very gentle, even horizontal slopes.

So far little has been said about the mechanism responsible for the initial slip which generates the whole flow. Bjerrum (1954*b*) and others have emphasized that this is probably associated with leaching out of salts from the porewater. In this process, not only is the sensitivity of the clay increased, but the undisturbed strength is also reduced. On a geologic time scale there is thus a gradual lowering of the shear resistance of the clay until eventually it is so low that instability occurs. This may account for the observation noted earlier that there is often no external cause readily associated with these flows. Sometimes, it is true, these flows do occur at times of prolonged heavy rains, and this may be important in increasing pore pressures at a depth sufficient to trigger off the initial slip. It should be emphasized, however, that heavy rains are not a necessary prerequisite to these earthflows; the natural water content is already high enough to convert the slip into a viscous flow.

Earthflows do occur in materials other than extra-sensitive clays (Peck and Kaun, 1948; Terzaghi, 1957), but, usually, only in special geologic conditions, and, often, where excess seepage of ground water occurs. The vast majority of these flows are restricted to quick clays; this is why so much emphasis has been put on them in this section, although not all failures in quick clays are, in fact, earthflows.

Micro-scale instability features

Many hillslopes show a regularly rippled surface appearance with miniature terraces or ridges extending across the slope usually normal to the direction of maximum slope. These terraces are rarely more than 0·5 m wide and deep. Such ripples have been termed *terracettes*; their origin is still very much a matter of debate.

It was thought, for a long time, that these small ledges were simply animal tracks, and they were disregarded as being indicators of a natural process of denudation on slopes. Admittedly, on some slopes, the

173

appearance of terracettes has been accentuated by the movement of sheep and other animals along them, but, since terracettes do occur in areas where animals are very rare, it would seem that some other mechanism is needed.

As long ago as 1922, Ødum (1922) suggested that these miniature terraces stem from instability of the soil mantle, and attributed them to a whole succession of very small rotational slips along the slope, similar to those described by Hutchinson (1965, 1967) in the London Clay, although on a much smaller scale. More recently Kirkby (1963) has also attributed them to surface instability pointing out that: 'the pattern of

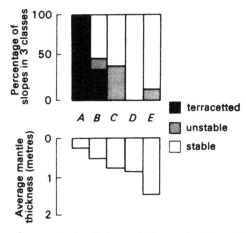

Fig. 7.14. Occurrence of terracettes in relation to thickness of soil mantle in the Pennine and Exmoor areas.
A Carboniferous Limestone (Pennines); *B* Exmoor Gritstone; *C* Exmoor Slates; *D* Shale Grit (Pennines); *E* Shales (Pennines).

cracks in the vegetation mat is exactly that to be expected in a uniform stretching of the soil layer below the grass, in such a way that the adhesion of the grass to the soil is greater than the mutual cohesion of the grass' (p. 291). This would be expected since the cohesion arises from predominantly vertical roots.

Why terracettes occur in some places and not in others is still a mystery. Some ideas were recently put forward by Carson (1967*b*) in an attempt to answer this, based on a study of slopes in the Exmoor and southern Pennine areas of England. The occurrence of terracettes in these areas appears to be very closely linked to the thickness of the soil mantle on the hillslopes. This is shown in fig. 7.14. In the area of Carboniferous Limestone and the Exmoor quartzite grits, the slope mantle is very shallow and most slopes are terracetted; it is difficult to describe any slope as either stable or unstable. In the other areas, the mantle is deeper, terracettes are very rare, and those slopes which are not clearly stable are distinctly *un*stable with pronounced scars and slides. The critical depth appears to be about 0·7 m. The impression is that these terracetted slopes are potentially very unstable and that, in some way, the shallowness of the mantle imparts extra strength to the slope which prevents major landslides, but still allows small-scale instability to occur and produce terracettes.

It seems very likely that shallow soil mantles (fig. 7.15a) overlying much stronger material will possess more strength than thicker mantles of the same material. One reason is that irregularities in the underlying

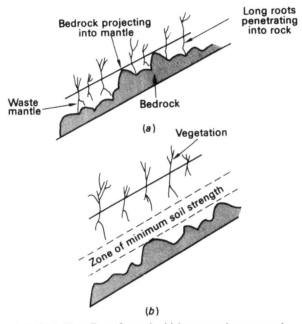

Fig. 7.15. The effect of mantle thickness on shear strength.

rock surface will tend to disrupt any large-scale sliding of the mantle. This is perhaps a minor element, however, in comparison with the effect of vertical root systems extending down into the underlying rock. The shear strength of the waste debris is now supplemented by the strength of the roots. In the case of thicker mantles (fig. 7.15b) few roots are capable of extending down into the stronger material and, as a result, there is a layer between the vegetation mat and the underlying rock surface in which the shear strength is at a minimum. Extensive shallow slides can take place in the deep mantle, but not in the shallow one. These ideas are supported by the observations of Gifford (1953) on Exmoor after the tragic 1952 storm; most of the landslides which occurred in the area involved the sliding of the complete vegetation mat over the thin layer of soil just above bedrock. It is thus suggested that terracettes occur where the shear strength of the waste mantle alone is insufficient to maintain the slope at its particular angle, and that pseudostability arises from the extra strength implanted by the vegetation mat. Whether this explanation is valid for terracettes in other areas is more dubious; it may well be that there is more than one type of terracette, and more than one mode of origin occurs.

Terracettes appear to be particularly common on slopes cut into the massive limestones of the Carboniferous Limestone where slopes are mantled by an extremely thin layer of waste debris. The limestone is usually so pure that only a small percentage survives as an insoluble residue and it is probably for this reason that the slope mantles are so thin. In turn, the thinness of the slope mantle prevents the occurrence of large-scale instability and produces terracettes.

THE RATE OF DENUDATION DUE TO INSTABILITY PROCESSES

In discussing briefly the rate of denudation associated with rapid mass movements it is perhaps worthwhile to distinguish between thin residual soils mantling stronger rock, and thick soil deposits, such as clay masses. The reason for this is that the rate of debris transport from the slope in the former case is still, to a large extent, *weathering-limited*.

Envisage a slope described by equation 6.18: a straight slope mantled by a very thin layer of scree-like debris produced by the retreat of a cliff wall under the process of rockfall. This slope is stable at the angle of repose of the scree which would be close to 35°. Over the course of geologic time, weathering alters the mantle. Suppose, for simplicity, that the ultimate weathered product is a cohesionless ($c' = 0$) silty-sand with ϕ' about 35°. Under conditions where the mantle never became saturated with water, the mantle would remain stable at the angle of the initial scree slope. In the case of the coarse talus mantle, it is unlikely that the material would ever become saturated because the voids between the rock particles are so great that water passes through the material very rapidly. When the talus has been weathered to a silty-sand soil, the pore spaces are then much smaller and are much more likely to become saturated at times of prolonged rainstorms. As a consequence we must expect positive pore pressures to reduce the strength of the mantle when it becomes wet, and shallow landslides would occur exposing the underlying bedrock. Most of the slides on Exmoor in the 1952 storm were probably of this type. Gradually the rock surface that has been exposed by the slide will disintegrate into scree-type rubble, and further weathering will convert the coarse mantle into a silty–sandy soil cover as before.

Saturation of the mantle and the development of positive pore pressures will then produce another slide stripping the mantle away from the underlying rock. Although the rate of this stripping process is partly affected by the frequency of storms big enough to saturate the mantle, the major element is clearly the rate of weathering, at least in temperate areas. In hot wet areas where weathering rates appear to be much greater, extremely thick residual soils exist (Lumb, 1962, 1965) and the limiting element appears to be the frequency of the development of large pore pressures in the mantle.

The thick residual soil mantles of the hot wet tropics lead conveniently into the general problem of thick soil masses, and introduce the second complicating element. This is the element of *feedback* between form and process which acts to slow down the rate of operation of debris removal. Assuming that undercutting is not taking place at the base of a slope, each rapid mass-movement generally acts to increase the stability of the hillslope and thus delay the lapse time until the next major slide. This is easily appreciated in the case of deep-seated rotational slips which, through backward rotation, contribute to lower the average angle of slope. A great deal depends however on whether the stripping process acts to produce parallel retreat or flattening of the slope, and this is discussed in more detail in the last section of this chapter.

So far we have neglected the case in which the slope is continually being undercut by a stream at the base or via another process. In this situation the rate of denudation will be much higher and to a great extent determined by the rate of undercutting. The actual processes operating on a hillslope will also be influenced by the rate of downcutting at the base. With a medium rate of stream erosion it might be possible for a rock to weather to produce a residual talus mantle and, with continued downcutting, to maintain a V-shaped valley with slopes at the angle of repose of the talus. With a more rapid rate of stream incision there might be insufficient time for weathering of the rock, and a vertical canyon is produced which will be converted to a V-shaped form only when the rate of downcutting has slackened or ceased. A similar point has been made by Hutchinson (1967, 1968) with respect to the coastal slopes cut in London Clay. Under conditions of moderate marine activity, weathering is able to keep pace with basal undercutting, and shallow landslides and mud-flows are able to maintain the slope at angles a little above 15°. Under more intense marine activity the type of mass-movement changes: weathering and shallow slides are unable to match the rate of basal cutting by the sea, and the slopes become steepened until a deep-seated rotational landslip is initiated. Throughout the coastal exposure of the London Clay, shallow landslides predominate

177

where marine activity is fairly moderate, and deep-seated slips only occur where marine erosion is really rapid. Under such conditions Hutchinson (1968) estimated that deep-seated slips would re-occur every 30–40 years comparable to a rate of marine erosion of 3 m per year.

MODELS OF HILLSLOPE DEVELOPMENT DUE TO INSTABILITY IN SOIL MASSES

In the last section of this chapter we shall neglect earthflows and concentrate on the other three major processes. This is because, firstly, earthflows are relatively rare phenomena, and, secondly, because earthflows have interested geomorphologists more through the particular landforms which they produce, than through a general contribution to the development of hillside slopes. The nature of this particular landform, the bowl-shaped depression, has already been described. In addition we shall not separate shallow and deep-seated slides, but for reasons which will become apparent in the text, we shall distinguish between thick soil masses and residual soil mantles.

Thick soil masses

An interesting model of the development of slopes produced by stream incision into a clay mass was provided by Skempton (1953), based on work in the Shotton boulder clay of County Durham. Skempton recognized three main stages in the sequence of slope development under mass-failure.

In the first phase, streams begin downcutting into the clay mass and steep, V-shaped valleys are produced. Side slopes are not vertical, as they would be if cut in strong rock, but occur at an angle which depends on the properties of the soil mass; in the sandy boulder clay at Shotton this angle was about 30–35°, while in more colloidal clays such as the Oxford and London Clays, the angle is nearer 15°. As streams continue to cut down, the valley grows deeper but not steeper, and shallow surface slides maintain a constant angle. No explanation was offered for the shallow slides but they probably arise from the loss of cohesion in the surface layers of the clay due to weathering. When cuts are made into material which behaves as though it is cohesionless, the maximum angle of stability is equal to the angle of internal friction of the material. This is easily shown by comparing the shear force and shear strength on a slope, as depicted in fig. 7.7. The shear force is given by

$$T = W . \sin \theta \qquad (7.16)$$

and the shear strength by

$$s = W \cos \theta . \tan \phi \quad (c = 0) \qquad (7.17)$$

The maximum stable slope corresponds to the value of θ when the safety factor F_s is equal to unity, since this is the state of limiting equilibrium for the slope. In this case we have the shear stress equal to the shear strength, and, therefore, from equations 7.16 and 7.17,

$$\tan \theta = \tan \phi$$

$$\text{or} \quad \theta = \phi \qquad (7.18)$$

No figures were given for ϕ' at Shotton, but a value of $\phi' = 30°$ would be very reasonable for a sandy boulder clay. In the London Clay, tests have shown that ϕ' is very close to 15°, agreeing with the concept put forward above. As streams continue to cut down, slopes are maintained at a constant angle by shallow landslides, provided that stream undercutting is proceeding fairly rapidly. The importance of this proviso will be appreciated from the third stage.

The second stage in this sequence occurs when deep-seated slips develop on the valley slopes. This occurs only when the valley depth attains a critical value. This limiting depth was previously given for vertical slopes (equation 6.8) and the general case linking critical depth to slope angle is given in the Appendix (equation A3.3):

$$H_c = \frac{4c}{\gamma} \frac{\sin i . \cos \phi'}{[1 - \cos (i - \phi')]}$$

where i ($= \theta$ in the present discussion) is the slope angle. In the Shotton area, field survey showed that H_c is about 45 m. This is inconsistent with i ($= \theta$) $= \phi'$, suggested previously, for which there is no restriction on slope height. Clearly, either $\theta > \phi'$ in the first stage, or ϕ' decreases with depth, or time, in the Shotton area. Once deep-seated slips occur, they become more prominent as the valley gets deeper.

Eventually streams cease to cut down rapidly, and floodplains develop protecting the side slopes from further undercutting by the streams; this marks the begining of stage three in the sequence. Slopes now flatten through shallow slides to a lower, more permanent, angle of limiting stability. In the sandy boulder clay at Shotton this angle is about 22°, and in the London Clay it is about 10°. As explained earlier, this angle in the London Clay is given approximately by equation 7.9 of the infinite slope case, and corresponds to the reduction of shear strength in the clay produced by the pore pressure pattern of equation 7.5 and fig. 7.5. It is interesting to note that application of the same equation to the Shotton clay would also produce a reasonable agreement with the actual limiting slope angle.

The flattening of the slope in this third stage is due mainly to the occurrence of high groundwater levels raising pore pressures and producing slides, until eventually the 'mature' angle of stability is reached. The important point is that these slopes have been protected from further undercutting by the stream, and have had enough time to flatten to this ultimate angle of stability. Undoubtedly high groundwater levels occur in stage one of this sequence, while the streams are cutting down rapidly, but, in this case, the continual undercutting by the stream allows insufficient time for slopes to flatten to the lower angle.

It is interesting that in the area of the Shotton boulder clay, Skempton (1953) observed that the initial slope at about 30° *flattened* to the stable angle at about 22°. Hutchinson (1967) believes that in the case of the London Clay, the comparable change, that is from slopes at about 13–15° to slopes at about 8–10°, occurs through the *retreat* of the degrading high-angle slope leaving the lower-angled slope as a basal slope on which material accumulates from above. He compares the geometrical changes in the slope with the mathematical model developed by Bakker and Le Heux (1947), discussed previously in chapter 6, where the rock slope angle β corresponds to the angle of the degrading slope, and the scree slope angle α is analogous to the slope of the basal area. The issue of flattening *vis-à-vis* retreat will be touched upon again.

The second stage of the model does not always occur. In situations in which downcutting proceeds only to a small depth, or alternatively, in rocks in which the cohesion is very high, the critical depth will never be attained and the sequence of slope development will pass directly from the first to the third stage. In the extreme case of downcutting in a very strong rock this marks the transition to the case of residual soil mantles. Before turning to a discussion of this case, it is useful to summarize the essential points of the model above.

1. Stream downcutting produces a V-shaped valley of constant side slopes, irrespective of depth, at an angle which is approximately the angle of internal friction of the material on the slope. This is a temporary angle of stability maintained only by continual undercutting at the base.

2. The development of a floodplain protects most slopes from further undercutting and allows the slope to be replaced by a lower-angled slope of more permanent stability; this may be considered the ultimate angle of stability.

3. The change from the high angle slope to the lower angle slope appears to be achieved through flattening or hinge-decline of the first slope, although under some circumstances it seems that retreat of the initial slope may take place.

This is perhaps the simplest model possible. In the case of residual soil mantles overlying stronger rock the situation is much more complex.

Residual soil mantles

In the situation where streams are downcutting into a strong rock mass, there is likely to be a number of major differences with the model for the clay mass described above. Consider a jointed rock mass. Stream incision will, assuming it is rapid, produce first of all a slope at any angle between the vertical and the angle of repose of the products of weathering, depending on the rate of undercutting. Eventually, however, a scree slope will develop standing at the angle of repose of the debris. This talus slope is able to stand at this high an angle because the voids in the mantle are large, and the mantle will never become saturated and pore pressures greater than atmospheric ($u = 0$) are unlikely to occur. Insertion of $u = 0$ in equation 7.4 along with equation 7.2 produces a θ value equal to the angle of repose as already given in equation 7.18. The talus slope is comparable to the straight slope produced in the initial stages of the model for a clay mass.

Continuous weathering, however, will progressively change the character of the mantle. The rock rubble will be broken down into soil and the mantle will pass through a sequence of talus, mixed talus and colluvium (*taluvium*), and soil. The exact change in c' and ϕ' will depend, of course, on the type of rock and soil. Consider the case in which sufficiently little clay is produced that throughout the weathering process the mantle remains cohesionless, and assume that a silty–sandy soil is produced. The angle of internal friction of this type of material would be about 35° and would thus differ little from that for the initial scree. The voids in this mantle are much smaller than on the initial slope, however, and positive pore pressures are likely to develop reducing the strength of the mantle and producing shallow landslides. Assuming, for comparative purposes, that the conditions given by equation 7.5 applied, the stable slope for this material would be given by equation 7.9:

$$\tan \theta = \tfrac{1}{2} \tan \phi',$$

so that θ is about 20–21°. This slope might be termed the ultimate stable slope and compared to the second angle of stability in the case of the model for clays above.

The picture presented above, comparable to the sequence of slope development envisaged by Skempton (1953) in a clay mass, is, however, very much over-simplified. It has implicitly made the assumption that the shear strength characteristics of the residual mantle do not change during the course of weathering from coarse talus to sandy soil. This is invalid. Although the angle of repose (ϕ') of loose dry sand is substantially the same as the repose angle of coarse scree debris, approximating to 35°, there is no justification for the assumption that the ϕ'

value remains constant during the progressive breakdown of the waste mantle. The evidence provided by many civil engineers, as given in table 4.2, is that mixtures of rock rubble and sandy particles possess higher ϕ' values than either pure scree or pure soil, due to the greater interlocking among the particles, and, in a loose state, the angle of shearing resistance is usually about 43–45°. Moreover the work of Vucetic (1958) and Holtz (1960), also noted in chapter 4, indicates that the change from a 35° ϕ' angle to a 43–45° ϕ' angle, and back to the lower value, is probably a relatively abrupt one. The conclusion is, therefore, that during the conversion of a rock rubble mantle into a sandy soil mantle, the mantle will for a long period of time be stable at an angle which is greater than the angle of 20–21° for the sandy soil cover. The actual angle of stability of this *taluvial* material (Wentworth, 1943) comprising mixed rock rubble and soil grains will depend on the pore pressures to which it is subjected. Since the loose mantle is much more permeable than the underlying bedrock it seems very possible that the pore pressure model given by fig. 7.5 and equation 7.5 may be applicable during times of perched water systems. Accepting this idea would produce an angle of stability of 25–26° for the taluvial mantle through substituting $\phi' = 45°$ in equation 7.9.

There is considerable evidence from geomorphologists (Savigear, 1956; Young 1961; Ruxton, 1958; Koons, 1955; Melton, 1965; Robinson, 1966) to indicate that, in a wide variety of climates, on rocks which weather to produce taluvial mantles of the type described above, angles of slope at 25–27° are very common. Only recently, however, has any attempt been made to explain this fact in terms of the soil mechanics properties of the material mantling the hillslopes.

The important point which emerges from this discussion is that, instead of just a temporary and an ultimate angle of stability as in the case of the clay mass model, we must recognize the possibility of a number of angles of limiting stability each associated with a particular stage in the breakdown of the mantle. The work by Carson and Petley (1970) on Exmoor and the Pennines provides an example of a trimodel distribution of straight slopes in the landscape associated with three stages of breakdown in the weathering of the waste mantle. There are strong grounds for suspecting that this pattern is probably very common wherever the underlying rock mass is jointed and weathers through the release and disintegration of these joint-controlled boulders. If this is so, it is not so surprising that angles of slope at about 25–26° are particularly frequent in the landscape. Nevertheless there is no reason why all rocks in all climates should pass through the three stages of disintegration noted above. The recognition of this point led Carson (1969b) to develop a general model of slope development under mass-failure applicable to most situations and summarized below. It contains three main points.

1. The effect of instability on a slope is to replace a steeper slope by a lower-angle one. In the case of the clay-mass model this involved one change in slope from a temporary angle of stability to an ultimate angle. This might be termed one phase of instability, and the model labelled a one-phase instability model. Many landscapes will undoubtedly experience more than one phase of instability. As an example a strong, well-jointed rock mass is likely to pass through three phases: an initial phase while the cliff face is replaced by a scree slope; a second phase with the change from a scree slope to a taluvial slope; and a third phase during which the ultimate soil-mantled slope emerges. The number of phases of instability which occur in a landscape depends very much on the history of disintegration of the rock material, although other elements such as the climate may also be important.

2. Each phase of instability is separated by threshold slopes on which, under the prevailing conditions, the mantle exists in a state of temporary stability. The angle of these threshold slopes depends on a number of elements, but, in particular, on the mantle character and the pore pressure pattern which occurs within it. We have already stressed that the variety of possible pore pressure patterns which occur in nature is immense, and, similarly, residual soil mantles may occur with a wide variety of different shear strength parameters. As a result it is inevitable that a complete range of threshold slopes must exist in the landscape. Nevertheless it seems that certain threshold slopes are especially common:

43–45°: slopes cut in rocks which are fractured and jointed so much that they are virtually cohesionless, but where, because of the high density of the packing the ϕ' angle is about 43–45°. Such slopes contain relatively large voids and probably do not develop pore pressures greater than atmospheric. The stable angle of slope thus corresponds with the ϕ' value of the material in the dense state. This angle is common in shale badlands. Strahler (1950) noted that many slopes in the Verdugo Hills occurred at this angle;

33–38°: slopes mantled by the same type of material as described in the 43–45° slopes, but this time in a much looser state of packing. The ϕ' value of this material is now much lower and approximates to the angle of repose. Slopes in this range, mantled by loose talus, have been described by many workers in a wide variety of climates (Bryan, 1922; Simpson, 1953; Terzaghi, 1958; Rapp, 1960; and Tinkler, 1966);

25–28°: taluvial slopes, described already in some detail;

19–21°: sandy slopes, again discussed previously, and based on the assumption that the pore pressure pattern corresponds to the distribution given by equation 7.5;

8–11°: slopes in clays.

It cannot be too heavily stressed, however, that the range of different types of soil along with the multitude of different groundwater patterns, and thus pore pressure patterns, make any angle of limiting slope possible. This is especially the case with clays, and although many clay slopes occur at about 8–11°, there are many that occur at angles outside this range.

3. The number of phases of instability, together with the steepness of the threshold slopes, determines most of the pattern of slope development. One major question remains unanswered. Is the replacement of one limiting angle of slope by a subsequent lower threshold slope achieved through a flattening process or through retreat of the steeper slope to leave the gentler slope as a basal slope? This point has been raised before and any answer to it must be very vague at this stage. Skempton (1953) indicated flattening in the case of the Shotton boulder clay; Hutchinson (1967) suggested retreat in the London Clay; and Carson and Petley (1970) argued that evidence existed for both decline and retreat. This is elaborated upon in connection with terracettes below.

Terracettes

The general model outlined above may be used to accommodate movements which produce terracettes. The first and second points are unchanged, and the same reasoning applies whether instability occurs in the form of rapid shallow slides or terracette movement. On the other hand, the limited evidence available to date (Carson, 1967b) suggests that the movement from one threshold slope to another occurs via flattening and not parallel retreat in the case of terracettes. This evidence is summarized below. It is based on work in the Carboniferous Limestone of the Pennines and the quartzite grits of northern Exmoor. The results offer an interesting contrast with those of the Shale Grit of the Pennines and the Devonian slates of Exmoor. The major evidence is summarized in fig. 7.16 and fig. 7.17.

In the lower histogram of fig. 7.16 the three peaks correspond to scree slopes at 33°, taluvial slopes at 25–27° and soil mantled slopes, predominantly sandy, at about 20°. Slopes are conspicuously rare in the range 28–31° and to a lesser extent 21–23°. These 'gaps' in the histogram might be interpreted as indicating retreat rather than flattening of slopes. In a retreat of the 33° slope to leave behind the 26° slope, no slopes should be produced at about 28–31° and similarly in a retreat of the 26° slope to produce the 20° slope, there should be a gap at about 21–23°. To some extent this is supported by examination of the slope profiles in these areas as illustrated in fig. 7.17*a*. These slopes are commonly composed of two segments; an upper one at about 33° appears to be retreating under the process of shallow landslides leaving behind a new slope at about 26°, the new angle of stability for the material. No slopes, however, were found with upper slopes at 25–27° and lower segments at 20–21° so that there is less evidence in this case of the last phase of instability.

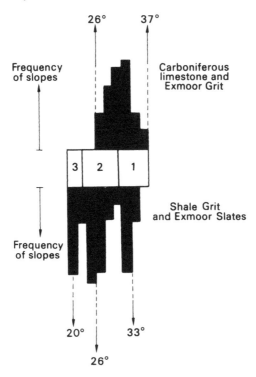

Fig. 7.16. The frequency distribution of straight slope angles in the Exmoor and Pennine areas in relation to angles of limiting stability (adapted from Carson, 1969*b*). Mantle: 1 scree; 2 taluvium; 3 sandy soil.

Terracettes are rare in the areas described above, and landslides are the major expression of instability. In the Limestone area of the Pennines and the grits of Exmoor the reverse is the case, and the effect that this has upon the course of slope development in the areas may be seen from the upper histogram of fig. 7.16. Slope mantles are less completely weathered in this area and only the first two stages of weathering are evident but the angles of stability are comparable to those in the other two areas. The distribution of slope angles relative to these limiting slopes is, however, very different. Instead of modes in the

185

histogram at about the angles of the slopes of limiting stability with a gap between, the reverse is the case, and the modal slope class occurs between the two stability slopes. It is tempting to conjecture that this is because straight slopes in these areas change from one limiting slope to another via flattening rather than through the retreat of the steeper slope. The majority of the slopes in these two areas have not yet accomplished the change from a talus slope to a taluvial slope and thus the modal slope class occurs between the two limits. This is supported by an examination of slope profiles. No two-segment slopes were found anywhere in the

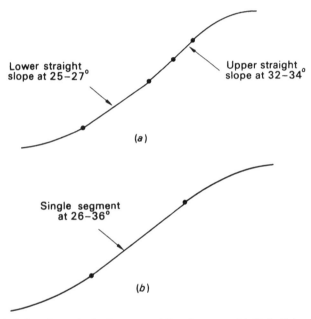

Fig. 7.17 Typical slope forms in the Exmoor and Pennine areas: (*a*) Shale Grit and Exmoor Slates; (*b*) Carboniferous Limestone and Exmoor Grit (adapted from Carson, 1969*b*).

area. Slopes usually possess one straight section with angles ranging from 26–36° and invariably the slopes are covered with terracettes except in the lower-angled slopes near 26° which are approaching stability.

It is not difficult to produce a hypothesis to explain why landslides should produce slope retreat and why terracettes should be associated with flattening of slopes. When shallow landslides occur on a fairly short straight slope the effect is to produce an almost instantaneous stripping of the whole mantle for the full length along the straight slope; inevitably the slope retreats. The instability which produces terracettes is presumably much slower with a gradual movement downslope only a little faster than soil creep. It is probable that *net* loss of soil takes place throughout the length of the slope, but presumably more is lost at the top than at the slope base since the moving mantle is continually exposing the upper slope. In this way the amount of net loss of debris would probably increase towards the top of the straight slope thus effecting a flattening of the profile. Clearly much of this is conjecture and needs to be substantiated by more evidence. The ideas do agree, however, with the meagre evidence so far available.

The net effect of instability in soil masses is to produce a landscape with gentler slopes upon which, since they are stable with respect to rapid mass-movement, debris movement is slow. Once a stable landscape has been produced, other processes such as surface and subsurface soil wash, solution, and soil creep, which may have been dwarfed into insignificance at the time of rapid mass-movements, now emerge as the major processes in moulding the landscape. The effects of these processes are discussed in the chapters which follow.

CHAPTER 8

SURFACE WATER EROSION

Falling and flowing water exert large forces on hillslope debris, which are resisted by frictional and cohesive forces. Hydrostatic forces may affect water erosion by helping to drive or slow down the flow, and by increasing or reducing the resisting forces of the soil. Different combinations of forces and resistances produce a number of rather distinct processes, most of which can be grouped under the broad heading of soil erosion. This includes the movement of particles which are detached by raindrop impact and may leap as much as a metre in the rebounding splash from the impact; and also of surface particles which are carried along in an overland flow. Where the flow is thin and relatively uniform, individual particles are removed from the surface without any marked concentration of erosion; but where there are marked fluctuations in flow thickness, some depressions may be eroded at a very much higher rate, initiating small channels in which the flow and transportation become progressively concentrated.

Rainsplash transport

A raindrop falling vertically onto a bare level surface may do two things; it may form a small crater on the surface with a raised rim, as can any projectile; and it will partially bounce back as a number of splashes in random directions and in doing so may carry soil particles with it for considerable distances. Debris will be transported outwards from the point of impact if the forces are great enough to move it, and the directions of movement will be completely random. Cratering will move large numbers of particles through short distances of the order of a millimetre, and splash will move a few particles for much longer distances of the order of 10 cm. These movements will combine to produce a diffusion of particles from areas of high particle concentration to areas of low concentration, so that a group of particles on a smooth surface will become progressively separated, as perhaps occurs with the fragments of a boulder which has been shattered *in situ*.

188

On a sloping surface the pattern of movement becomes asymmetric and there is a net transport of debris downslope. The asymmetry is produced because (1) the downslope component of raindrop momentum acts directly to move debris and (2) the same set of splash trajectories will produce longer jumps when directed downhill rather than uphill. Both of these effects produce a net downslope transport, and outweigh any asymmetric tendency that may exist in the direction of rainfall. When the surface has a layer of water standing or flowing on it, raindrop impact directly on the soil, and therefore the effect of cratering and splashback, is considerably reduced, but raindrops falling on the layer of water increase turbulence and so indirectly increase the rate of surface erosion by the flow.

Rainsplash can directly move debris of up to at least 10 mm diameter and indirectly it can move very much larger stones. One way in which this is achieved is by undermining the downslope side of a stone by removing the surrounding fine material until the stone overhangs and falls over. Undermining by rainsplash may also play a part in the transport of stones on bedrock slopes. A sequence has been observed where fine material is washed across the bedrock until it is trapped beneath a fresh stone; plants then begin to grow in this soil until eventually their roots wedge the stone up from the surface and allow more soil to accumulate until the stone is supported by a pedestal of fine material. Rainsplash then undercuts the downslope side of the pedestal and a shelf is formed on the upslope side by deposition from above. Finally the stone is undermined and rolls downhill again leaving the pedestal of soil unprotected and subject to fairly rapid removal although the binding action of plants and their roots may delay its destruction for some time.

However, the main effect of vegetation on rainsplash movement is by shielding the surface from direct raindrop impact. Some of the rain collected on the plant surfaces does eventually fall to the ground, often in drops of larger size, but they are moving at much less than their terminal velocity, particularly under grass or other low cover, so their erosive power is greatly reduced. Under temperate conditions, the plant cover almost completely prevents rainsplash transport, so that small breaks in the cover assume great importance as points of accelerated erosion (Slaymaker, 1968). Under semi-arid conditions the exact amount of the vegetation cover, and the extent to which it is in phase with the seasonal rainfall, are critical in determining overall erosion rates (Williams, 1969). Perennial shrubs produce local differences in surface erosion which commonly results in their being elevated on mounds. Rainsplash contributes to mound formation because it can only produce splash transport inwards from the edges of the mound towards the centre and thereby has a heaping effect; and driving rain, such as might erode the mound, can only penetrate with full effect on the windward side.

Unconcentrated surface wash

Flows over the surface of a hillside vary greatly in the extent to which the flow is concentrated into depressions or channels. A more or less continuous transition of types of flow exists, from flows of uniform depth; through types of flows where most of the water moves in discontinuous depressions; through flows in definite channels which nevertheless move in position; to flows in permanent channels. In order to discuss slope processes a somewhat arbitrary division must be made between slope and river processes. In this chapter flows in channels which periodically move in position and so act on the whole of a hillside are considered as a part of the *slope wash* process group; whereas flows in channels which remain in essentially the same position and degrade or aggrade a valley bottom but not the whole slope are considered as contributing to *river* processes. Transport of debris on a broad floodplain is marginal to both groups of processes according to this definition, and will normally be taken as fluvial.

Flow in a thin uniform sheet is only possible on rather smooth surfaces, and will generally be very thin in depth so that its erosive effect is slight. Because only fine material can be transported in this way, the force required to erode the debris may be very much larger than the force required to transport it. It is reasonable therefore to consider such removal as partly *weathering-limited* because the thin skin of surface material which is dependent on wetting and drying, and freeze–thaw processes to loosen it, can readily be carried away. Over a period, therefore, the rate of this removal depends not on the erosive force of the flow but on the rate of break-up of the surface between flows. During the flow itself, further removal can be effected by rainsplash impact, by direct hydraulic shear on the roughness elements of the surface, by abrasion of the surface with coarser debris, and by wetting the surface and so lowering the cohesive forces which other forms of removal must overcome. For very thin flows over silts and clays removal of material previously loosened or detached by rainsplash is the most important factor, but in deeper flows the erosive power of the flow itself and also its transporting capacity become increasingly important. In sand and gravel, if the flow is deep enough to transport any debris at all, then the transporting capacity is always the limiting factor and the removal is strictly *transport-limited*.

Almost all natural surfaces are too uneven to produce uniform flow, and instead most of the water is concentrated in small discontinuous depressions and in smoother patches on the surface. The variations in flow thickness produce variations in the ways in which debris is being carried, so that there may be surface erosion at some points of high concentration, while other points show transportation of sand grains without erosion and others show transport only of fine material. Flow concentrations are too local and flow durations too brief to produce true channels. Under a vegetation cover the organic-rich topsoil is very permeable, so that overland flow is less common and less intense than where vegetation is absent. Breaks in the vegetation cover become very important as sources for slopewash debris, because they commonly expose less organic and less permeable subsurface horizons which encourage overland flow and often cause seepage of throughflow water, as well as allowing rainsplash to be more effective (Hadley, 1955; Slaymaker, 1968).

Concentrated surface wash

If flows are sufficiently intense they will generally erode a number of small channels, and if these channels remain fairly shallow they tend to move in position from time to time so that, over a period, the hillside is lowered evenly. There are two common types of semi-permanent channel pattern; braids and rills, and these are associated with sandy and silt-clayey soils respectively. In sands or gravels material can be readily eroded but transportation is limited to the capacity of the flow. Coarse debris is frequently deposited as bars in the centre of the flow and these divert and subdivide the flow into many channels which join and separate over the whole surface. Each channel shifts its course periodically, and during each flow a few channels will move position. There is little tendency for channels to be destroyed *between* successive flows, so that there is always a channel pattern on the surface. In fine debris, erosion is difficult and transportation easy, so that channels, once formed, tend to be maintained by the flow, and form a set of sub-parallel rills, each a well-defined small channel running down the slope. Between flows or between rainy seasons freeze–thaw and moisture changes, which are most effective in fine debris, tend to produce irregularities and movement in the surface material which obliterates the channels. Successive flows or seasons therefore show rills in different positions, so that, as with braids, the whole hillslope is evenly lowered.

A single braided channel is typically 1–10 cm deep and 5 cm to 5 m wide. Typically such channels occur on almost smooth desert surfaces sloping at 1–5°, and together they may occupy 5–20 per cent of the surface. In a braid the debris which cannot be carried at any time (usually coarse material) is deposited as bars and islands, but this deposition need not be permanent, but is simply the result of temporarily low discharge. Because deposition is essential to the process of channel shifting, braids are most commonly associated with aggradation but they need not be, and they may be an equilibrium form (Leopold, Wolman and Miller, 1964, pp. 284–95). The tendency to braid increases as the debris becomes more heterogeneous, as the discharge becomes more variable and as the probability of aggradation increases. The first two of these factors refer to the conditions of transportation at a point, but the third refers to the way in which sediment discharge is varying downstream, and so ultimately to the slope profile.

Rills are usually only a few centimetres wide and deep, and their dimensions are controlled by the erodibility of the soil (usually fine grained) which they cut into. They are prevented from becoming large by the short duration of most flows and also by the tendency of the soil to become tougher with depth, because it is more compacted so that differences in erosive power of neighbouring rills is minimized by the increasing difficulty of cutting into lower layers. The small channels formed in this way can be obliterated by a number of processes, but some 'master' rills may grow large enough to escape destruction. Schumm (1956 *a, b*; 1964; Schumm and Lusby, 1963) has described how frost heaving in the winter can deform the soil surface to such an extent that rill channels formed during summer rains are completely destroyed. Observation and measurement at sites with sparse vegetation and substantial winter frosts, in the Colorado Plateau and in New Jersey, show a marked seasonality. Surface markers move steadily downslope, but move more during the winter than in summer. Stakes in the ground show fluctuations in the ground level (shown by stake exposure) resulting from periods of freeze–thaw and rain-beat.

These movements correspond to seasonal changes in the character of the soil surface. During the winter months frost action breaks up the soil creating a highly permeable surface composed of polygonal aggregates separated by a network of shallow cracks. With the onset of spring rains a marked transformation occurs: the edges of aggregates are destroyed as they crumble into the cracks and partly close them. Although rain-splash never really succeeds in obliterating the aggregated structure, its compacting effect tends to seal the surface into a highly impermeable crust. As a result, runoff becomes extremely important and the runoff–precipitation ratio in August–October is substantially higher than in April–June. Increased runoff during the late summer initiates rill erosion (which increases stake exposure) but these rills are only temporary features. The return of frost activity in the winter once again breaks up the surface and obliterates all evidence of the rill pattern.

During the winter months on such badland slopes, when infiltration capacity is high, runoff is negligible and heave is important, the effect of soil wash is minimized and many of the surface properties are then characteristic of humid temperate areas. The late summer period, when infiltration is much reduced, runoff much higher and surface wash effective corresponds more typically to semi-arid areas. These slopes with little vegetation and frequent frost are therefore hybrid in their characteristics, but other factors effectively destroy rills, even in the absence of frost.

For example, Schumm (1964) describes how disturbance of the surface by animals can move material downslope and fill rills, although this process is most effective in the spring thaw period. A more widespread effect is that, during a rainstorm on unvegetated slopes, a surface layer 1–5 mm thick is wetted and subsequently dries into a thin mud-cracked cake which requires little or no disturbance to slide off in pieces several metres across which fill rills and other surface depressions with a talus of small blocks of dried soil. This is similar to processes of infilling described by Bryan (1940a) and Beaty (1959). Rills also form in cultivated fields, and it is normal practice to plough them in seasonally, so destroying them. In all of these ways rills are destroyed, and rills forming during a subsequent overland flow will have no reason to select the same courses; indeed if destruction is achieved by infilling, the fill is usually more permeable so that new rills will preferentially form elsewhere. The changing location of rill erosion thus produces more or less even lowering of the hillslope surface by channel erosion, and if rills are destroyed by infilling, there is also a recurrent downslope movement from inter-rill areas.

If more successful master rills do form then the hillslope surface will gradually form a valley around them so that rills in successive storms will change in direction; starting off parallel to the master rill, but progressively turning in towards it as it develops a small valley (fig. 8.1). This process of cross-grading may also produce changes within the time limit of a single storm, as has been described by Horton (1945). The cross-grading process clearly tends to form a permanent channel which cannot be destroyed between flows, and shows one way in which slope processes can begin to form a river.

Where there is a good vegetation cover, rills are less common than on bare soil because overland flow is less frequent and less intense, so that sufficient erosive force to produce a rill is rare. When the vegetation layer is locally removed, however, overland flow and seepage are greatly increased so that the water is able to erode the soil and produce a channel that grows large enough not to be filled in or obliterated before another storm occurs. Under these conditions, the alternation of incision and obliteration has a time scale of centuries rather than a single season. On both vegetated and unvegetated hillsides, rills are related to gullies and headward extensions of existing channels, but a near-permanent channel extension differs from a rill because it is generally along the axis of a well-defined depression so that even where the channel is periodically obliterated, it always re-forms in roughly the same position.

Rills and braids are also related to mudflows, which are shifting flows with definite channels and levees at their sides. Typically they

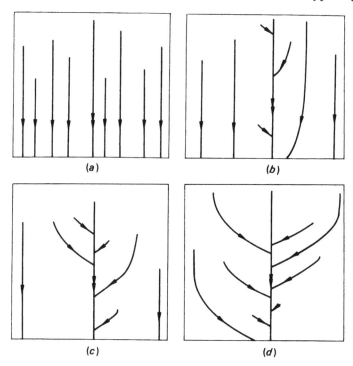

Fig. 8.1. Hypothetical stages in the development of a master rill (double arrow) by cross-grading as it enlarges its valley. Diagrams show successive periods of rilling, with only the master rill reforming in the same position (derived from Horton, 1945).

form on steep slopes in very heterogeneous material, containing both fine material and boulders. The flow advances in a series of surges, each of which is contained by an arcuate dam of deposited coarse debris. The flow breaks through the front of this dam and surges forward only to form another dam. The sides of these dams are left behind to form continuous levees. The widths and depths of these channel and levee systems may be anything from a centimetre up to tens of metres, but they are most typically several metres across, and found on high gradient alluvial fans (Hooke, 1967).

194

METHODS OF MEASUREMENT

The ideal way of measuring the rate of a slope-process is to measure the sediment discharge of that process; that is, the quantity of soil material passing a unit length of contour in unit time. One may attempt to substitute space for time and measure the distribution of material over the ground surface, but in this case it is also necessary to measure the mean velocity of each size-class of material. At the soil surface itself another method is possible; to measure the change of elevation of the surface over a period. This change in elevation is a direct measure of landscape development, and is related to the rate of change of transport downslope by the continuity equation (5.1). However, it is usually much smaller in magnitude than the sediment discharge and therefore more

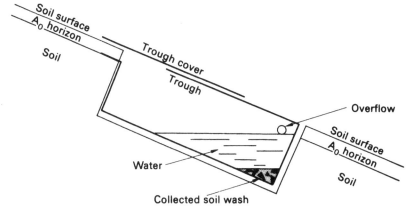

Fig. 8.2. Soil wash trough (after Young, 1960).

difficult to measure accurately. In processes involving a diffusion exchange of material, it may also be relevant to measure the dispersion of material initially together, even when this produces no net transport downslope; a special case of this approach is in the dispersion of material up from the soil surface (as well as laterally) during rainsplash.

Sediment discharge

The direct measurement of sediment discharge carried by running water consists of catching all or part of the flow, in a way which does not modify the flow above the point of measurement, and analysing the sediment in the water before it is able to change its electro-chemical state or mechanical size. These ideal requirements are never fully met. For measurement of surface wash, short troughs, usually 20–50 cm in length have been inserted in holes in the soil, with a lip on the upslope side projecting 1–2 cm into the soil at the base of the A_0 horizon (Young, 1960). The trough is covered and provision made for overflow on the downhill side (fig. 8.2). Disadvantages of this method are principally the loss of material in fine suspension or solution in overflow water, and the tendency of material to wash in at an accelerated rate from the small free face exposed in the A_0 horizon. The former problem can be reduced by increasing the size of the container and the frequency of emptying it; and the latter by attempting to get a smooth transition from the soil surface to the trough with an apron of plastic sheeting or concrete. Similar troughs, 1·8 m (6 ft) long, have been used in U.S. Soil Conservation Service experiments at the base of artificially sprinkled runoff plots. With this equipment all runoff and sediment produced in an experimental run is drained off and analysed.

For surface flow the interruption of the flow by a collecting trough does not materially influence the flow, because it is driven by the surface gradient only. To attempt a similar measurement for sub-surface wash is much more difficult. Any pit or trough will modify the hydraulic gradients which are important in determining the rates of water flow, except at a natural stream bank. What is worse, the introduction of a free face will cause vastly accelerated erosion over the whole trough face, so that the measurement will record not the true sub-surface transfer of material, but merely the conditions at the free face, whether this is left open, or shielded with some sort of gravel filter. Troughs have been used to record throughflow in the soil (Whipkey, 1965), and may be satisfactory for measuring dissolved material in the soil, but are useless for measuring solid sediment discharge. In principle, what is needed to measure solid sediment discharge is a block of material of identical pore size distribution to the soil, but which is readily distinguishable. Hydraulic gradients and flows will not then be interrupted, and soil particles washing into the block can be distinguished and their total measured. Suitable materials might include ground glass, fine glass beads, gypsum or layers of fabric. In this type of measurement the block will not generally be chemically identical to the soil, so that there may be difficulties caused by re-deposition of clays and solutes.

When flows become large, as in rills or small streams, the flow must either be divided or sampled. In a divider, water flows through an almost rectangular flume (slightly wider at the top to compensate for the frictional effects of the walls) and is allowed to fall freely from a sharp edge. A trough catches a proportion of the falling water and sediment equal to the ratio of trough width to flume width. If the flow fraction is still too large to store, several dividers can be installed in series, each dividing the overflow from the previous storage tank (fig. 8.3). Alternatively normal stream sampling equipment can be used, consisting of

196

integrated sediment samplers in combination with a continuous discharge or stage recorder.

Long period measurements of sediment discharge are rare, and usually rely on the existence of natural or artificial breaks in the land surface which trap a part of the sediment (mainly solid) moving past it. Reliability of the results depends greatly on the estimates made of trap efficiency. Under natural conditions, raised beach benches form steps which can act as traps for accumulating some debris. Suitable artificial structures are much more common, though generally providing shorter periods of record. They include, in order of increasing trap efficiency agricultural terraces, trees (Lamarche, 1968; Ruxton, 1967), fences and

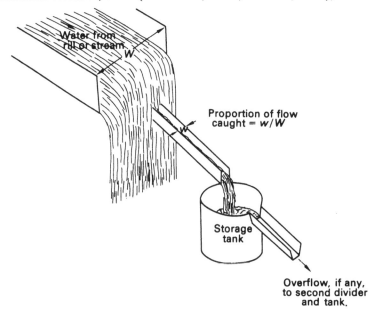

Fig. 8.3. Flow divider for measuring water and sediment discharge.

walls at field boundaries, contouring irrigation canals, erosion check dams and water storage dams (Federal Inter-agency River Basin Committee, 1953). Material in the last two of these may of course be derived from channel erosion in addition to slope erosion.

The use of tracers provides an effective indirect method of measuring sediment discharge. The ideal measurement consists of labelling every particle beneath a unit area of slope surface and recording its movement over a period. The total sediment discharge is the sum of these velocities, each multiplied by the mass or volume of the particle. Clearly this is impractical, and real measurements are based on a sample, usually a stratified sample. For grain movement, grain size is a normal basis for stratification, and also depth in the soil if sub-surface movement is considered. For movement as colloids or in solution, chemical composition is also a relevant basis. For surface particle movement of coarse sand and gravel a sample of stones of each size-class may be painted, or particles of the correct size introduced (Schumm, 1964), and placed in known positions. Over a period, mean rates of movement are obtained for each size-class, and weighted with the volume of this size present per unit surface area to obtain its contribution to the sediment discharge. For periods of several years masonry paints or resin and dye mixtures are suitable coatings for stones, but over a period of a few months aerosol spray gloss paints applied to dry stones are sufficiently durable: on unvegetated surfaces they can be applied in the form of a continuous line. For movement of fines either on or below the surface, radioactive tracers are ideal where they may safely be used.

TABLE 8.1. *Examples of suitable isotopes for radioactive labelling of compounds moving through the soil* (data from I.A.E.A., 1967)

Labelling isotope	Compounds to which label is attached
^{45}Ca	$CaCl_2$
^{36}Cl	$CaCl_2$
^{128}I	KI
^{131}I	KI, NaI
^{35}S	H_2SO_4
^{55}Fe	$Fe_2(OH)_3$, Fe–kaolinite, Fe–montmorillonite
^{59}Fe	EDTA, EDDHA, KCN, Na-citrate, NaCNS, $FeSO_4$
^{51}Cr	$K_2Cr_2O_7$, EDTA, Na-oxalate
^{54}Mn	$MnSO_4$, Mn–EDTA
^{60}Co	EDTA, KCN
^{65}Zn	EDTA
^{14}C	Humic acid

Examples of suitable radioactive labelling materials and the compounds to which they may be attached are shown in table 8.1. (I.A.E.A., 1967). For particle movement it may also be possible to use fine glass beads, iron filings or spores, but the density differences between such tracers and soil particles may invalidate their use.

Changes of surface elevation

Repeated accurate surveys show net changes of elevation over a period. Hillslope measurements generally differ from normal methods of topographic survey in order to achieve two aims: (1) by increasing the frequency of benchmarks, it is intended to reduce measurement errors, even using simple instruments; and (2) by modifications of the technique, it is intended to obtain additional detail about the sequence of elevation changes between readings, especially in the case of alternating erosion and deposition at a point. The ideal benchmark will be firmly fixed at depth in a zone of soil or bedrock which is not liable to vertical heaves or lateral movement. Any parts of the bench-mark protruding above this zone must be prevented from moving by a combination of rigidity and protection from the moving soil. Benchmarks in actual use rarely satisfy these criteria, and generally, the more there are in an area the farther each one is from the ideal.

Steel rods, 50–100 cm long and 1–2 cm in diameter, have commonly been used as bench marks. They are driven into the soil with their tops just above or below the surface of the soil (if below, their positions must be roughly located relative to surface features). Two such benchmarks can form a datum for levelling and re-levelling the line between them, and this method is suggested for the measurement of changes in channel cross-section at 'Vigil' network sites (Leopold, 1962). A second method consists of placing a stake at each measurement point and measuring changes in the exposure of the stake. Schumm (1964) has used straight metal stakes, 45 cm long, and Leopold, Emmett and Myrick (1966) have used 15–25 cm nails with a washer on the shaft of the nail. The purpose of the washer is to record the minimum level of the surface between readings, in addition to the net change in level shown by the change in exposure of the nail. This method may lead to false readings if rainsplash removes material around the washer, so that it stands on a pedestal. With relatively short stakes it is also important to ensure that they penetrate well below the depth of soil heaving. These simple methods by no means exhaust the possibilities of accurate measurement using either normal surveying instruments, or special purpose-built devices, such as point gauges.

Dispersion and miscellaneous methods

In diffusion movements, which include rainsplash and ionic exchanges, the tracer methods described above are appropriate for recording changes in positions of particles. What differs is the computation: in a complete analysis it may be desirable to treat each movement as a vector and analyse the distribution of particles with distance in each direction. If the movement is a simple diffusion, then it may be sufficient to calculate the diffusivity as the mean square movement per unit time (relative to the mean, if it moves). In the case of rainsplash, dispersion is accompanied by movement up from the surface, and this separation provides material which can readily be carried away by overland flow.

199

Ellison (1944) designed a splash board to catch material lifted from the surface, consisting of a very narrow trough sunk flush with the ground surface. Two of these were placed side by side, with a vertical wall between them, and a small canopy over the troughs to prevent material splashing out again (fig. 8.4). This allows material coming from the upslope and downslope directions to be distinguished. In using this device, it is important to ensure that the trough remains flush with the ground surface throughout the period of measurement.

Other methods of observing sediment transport by running water relate mainly to the examination of forms which are assumed to result from these processes. In some cases careful repeat observations or the

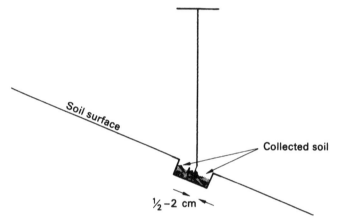

Fig. 8.4. Splash board to measure material lifted from the soil surface during rainsplash (based on Ellison, 1944).

existence of datable starting points can allow rates to be computed. Overland flow can be seen in the partial flattening of grassy vegetation and perhaps by detached leaves and stems trailing downslope around living stems. If there has been debris transport, some of the material may pile up in small dams above small and large stems. Where debris transport is considerable, transported material may lie on top of living vegetation or over a more compacted former surface soil. On the downslope side of tree trunks stem-flow may move material even without widespread overland flow (Ruxton, 1967), and leaf litter may be cleared from a small area in this way. Rainsplash may be shown by the formation of small pedestals, each capped by a stone. Within the soil, thin sections examined under a petrographic microscope may show accumulations of oriented clay in the B_T horizon in laminated coatings around soil particles and linings inside pores (Cady, 1965). The relative clay depletion of A horizons probably shows that some form of clay transfer has occurred (see also chapter 9).

EVIDENCE OF RATES OF ACTION

Rainsplash transport

Ellison (1944) first demonstrated that rainsplash could be a major process of sediment release and transport on the soil surface, where it is unshielded by vegetation. In splashboard experiments he showed that small particles could move up to 1·5 m, and that even 4 mm stones could move as far as 20 cm. Finer material is lifted from the surface in preference to coarse so that the material removed is different in composition from the original soil material (fig. 8.5), and the coarser fractions tend to remain on the surface forming a lag deposit if the rainsplash material is

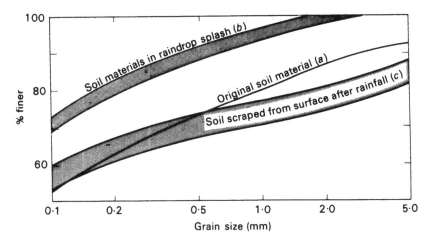

Fig. 8.5. Grain size analyses of (*a*) original soil material used in rainsplash experiments; (*b*) material lifted from the surface by rainsplash and (*c*) material scraped from soil surface at end of experiment (after Ellison, 1944). Bands in (*b*) and (*c*) indicate range of results for differing drop size and impact velocity.

removed. The rate at which material is lifted from the surface was found to be proportional to the fourth power of raindrop velocity, the first power of raindrop diameter, and the two-thirds power of rainfall intensity, under conditions where these factors could be separately controlled (Ellison, 1944). For normal raindrop distributions (Laws and Parsons, 1943) this relationship is equivalent to an increase proportional to the 1·0 to 1·5 power of rainfall intensity.

The asymmetry of movement on a sloping surface leads to a net downslope transport from rainsplash on its own, so that Ellison (1944) found three times as much material moving downslope as upslope in experiments on a 6° (10 per cent) slope. During a single storm particles of the same size do not all move the same distance, and their behaviour is best described by a probability distribution. Fig. 8.6 shows the movements experienced by small stones in a single storm, on slopes of 6° and 17°, and it can be seen that the distributions approximate to the log-normal form. Mean distances moved decrease with grain size and increase with slope gradient. Fig. 8.7 shows a series of measurements, each point representing a mean for a 5° range of slope angles. The solid lines show movements on natural hillslope surfaces during a single

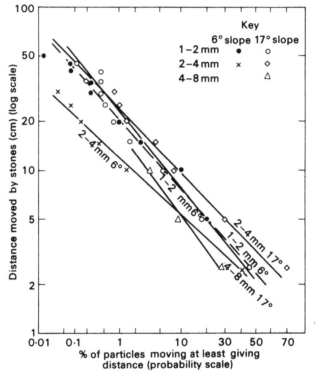

Fig. 8.6. Cumulative frequency distributions of distances moved by rainsplash in a single storm of 16 mm rain on bare ground near Tucson, Arizona (data from Kirkby and Kirkby: in preparation).

month of intense rainstorms, falling at about 50 mm per hour. Data is somewhat scattered, especially for the finer material, but there appears to be an increase in the distance moved as gradient increases, for all grain sizes. However, for the larger grades, the rate of increase with slope is much higher. Schumm's (1964) data for creep and rainwash in Colorado is only partly comparable, as a large portion of it occurred during cold winters, but seems to suggest that this trend continues for coarser material. In contrast, laboratory data for fine (1–2 mm) material on a uniform surface of the same material shows a much steeper rate of increase with slope than with same-sized material on a natural slope. Differences in surface roughness may be responsible for these different behaviours: on natural slopes, steeper gradients are associated with larger amounts of coarse material, and it seems possible that on these steeper and rougher surfaces the movement of fine material is much more impeded than that of coarser debris.

If the mean distances of travel in fig. 8.7 are multiplied by the grain

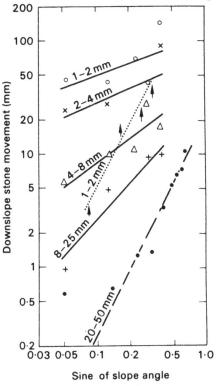

Fig. 8.7. Relationships between rainsplash particle movement and slope gradient for different grain sizes and surfaces. (*a*) Natural hillslopes near Tucson, Arizona for one month (July/Aug. 1964) (Kirkby and Kirkby, in preparation). (*b*) Schumm (1964) data for creep and rainwash in Colorado. Data multiplied by $\frac{1}{10}$ to reduce to approximately same basis. (*c*) Lab movement on surface of uniform grain size (Kirkby, unpublished data).

Key:

(*a*) Arizona data for July–Aug. 1964
    ○ 1–2 mm stones   △ 4–8 mm stones
    × 2–4 mm stones   + 8–25 mm stones

  (*b*) Schumm (1964) data showing annual rates × $\frac{1}{10}$
    ● 20–50 mm markers

  (*c*) Lab data for stones forming surfaces of uniform grain-size (relative values)
    1–2 mm stones.

diameters, we obtain values of mass transport for each grain size, per unit area covered by that grade. If this mass transport per unit area is plotted against grain size, a peaked curve is obtained, the peak corresponding to grain size which is carried most effectively, and this 'most effective grain size' increases from about 2 mm on a 3° slope to about 12 mm on a 25° slope. To obtain the *actual* mass transport at a point,

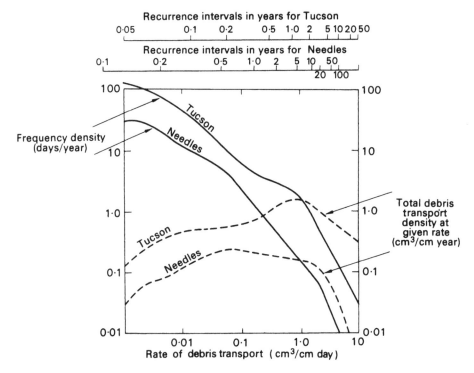

Fig. 8.8. Calculated frequency distributions for rates of debris transport by rainsplash on unvegetated ground (solid lines); and total transport effected at each rate (broken lines); for Tucson and Needles, Arizona.

Tucson: total annual rainfall = 325 mm: total transport = 4·2 cm³/cm year.

Needles: total annual rainfall = 125 mm: total transport = 0·5 cm³/cm year.

(Kirkby and Kirkby, unpublished material in conjunction with U.S. Weather Bureau rainfall frequency data.)

values of mass transport per unit area must be multiplied by the percentage of the slope area covered by the corresponding size grade. This transport rate shows no systematic increase with slope gradient because the increases in gradient are normally offset by increases in grain size, which counteract the effect of slope so that similar rates of transport can exist above and below sharp breaks in slope. Rates of transport are, however, related strongly to rainfall; storm transport being proportional to storm rainfall to the 2·5 power in Arizona studies (Kirkby and Kirkby, *in preparation*). Combining mass transport rates on unvegetated areas with meteorological data on storm frequency, magnitude and frequency curves for transport on unvegetated areas can be synthesized (fig. 8.8). The solid-line curves show the relative frequencies of different rates of transport, with most frequent rates of 0·001 cm³/cm day or less, corresponding to storms which are exceeded on seven to twenty days per year. The broken lines show the total transport which occurs at the various rates in year. Dominant event sizes differ considerably if they are defined in terms of the daily rate of debris transport (Tucson at 0·8 cm³/cm day; Needles at 0·06 cm³/cm day), but somewhat less in terms of recurrence intervals, which are in the range 0·5 to 1·0 years. These values for the return time of dominant-sized events are somewhat lower than those found for dominant sediment discharge in rivers (Wolman and Miller, 1960), but again show that it is the storm of moderate size, in which adequate carrying capacity is combined with adequate frequency, that is the most effective. If rainsplash alone is considered, there is no reason to suppose that the erosion will change its character qualitatively during more intense storms, since all rains contain a spectrum of raindrop sizes, including some of maximum size, so that the largest particles may still be moved during comparatively gentle rains. It is clear, however, that other processes of surface erosion will become increasingly important in more intense and prolonged rainstorms, so that changes in the character of the movement will be associated with a progressive change in the dominant process.

Summing the areas under the broken line curves in fig. 8.8, we obtain the mean annual rate of rainsplash transport on unvegetated ground for Tucson and Needles. Repeating this calculation for other points gives the approximate relationship:

$$S = 0.3 \left(\frac{R}{100} \right)^{2.2} \tag{8.1}$$

where S is the mass transport in cm³/cm year on unvegetated ground, and R is the mean annual rainfall in mm (fig. 8.9). This relationship can only be valid for conditions of rainstorm distribution and ground stoniness similar to those in Arizona, but forms a basis for examining the influence of a vegetation cover on the rate of transport. It is apparent in the field that a low grassy or shrubby vegetation cover almost completely suppresses rainsplash by breaking the impact of the raindrops, so that an approximate value for the actual transport is the product of the rate of transport on bare ground and the proportion of unvegetated area. Curves for these two components, generalized for southern U.S.A., are shown in fig. 8.9 in terms of mean annual precipitation; and their product, the actual net transport rate, is also shown. The two main effects of increased rainfall – that of increasing the rate of bare-ground transport and that of decreasing the amount of bare ground available by producing denser vegetation – tend to oppose each other. The resulting net transport rate has a peak value at annual precipitations of 300–400 mm for it is at this rainfall that the optimum combination of sparse vegetation and high transport rate occurs. This

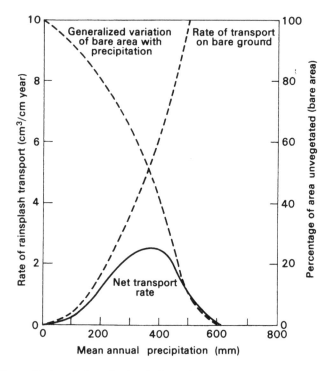

Fig. 8.9. Generalized variation of rainsplash erosion on unvegetated ground; percentage bare (unvegetated) area and net transport rate, calculated as the product of the other two: all expressed in terms of mean annual precipitation in southern U.S.A. (from Kirkby, 1969).

rainfall corresponds to moist semi-arid conditions in the United States. In other parts of the world these curves must be modified to allow for differences in the distribution of intense storms and in the seasonality of rainfall and its influence on the vegetation cover. At high annual rainfalls, under primary rain forest, another factor affecting rainsplash becomes important; that is, the long height of fall (about 8 m) from canopy trees on to a ground surface which is not well covered by low vegetation and leaf litter, so that drops strike the surface with almost full terminal velocities after accumulating to large diameters on the leaves,

206

even during fine drizzle (Ruxton, 1967). Under these conditions, the rate of rainsplash erosion may increase again at extremely high annual rainfalls.

Soil surface erosion

Except in the case of rainsplash, different processes of surface erosion by water are seldom distinguished in measurements. At low intensities of erosion, removal is generally more or less uniform, and at higher in-

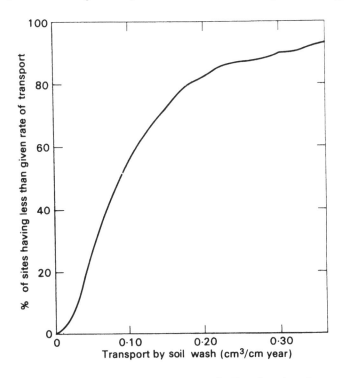

Fig. 8.10. Cumulative frequency curve showing rates of soil wash under a dense grass cover in Great Britain, on slopes of 5–30°, at 118 sites. Data from Young (1958) and Kirkby (1963).

tensities braids, rills or small gullies play an increasingly important role, but their separate contribution will not be discussed in this section. Factors which have a demonstrable influence on the rate of erosion of surface soil are the topography, the vegetation cover, the rainfall characteristics, and the soil texture, structure and micro-topography. These factors are examined below, starting with some actual rates of transport and their variation within the area of a single experiment.

207

Under humid temperate conditions, with moderate rainfall intensities and total vegetation cover, measured rates of soil wash are exceedingly low. Figure 8.10 shows the distribution of values found at 118 sites in Great Britain, 49 on Millstone Grit in the Pennines (Young, 1958) and 69 on Silurian shales in Galloway (Kirkby, 1963). The separate sets of data show very similar distributions, and each covers a considerable range of slope gradients and positions on the slope profile, but no consistent correlation with these topographic factors can be detected. These

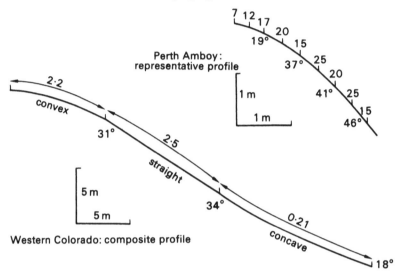

Fig. 8.11. Hillslope profiles associated with graphs in fig. 8.12. Figures above the profile show rates of erosion at measurement stakes in mm/year. Figures below the profile show slope gradients in degrees (data from Schumm, 1956a and 1964).

measurements, made with soil wash collecting troughs, are liable to be over-estimates of the transport by soil wash, but they show an extremely small median value of 0·09 cm³/cm year.

Where vegetation is sparse or absent, as occurs naturally in arid or semi-arid areas, but may also occur in more humid areas as a result of human activity, values of soil erosion are very much higher. Fig. 8.11 shows rates of exposure of stakes on hillslope profiles. One of these is a composite profile, averaged from four representative profiles from areas of Western Colorado on Mancos shale (Schumm, 1964; profiles 6, 13, 16 and 20), with mean rates of erosion calculated for convex, straight and concave slope segments. The other is a representative profile from badlands near Perth Amboy, New Jersey (Schumm, 1956a), where natural slopes have developed on a fine-grained artificial fill with no vegetation cover. The variations in stake exposure on these profiles can be accumulated to obtain the rate of sediment transport at each point, and the variations in transport rate can be correlated with gradient and distance from the divide as simple power laws (fig. 8.12). Since gradient and distance from the divide are related to each other, however, the exponents in these power law expressions are not critically determined. It can be seen that the rates of sediment transport on the bare slopes are many times larger than in the vegetated case, with values of 200–500 cm³/cm year at the base of the slopes. Hillsides in both of these examples are rilled. Equally high transport rates have been ob-

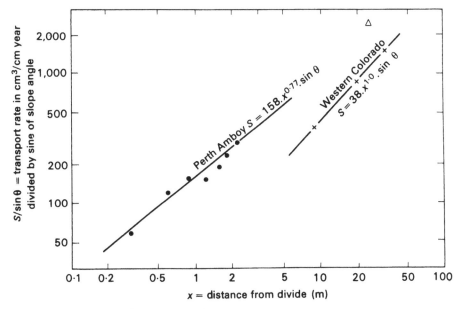

Fig. 8.12. Relationships between rates of soil wash transport and slope length and inclination for unvegetated and sparsely vegetated hillslopes.

● Data from Schumm (1956*a*) for badlands on artificial fill material near Perth Amboy, New Jersey.

× Generalized data from Schumm (1964) for badlands in Western Colorado on soils derived from Mancos Shales.

△ Sangre de Cristo Mtns, New Mexico (Leopold, Emmett and Myrick, 1966).

tained under semi-arid conditions on natural slopes in the Sangre de Cristo Mountains of New Mexico (Leopold, Emmett and Myrick, 1966), with values of 1,050 cm³/cm year on 20–30° slopes about 25 m long; and for braided pediment-like surfaces in the Sacaton Mountains, Arizona, with about 1,500 cm³/cm year on 2° slopes at about 300 m from the divide (Kirkby and Kirkby, in preparation). Both of these values lie close to the best-fit lines in fig. 8.12.

The effect of slope length and gradient can also be seen in studies on soil erosion plots and in cultivated fields, although the rates are less relevant as measurements of natural erosion conditions. Relationships are usually expressed as power laws, but exponents vary. Zingg (1940) analysed data from five Soil Conservation Experiment Stations and obtained

$$S \propto x^{1 \cdot 6} . \tan^{1 \cdot 4} \beta, \tag{8.2}$$

where S is soil transport (cm^3/cm year), x is length of slope (m) and tan β is the slope gradient. Musgrave (1947) obtained, for U.S. Dept. of Agriculture data:

$$S \propto x^{1 \cdot 35} . \tan^{1 \cdot 35} \beta, \tag{8.3}$$

and Kirkby (1969), for a ploughed field in Maryland:

$$S \propto x^{1 \cdot 73} . \tan^{1 \cdot 35} \beta. \tag{8.4}$$

These values, all for unvegetated field soils, show a reasonable uniformity which suggests that power laws of this sort provide suitable empirical models for the variation of soil wash transport with topography. The difference between these cultivated-field values and the exponents obtained for natural slopes (fig. 8.12) may be due to the differences in soil and vegetation which occur under natural conditions, but which are minimized by experimental procedures: for example soils tend to become moister and vegetation somewhat denser downslope under natural conditions, so that the influence of slope length is reduced, leading to lower slope length exponents under natural conditions. All of the studies quoted, however, refer to fine-grained soils in which steep slopes do not develop much coarser surface material than neighbouring low slopes. The influence of gradient has been explored in a somewhat different way by Renner (1936). He examined soil surfaces within 5° slope classes, and calculated the percentage of area in each slope class which showed signs of erosion, either as a sheet or in gullies. The results (fig. 8.13) show an increased incidence of erosion up to 35° and a decline at higher gradients, which may be attributed to the high resistance of slopes which are able to stand at such high gradients, and also to the reducing collecting area per unit slope surface area on steep slopes (Horton, 1945).

The massive difference between rates of soil wash on vegetated and unvegetated slopes, of the order of 1,000 to 10,000 times, shows that the presence or absence of vegetation is the largest single factor in determining the rate of transport. Vegetation acts by (1) intercepting precipitation, (2) breaking the impact of raindrops and so preventing dispersion of soil particles by rainsplash and the formation of a compacted surface layer with a very low infiltration rate, (3) improving soil structure and infiltration within the soil around roots, (4) interrupting and subdividing overland flow around plant stems and leaves, and (5) physically binding the soil with the root mat.

Weaver (1937) exposed soil containing plant roots to erosion from a water jet from a hose, and found that the presence of annual crop plants extended the period before erosion began by a factor of one to two times; native big bluestem grass and bluegrass extended it up to five times, and Hungarian brome grass, which has a dense mat of interwoven rhizomes, delayed the onset of erosion by fifteen to twenty times. It appears that while root systems, especially rhizomes, do protect the soil from erosion, they are not able to explain the differences in erosion rate associated with vegetation.

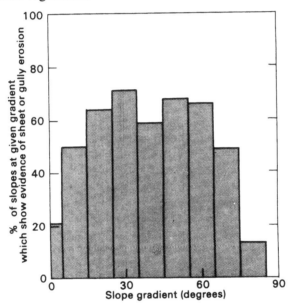

Fig. 8.13. Percentage of slopes in each gradient class which show evidence of sheet or gully erosion. Boise River Basin, Idaho (after Renner, 1936).

All of the other ways in which vegetation influences surface erosion (nos. 1–4) act by reducing the quantity and velocity of overland flow, mainly by increasing the infiltration rate of the soil. Fig. 8.14 shows two sets of data, each set for comparable conditions of slope length and gradient, but with a range of different vegetation covers. It can be seen that increasing density of vegetation leads to a considerable reduction in both overland flow runoff and in erosion, and that the erosion loss is closely related to the rate of runoff. These data appear to support the evidence of root erosion studies that the main influence of vegetation

211

lies elsewhere – that is in its control of overland flow *runoff*. Point 10 in fig. 8.14 shows the effect of mosquito gauze spread just above the soil surface, which has a very similar effect on runoff and erosion to the dense grass cover (point 11). At the end of the rainy season, the bare soil plot has 'a dense compacted layer on the surface with no visible structure, looking and feeling like a road, but still eroding rapidly under high speed runoff'. Under the mosquito gauze however, the soil 'remains cloddy and loose throughout the season, and only exceptional rain pro-

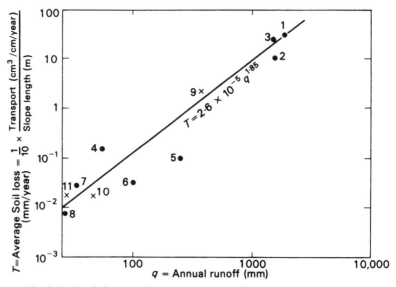

Fig. 8.14. The influence of vegetation on runoff and on soil erosion.
● Data from Meginnis (1935) showing different vegetation covers in an area of south-east United States having 1500–1800 mm annual precipitation.
× Data from Hudson and Jackson (1959) showing different vegetation covers in Rhodesia, with 1,000 mm annual precipitation.
1. Cultivated cotton: rows downslope. 2. Cultivated cotton: rows on contour. 3. Barren abandoned land. 4. Black locust–osage orange plantation. 5. Scrub oak woodland. 6. Bermuda grass pasture. 7. Broomsedge field. 8. Oak forest. 9. Bare soil. 10. Mosquito gauze over bare soil. 11. Dense grass.

duces any surface movement' (Hudson and Jackson, 1959). It may tentatively be concluded that the most important effect of the vegetation is in protecting the surface from raindrop impact and the consequent dispersion and compaction of fine surface material.

The distribution of rainfall intensities is also an important factor in determining the rate of soil erosion. Intense rain contains more large drops which are more effective in compacting and dispersing the soil, and intense rain is also more likely to exceed the infiltration capacity of the soil and so produce Horton overland flow (fig. 3.13). Wischmeier and Smith (1958) have related total soil loss on moderate slopes with sparse vegetation cover, to rainfall intensity and energy by the empirical relation:

$$T = (0.00094 \pm 0.00006) . E . i_{30} \quad \text{with 95 per cent confidence,} \quad (8.5)$$

where T is the total soil loss in millimetres depth during the storm; i_{30} is the maximum 30-minute intensity during the storm (mm/h); and E is the total storm energy in J/m^2, which is obtained by summing the instantaneous rate of energy production, e, over the storm, where

$$e = 13.32 + 9.78 \log_{10} i \quad J/m^2/\text{mm rain,} \quad (8.6)$$

and where i is the instantaneous rainfall intensity (mm/h.). These linear regressions gave good agreement with data from a single intense storm in a Mancos Shale basin of Western Colorado (Hadley and Lusby, 1967).

Musgrave (1947) has also related erosion loss to rainfall intensity specifically to the 1·75 power of the 2 year-30 minute rainfall, but where vegetation is relatively dense and infiltration more variable, runoff is less closely related to rainfall alone, and it is more productive to correlate the soil wash transport with the overland flow discharge. Fig. 8.14 shows a correlation with runoff under comparable conditions of rainfall, and this shows erosion proportional to the 1·85 power of the annual overland flow runoff. Horton (1939) has analysed runoff plot experiments to obtain the empirical relationship:

$$T = 0.04 \, q_s^2 . \tan^{0.62} \beta, \quad (8.7)$$

where T is the soil removal rate in mm/h; q_s is the runoff intensity in mm/h; and $\tan \beta$ is the surface slope. Compared with fig. 8.14, this value is about 1,000 times greater, but this is mainly due to the difference between instantaneous runoff rates and annual rates. Horton's equation may also differ in that it applies mainly to erosion in which there is some gullying.

Soil conditions influence erosion rates mainly by influencing runoff, which they do through variations in infiltration capacity with texture, structure, and antecedent moisture content; and also by variations in depression storage which may be modified considerably by cultivation practices. There are also differences in erosion rate which are caused by the inherent erodibility of the soil material and by the condition of the soil surface particles at the start of a rain. Variations in infiltration rate with soil and soil moisture properties have been discussed in chapters 3 and 4. The resistance of soil to erosion under given runoff conditions depends primarily on the size of its aggregates, rather than on its textural components (Diseker and Yoder, 1936) and the cohesive forces holding these aggregates together. Loose, granular soils are resistant to small amounts of runoff, because the aggregate size is relatively large, but if runoff becomes more intense the soils can erode rapidly down to a level which is compacted (Baver, 1956). Fine-textured soils, on the other hand, may have a thin surface layer of loose, fine-textured material which is produced by drying of the soil between rains. This layer is very readily removed by the first runoff, and continued runoff

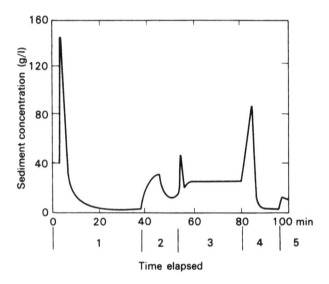

Fig. 8.15. Variations in soil loss during the course of an experiment, applying rainfall and overland flow to Keene silt loam in differing combinations (Ellison, 1945).
1. 0–38 min: overland flow only at 50 cm/h. 2. 38–55 min: overland flow plus rain at 82 mm/h. 3. 55–80 min: rain at 82 mm/h only. 4. 80–96 min: overland flow at 50 cm/h only. 5. 96–100 min: overland flow plus rain.

will produce less erosion, even if it is more intense (Baver, 1956; Emmett, 1970). Fig. 8.15 shows this kind of behaviour in a continuing experiment during which overland flow and raindrops are applied in different combinations. At each change of the combination, there is an initial peak of the sediment concentration curve, as material which is available to the new combination is removed, before settling down to a new steady level (Ellison, 1945). It should be noted that the total sediment production (as opposed to concentration) is greatest when raindrops and overland flow are present; and also that at each succeeding change-over there is less freely available material, so that the peak at 85 minutes, for

214

example, is lower than the peak at 5 minutes. Once the initial loose material has been removed from fine-textured surfaces, the moist compact surface is relatively resistant, and its resistance increases with clay or organic colloid content (Baver, 1956) and usually with antecedent moisture content (Grissinger, 1966).

Under natural rainfall conditions, the inherently greater resistance to erosion of clay soils is balanced against their greater runoff. Indices of erodibility usually refer to the amount of silt and clay in soils, for example in the ratio:

$$\frac{\text{surface area of particles} > 50\mu \text{ in cm}^2/\text{g.}}{\text{Total \% silt} + \text{clay in dispersed soil} - \% \text{ in undispersed soil}}$$

(André and Anderson, 1961). Lutz (1934) obtained a measure of erodibility which increased with ease of dispersion and decreased with infiltration capacity at the surface, permeability within the soil profile, and grain size. These indices combine soil properties in somewhat opposing ways.

The control of rapid soil erosion is of great economic benefit, and has been achieved mainly by reducing overland flow. Where practicable, a complete vegetation cover is the ideal control measure, but this conflicts with the needs of farming, and it may only be possible to leave a vegetation cover on some contouring strips, though these need not always be the same. Soil structure and texture may be controlled, especially by maintaining a high organic content, which improves infiltration and limits dispersion during erosion. Perhaps most important, however, is the modification of topography, both on the large scale by terracing which reduces slope gradients and concentrates erosion risks on the terrace outlets and banks, which must therefore be fully protected; and on the small scale by contour ploughing and crosstying furrows, which can lead to a tenfold increase in depression storage on a slope.

There are no effective magnitude and frequency studies for soil erosion, but the evidence reviewed allows some generalizations to be made about the form of the curves. Daily rainfall amounts are widely available and commonly approximate to an exponential frequency distribution:

Cumulative frequency, $\qquad J = J_0 \cdot e^{-r/r_0};$ $\qquad\qquad$ (8.8)

or frequency density, $\qquad j = \dfrac{J_0}{r_0} \cdot e^{-r/r_0}$ $\qquad\qquad$ (8.9)

where J_0 is the total number of rain-days per year, r_0 is the mean rainfall (in mm) per rain-day, and r is the daily rainfall (in mm), the independent

215

variable. Then the total rainfall, R, is equal to $J_0 . r_0$ (Values of J_0, r_0 in this model should strictly be best-fit values to the exponential distribution rather than exact values.) Under given conditions of soil and vegetation, it is supposed in this model that a daily quantity of rainfall equal to r_c is able to infiltrate and evaporate during rainfall, so that daily rains in excess of r_c will cause overland flow on the hillside, equivalent to a runoff of $(r - r_c)$ mm. Then the annual overland flow runoff is equal to

$$R . e^{-r_c / r_0} .$$

Using this estimate of slope runoff for comparison with the data presented in figs. 8.12 and 8.14, an estimate for the total annual soil erosion, on a slope ten metres long, inclined at $10°$ is:

$$S = 0.003 \ R^2 \cdot e^{-2r_c / r_0} \text{ cm}^3 / \text{cm year.} \qquad (8.10)$$

The corresponding daily rate of transport in a day with rainfall r, is:

$$s_r = 0.0015 J_0 \cdot e^{-r_c / r_0} \cdot (r - r_c)^2; \qquad (8.11)$$

and the total amount of transport per year, in days during which the rainfall is r, is the product of this rate and the frequency density, giving:

$$S_r = 0.0015 \frac{J_0^2}{r_0} \cdot e^{-(r + r_c)/r_0} \cdot (r - r_c)^2. \qquad (8.12)$$

If this value is plotted against the corresponding rainfall rate, then the peak value of S_r is at the geomorphically most effective rainfall. This peak value of the transport rate is equal to:

$$S_r^* = 0.0008 J_0^2 . r_0 . e^{-2r_c / r_0}. \qquad (8.13)$$

It occurs at a daily rainfall of $r = 2r_0 + r_c$, which has a cumulative frequency of $0.136 J_0 . e^{-r_c / r_0}$. These results are obtained mathematically from the assumptions. They can be interpreted to show the effect of rainfall and vegetation on soil wash transport.

Fig. 8.14 can be used to calculate suitable values of the constant r_c for different vegetation covers. Both the sets of data are consistent with the values $r_c = 10$ mm for bare ground; $r_c = 40$ mm for a good grass cover and $r_c = 100$ mm for an oak forest. Fig. 8.16 shows the effect of using these values to estimate magnitude and frequency curves, in this case for south-west England. Under a given rainfall regime, increasingly dense vegetation cover produces progressively lower peaks at higher and higher rainfalls. This means that major storms play an increasingly important role under denser vegetation, and that longer periods of measurement are needed to obtain valid mean transport rates. Whether a hundred-year period is needed for grassed surfaces, as is implied by fig. 8.16, or whether a shorter period will contain an adequate proportion of high-intensity short-period rains is an open question.

In fig. 8.17, annual totals are plotted in terms of annual rainfall and the ratio of critical rainfall, r_c, to mean rain per rain-day, r_0. It can be seen that where the rain per rain-day is large, there is a much smaller effect produced by removal of the vegetation. If there are 5 mm of rain per rain-day, as in south-west England, then removal of grassland will increase soil transport by a factor of 400 times: at the other extreme the monsoon in India commonly produces average rains of 100 mm or more per rain-day and under these conditions removal of grassland will

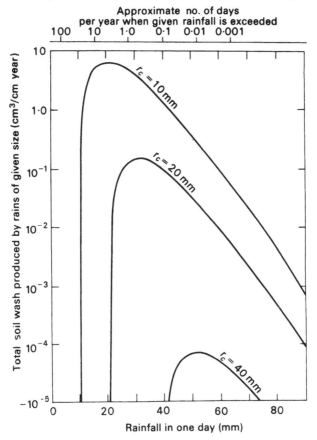

Fig. 8.16. Estimated total rates of soil wash for conditions in south-west England for different values of one-day rainfall.
Conditions for 10° slope; 10 m long.
Total annual rainfall = 960 mm.
Total number of rain-days = 180 per year.
r_c = daily rainfall required to produce surface runoff.

only increase the soil wash transport by 1·3 times. Under the latter conditions the actual rates of removal will be very much higher both for bare and vegetated ground, and maximum removal appears likely to occur under conditions of high rainfall concentrated into a small number of rain-days, so that vegetation is relatively poor and the mean rain per rain-day is high. Similarities can be seen between this model and Fournier's (1960) empirical correlation of world-wide sediment yield (in rivers) with a factor:

$$\frac{(\text{rainfall of wettest month of year})^2}{(\text{mean annual rainfall})}.$$

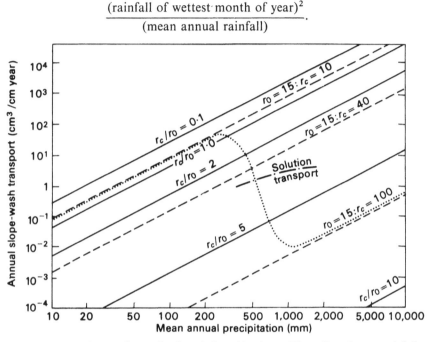

Fig. 8.17. Estimated rates of annual soil wash for a 10 m long, 10° gradient slope as rainfall characteristics vary. Rate of solution transport in U.S.A. shown for comparison (dot-dash line).
Solid lines: relationship between transport rate and annual rainfall for specified values of r_c/r_0. (r_c = daily rainfall required to initiate surface runoff: r_0 = mean rainfall on rain-days.)
Broken lines: relationship between transport rate and annual rainfall for r_0 = 15 mm (approximate value for southern U.S.A.); and r_c = 10 mm (bare surface) = 40 mm (dense grass) and 100 mm (oak forest).
Dotted line: approximate relationship between transport rate and annual rainfall as natural vegetation varies with rainfall across southern U.S.A.

Fig. 8.17 also shows (broken lines) the relationships for bare soil (r_c = 10 mm), grass (r_c = 40 mm) and forest (r_c = 100 mm), when the mean rain per rain-day = 15 mm, which is approximately equal to the best-fit value across southern U.S.A. Superimposed on these three lines is a dotted line which follows the transition from one vegetation type to the other as rainfall changes, under natural conditions in the United

States. This line shows the net variation with rainfall under natural conditions, and is characterized by a peak at 250–350 mm of rain, a minimum at about 1,000 mm, and the indication that a further increase would take place in wetter areas with similar rainfall distributions. This curve may be compared with the data of Langbein and Schumm (1958) for fluvial sediment in the United States: both the forms of the curves and the quantities of debris involved imply that soil wash erosion is a major part of the sediment budget in semi-arid North America. Fig. 8.17 may also be compared with the much smaller rate of erosion associated with rainsplash (fig. 8.9).

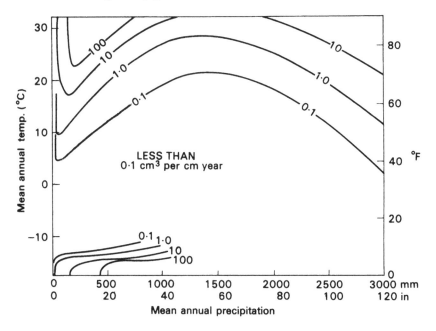

Fig. 8.18. Generalized rates of soil erosion in cm³/cm year* as climate varies around the world. Rates are calculated on the basis of fig. 8.17, and may be too large in cold areas where much of the precipitation falls as snow.

* On a 10 m long, 10° slope. Equivalent to lowering in one-hundredths of a millimetre.

The estimates shown in fig. 8.17 may also be combined with data for world-wide meteorological stations, plotted on a rainfall–temperature diagram (fig. 8.18), to show morpho-climatic regions in which surface soil transport is most intense. Values have been calculated for a slope at 10° on the basis of gross precipitation values, so that cold region rates are probably over-estimates because much precipitation falls as snow. Maxima may be seen to occur in warm, near-arid areas, in tropical forests and in bare arctic areas.

PROCESS MECHANISMS AND PROCESS–RESPONSE MODELS

Rainsplash erosion

A raindrop of diameter l, falling at velocity v, has a momentum of $\rho_w \pi l^3 v/6$, where ρ_w is the density of water. The downslope component of this momentum is transferred in full to soil particles, while the component normal to the surface is reflected, and only a proportion q (<1) is transferred to soil particles. Suppose this momentum is transferred to a soil particle of diameter d (initially assumed $>l$). This particle will receive a velocity which has a component out from the surface and will be launched into a trajectory, which has an average range, measured horizontally, of:

$$\Delta x = 2 \left(\frac{\rho_w}{\rho}\right)^2 \left(\frac{l}{d}\right)^6 \cdot \frac{q \cdot v^2 \sin \beta \cdot \cos \beta (1+q)}{g}, \qquad (8.14)$$

where ρ is the density of the soil particle, β is the surface gradient angle, and g is the gravitational acceleration constant.

The particle will then land, with a downslope velocity component of:

$$\frac{\rho_w}{\rho} \cdot \left(\frac{l}{d}\right)^3 \cdot (1+q) \, v \sin \beta,$$

and may then repeat the process, in a series of bounces. In this model, particles moving on a uniform surface will move a distance proportional to $\sin \beta \cdot \cos \beta$.

An alternative simple model ignores the reflected normal momentum component, and considers the movement of a particle sliding on a rough surface, after being given an initial downslope velocity. In this case the distance of travel is:

$$\Delta x = \left(\frac{\rho_w}{\rho}\right)^2 \cdot \left(\frac{l}{d}\right)^6 \cdot \frac{v^2 \sin^2 \beta}{2g (\tan \phi - \tan \beta)}, \qquad (8.15)$$

where ϕ is the friction angle for the stone on the surface. In this second model the distance moved on a uniform surface is proportional to the (sine)^2 of the slope angle at moderate gradients, and increases more rapidly than this as the angle of friction is approached. Clearly both of these models are a simplification, each emphasizing one aspect of the process at the expense of others. It appears however, that the latter model is a closer approximation to reality, as presented in fig. 8.7, where uniform surfaces show sine-square laws of movement; and non-uniform surfaces, roughening at steeper gradients, show less steep rates of increase with gradient. The exponential distributions of movement distance shown in fig. 8.6 are consistent with this model, and indicate that the friction corresponds to a situation in which the probability of coming to rest is constant with distance.

In both of these models the distance moved by a single particle is proportional to d^{-6}, so that the mean distance moved by particles which are larger than the raindrops (completely covering a uniform surface) is proportional to $d^{-6} \times$ particle area, that is to d^{-4}; and the total transport for a surface of a single size-grade is proportional to d^{-3}. For particles which are smaller than the raindrops, the momentum of a single drop is split between a number of particles, and mean distance moved is proportional to d^{-2}, total transport to d^{-1}. Only on surfaces of mixed grain-size will coarser particles be able to move farther than fine, because to the coarse particles the surface is relatively smoother. In both models, the total transport is proportional to about the square of the rainfall momentum (empirically therefore, transport $\propto$ intensity $^{2 \cdot 2}$).

Process–response models for rainsplash are generally inapplicable, because rainsplash is almost always accompanied by much larger rates of soil wash in occasional overland flows. The one location in which rainsplash may be more important is close to divides, where its independence of a collecting area, such as is needed to provide overland flow, may allow it to dominate. In this location, a sine-square law will

$$(y_0 - y) \propto x^{3/2}, \tag{8.16}$$

$(y_0 - y)$ is the vertical fall from the divide at horizontal distance x from it. A sine law leads to an exponent of $2 \cdot 0$; and these forms are approached if the hillslope base is not being down-cut too rapidly. Both are convex profiles, the latter more broadly rounded near the divide. This process is thought to be responsible for divide convexity in semi-arid and arid regions, and Gilbert's (1909) argument (chapter 10, p. 299) applies to explain convexities produced by rainsplash as forcibly as to convexities produced by soil creep.

If rainsplash were responsible for whole hillslope profiles, then they would have approximately the above forms if stream position at the slope base were fixed, and they would be convex throughout. If the stream cuts down or moves laterally into the slope, the profile will become more convex; if the opposite, basal concavities will form, associated with deposition. All of this assumes that grain-size is constant, and it may well not be so. The extreme example is a semi-arid break in slope, at which grain size changes, but transport rate remains the same. If this form were in equilibrium with rainsplash forces alone, then grain size would increase as the first power of gradient for small material (smaller than rain-drops) and as the 1·5 power of gradient for

221

the larger grades, as can be seen from the equations above. In fact fig. 13.4 shows an increase with the *square* of gradient.

Surface erosion

Wash erosion is essentially a special case of fluvial transport, in which thin flows occur over very rough surfaces. To what extent do these special features limit the application of sediment transport models derived from flume and river data? Comparisons can be made both for the relationships between water discharge and depth of flow; and for the relationship between sediment discharge and water discharge.

The basic equation (3.2) for mean velocity of a water flow is:

$$v^2 = \frac{2g}{f} . r . s, \tag{8.17}$$

where v is the mean velocity, in a downslope direction, r is the hydraulic radius (mean depth), s is the slope, in the form $\sin \beta \cos \beta$, and f is the dimensionless friction factor, which is related to the equivalent grain roughness diameter, k, by the relationship:

$$\frac{1}{\sqrt{f}} = 4 \cdot 07 \log_{10}\left(\frac{r}{k}\right) + 2 \cdot 0, \tag{3.3}$$

where k is to be interpreted as the size for which 84 per cent of the grains or surface irregularities are smaller. (These equations have been discussed more fully in chapter 3.) This relationship can be approximated by a power function in r over limited ranges of r/k, so that

$$\frac{1}{\sqrt{f}} = A . \left(\frac{r}{k}\right)^b, \quad \text{and} \quad v = A . \sqrt{(2g)} . k^{-b} . r^{0 \cdot 5 + b} . s^{0 \cdot 5} \tag{8.18}$$

where A, b are constants. The water discharge per unit width, q, is then given by:

$$q = v . r = A . \sqrt{(2g)} . k^{-b} . r^{1 \cdot 5 + b} s^{0 \cdot 5} \tag{8.19}$$

Over limited ranges of r/k, approximate values of the exponent b are:

$$
\left.
\begin{aligned}
40 &< \frac{r}{k} < 5 \times 10^4; \; b = \tfrac{1}{8} \\[2mm]
10 &< \frac{r}{k} < 100; \quad b = \tfrac{1}{6} \\[2mm]
1 &< \frac{r}{k} < 10; \quad b = \tfrac{1}{2} \\[2mm]
\tfrac{1}{2} &< \frac{r}{k} < 2; \quad b = 1.
\end{aligned}
\right\} \tag{8.20}
$$

These values give a range of possible exponents in the equation for discharge, between 1·62 for deep flows, 1·67 for moderate flows (Manning's equation), and 2 to 2·5 for shallow flows. In the limiting case a value of 3 is obtained for fully laminar flows. Comparing these depth exponents with those obtained in runoff plot experiments, the latter show a range of values between 1 and 3·7 (Horton, 1945; Emmett, 1970), the lowest values being associated with flow through vegetation as 'subdivided' flow (see table 3.2). Horton has interpreted this flow as a super-turbulent condition, but it may also be considered to depend on the arrangement of the roughness elements in a flow through vegetation, which is totally different to the pattern of a grain roughness on the bed.

In the simplest case, grass leaves may be thought of as constant linear shapes projecting right up through the flow, so that the ratio of water depth to roughness is constant (i.e. r/k constant) until the grass is completely overtopped. Before overtopping, the depth exponent in the discharge equation will then be 1·5 and the exponent $b = 0$. Two effects may even produce *negative* values of the exponent b; that is, conditions in which the effective depth to roughness ratio, r/k, actually decreases as water depth increases. First, this can occur if plants branch, so that the number of immersed linear roughness elements increases as depth of flow increases; and, second, deep flows begin to drag plants down into the flow through the force of the water and again increase the density of roughness elements. In these cases the exponents of depth in the discharge equation can be less than 1·5, and reach the values of 1·0 quoted by Horton (1945) without any necessary change in the *type* of flow.

Sediment discharge in rivers and flumes has been expressed in dimensionless terms above (chapter 3) as a relationship between the dimensionless sediment discharge:

$$\Phi = \frac{S}{(g.\Delta.d^3)^{\frac{1}{2}}} \tag{3.13}$$

and the dimensionless tractive stress:

$$\theta = \frac{r.s}{\Delta.d}, \tag{3.10}$$

where S = sediment discharge on a volume basis per unit width, Δ = submerged sediment density/water density$\approx 1\cdot65$ for quartz, d = grain size or equivalent grain roughness (k above).

In chapter 3, suspended load was expressed in the alternative equations:

$$f\cdot\Phi = 0\cdot1\ \theta^{5/2} \quad \text{or} \tag{3.11}$$

$$f\cdot\Phi = 0\cdot077\ \theta^2(\theta^2+0\cdot15)^{\frac{1}{2}} \text{ (Engelund and Hansen, 1967).} \tag{3.12}$$

223

These forms suggest that a power function is a good approximation, with the power rising from 2 when most of the movement is as bedload, up to 3 when most is suspended. Pure bed-load formulae suggest an even lower exponent of θ; for example:

$$\Phi = 8(\theta - 0.047)^{3/2} \text{ (Meyer-Peter and Müller, 1948).} \qquad (3.14)$$

Considering first of all soil wash movement on fine-grained surfaces, it seems appropriate to use the suspended load formulae, and replace the grain size d in the expressions for Φ, θ by the equivalent grain roughness of the surface, k. Expanding the first expression for suspended load discharge above:

$$\frac{10 . \Delta^2 . S}{g^{1/2} . k^{3/2} . s^{5/2}} = \left(\frac{r}{k}\right)^{5/2} \left\{4.07 \log_{10}\left(\frac{r}{k}\right) + 2\right\}^2. \qquad (8.21)$$

While water discharge per unit width, q, is given by:

$$\frac{q}{(2g)^{1/2} . k^{3/2} . s^{1/2}} = \left(\frac{r}{k}\right)^{3/2} \cdot \left\{4.07 \log_{10}\left(\frac{r}{k}\right) + 2\right\}. \qquad (8.22)$$

In each of these expressions, the value of the roughness, f, has been written out in terms of r/k, and all terms in r/k have been separated on the right-hand side of the equation. If these two expressions in r/k are plotted against one another, then the former is approximately equal to 1.0 times the latter, raised to the 1.75 power, and the errors involved are very small ($<1\%$ for $0.4 < r/k < 100$). In other words,

$$\frac{10 . \Delta^2 . S}{g^{1/2} . k^{3/2} . s^{5/2}} = 1.0 \left(\frac{q}{(2g)^{1/2} . k^{3/2} . s^{1/2}}\right)^{1.75}, \qquad (8.23)$$

which can be simplified to:

$$S = 0.0158 \, q^{1.75} . k^{-1.11} . s^{1.625}.$$

If, in place of

$$f \cdot \Phi = 0.1 \, \theta^{2.5},$$

the latter form of the suspended sediment equation is used that is, approximately:

$$f \cdot \Phi \propto \theta^n \text{ for } 2 < n < 3.$$

then the following forms for sediment discharge, S, are obtained:

$$\left. \begin{array}{ll} n = 2.0 & S \propto q^{1.5} . k^{-0.75} . s^{1.25} \\ n = 2.2 & S \propto q^{1.6} . k^{-0.90} . s^{1.40} \\ n = 3.0 & S \propto q^{2.0} . k^{-1.50} . s^{2.00}, \end{array} \right\} \qquad (8.24)$$

and these exponents of slope and of water discharge (proportional to distance, x, to the power of 0·7–1·2) compare very favourably with the empirical values obtained from runoff plot experiments (equations 8.2–8.4 above).

Where low vegetation is important, it has been suggested above that the effective ratio r/k may fall as depth increases, so that

$$\frac{1}{\sqrt{f}} \propto r^b \text{ for } 0 > b > -\tfrac{1}{2}. \tag{8.25}$$

For $b = 0$, that is a constant number of linear roughness elements,

$$S \propto q.s^2, \tag{8.26}$$

and for $b = 0·5$, the limiting case found by Horton for the water flow equations,

$$S \propto q.s^2 \text{ again } (\tfrac{1}{2} < r/k < 2). \tag{8.27}$$

In other words, flow through dense low vegetation (e.g. grass) seems likely to produce lower exponents of discharge and higher exponents of

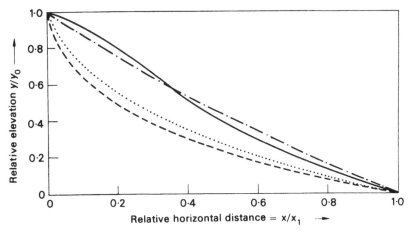

Fig. 8.19. Approximate profiles which are characteristic of wash processes, according to a range of possible laws; under the conditions of a fixed divide at $x = 0$; $y = y_0$ and a stream removing all debris at $x = x_1$; $y = 0$

(a) . . . . . . $S \propto x^{1·5}.s^{1·25}$
(b) – – – – $S \propto x^{2·0}.s^{2·0}$
(c) . . . . . $S \propto x^{1·0}.s^{2·0}$
(d) ——— $S \propto (0·04 + x^{2·0}).s^{2·0}$
(a), (b) show the range of equations governing movement on fine-grained surfaces without vegetation
(c) shows the influence of low vegetation in making the profile less concave
(d) shows the influence of rainsplash in producing a divide convexity.

gradient that might be found on bare ground. These differences might have an influence on slope forms in humid tropical areas if there is some low vegetation cover (though most vegetation is above the flow in primary forest) because wash erosion is important in very humid areas.

In fig. 8.19 the method of characteristic forms, described in chapter 5 and in appendix B, is applied to the several slope transport equations obtained above. Each curve describes the form of hillslope profile obtained after long-continued application of a given process under conditions of a fixed divide and, for simplicity, basal removal at a fixed point. Each profile approaches a form $y = y_0 \cdot f(x)$; in which the lowering through time is applied equally all over the profile, so that only y_0 is time-dependent; and the function $f(x)$ is dependent on the *process* alone. Derivation of the curves shown in fig. 8.19 is described in appendix B. The curves (*a*) and (*b*) correspond to the bare-surface extreme forms of equation (8.24), showing similar moderately concave profiles without summit convexity. Curve (*c*) shows the modification which is predicted above if soil wash is effective under a low vegetation cover ($S \propto q.s^2$), and which produces an only very slightly concave profile. Curve (*d*) shows the effect of adding a term for splash transport to the form producing curve (*b*). The splash introduces a term in the transport equation which is independent of distance, and this is matched in the profile by a summit convexity. The magnitude of this convexity is approximately correct for a slope with a total horizontal length of 25 m: on a slope twice as long the convexity will have approximately the same absolute size, and so occupy a smaller proportion of the total slope length. In this simple model, the splash transport has been assumed proportional to the square of the slope: this has been seen to be most accurate for gradients well below the angle of limiting stability of slope debris. On steep slopes an almost straight slope segment will form, at approximately the angle of limiting stability.

In these models no distinction has been made between unconcentrated wash and slope erosion in rills or braids. This approach appears to correspond with the evidence, in which distinctions are rarely made; and also with the hydraulic models where a single model appears to be appropriate for thin flows and for rivers. Clearly areal concentrations of flow will produce an areal distribution of effective roughness diameters and of rates of erosion, but over a period of years the concentrations move around so that the whole slope is lowered evenly. Where this is not so, and erosion becomes concentrated, streams or gullies will develop and consistently change the distribution of erosion, and also tend to produce changes in overall drainage density. In temperate climates, however, under a vegetation cover, soil wash is negligible except along lines where the vegetation has been stripped (pp. 189, 193) in rare storms, producing semi-permanent gullies which fill in very

226

gradually over periods of centuries, if at all. The formation of such gullies produces very much more runoff, and therefore sediment, than would be produced if the vegetation remained intact, and the cutting of such gullies may be geomorphically dominant events (Hack and Goodlett, 1960), with recurrence intervals of 100 years or more. As such they may also be important in interpreting slope development. The concepts of critical erosion distance and of the belt of no erosion (Horton, 1945) therefore appear to have possible relevance only to temperate slope development and not to areas where vegetation is sparse, because it is only under a continuous vegetation mat that the formation of a gully may produce a marked acceleration in the rate of sediment production.

Horton (1945) determined the distance from the divide at which erosion begins by balancing the tractive force of the flow against the resistance required to erode the soil and vegetation mat. In this balance the flow was described in terms of Manning's equation, and the resistance was considered a constant (i.e. a pure inter-granular cohesion, neglecting normal stresses for surface debris). Under steady state conditions of overland flow, which are probably appropriate to the very intense storm conditions necessary to produce gullies on vegetated slopes:

$$\gamma_w . r . \sin \beta . \cos \beta = c_0 \qquad (8.28)$$

where γ_w is the unit weight of water, β is the slope gradient angle, and c_0 is the critical cohesive strength of the soil and vegetation mat at the surface. Eliminating the hydraulic radius r, using Manning's equation, and substituting

$$q_s = i . x . \sec \beta, \text{ measured downslope}, \qquad (8.29)$$

where i is the excess rainfall intensity (after infiltration, etc.); then the critical erosion distance, or width of the belt of no erosion, is

$$x_c = \frac{1}{n . i} . \left(\frac{c_0}{\gamma_w . f(s)} \right)^{5/3}, \qquad (8.30)$$

where n is the Manning roughness, x_c is the critical distance in metres, and $f(s)$ is a function of slope $= \sin^{0.7} \beta . \cos^{0.4} \beta$; this last expression is very similar to that found by Horton,

$$\left(\frac{\sin \beta}{\tan^{0.3} \beta} \right).$$

It has been assumed above that overland flow discharge increases linearly with distance from the divide. This is only a reasonable approxi-

227

mation when the overland flow is of the Horton type (see chapter 3), and is produced because rainfall intensity exceeds infiltration. On temperate vegetated slopes, however, much of the overland flow is produced because the soil is saturated, and this saturation overland flow is produced much more at the base of slopes (chapter 3), especially at lower rainfall intensities, and produces quantities of overland flow which are smaller than the Horton type, but increase much more rapidly than linearly with distance from the divide. This property results from the maintenance of good soil texture by the vegetation, with consequent high infiltration rates. Its effect on slope profiles would be to make them more sharply concave, and so counteract the effect produced by the unusual roughness properties of grass and other low vegetation (described above), but temperate vegetated slopes have such low rates of soil wash that this influence will rarely be expressed in the topography.

For coarse-grained and relatively non-cohesive surfaces, such as are common in semi-arid areas, it is appropriate to consider debris movement as comparable to bed-load rather than suspended load, and the roughness as that introduced mainly by the grains themselves rather than by surface irregularities. The bed-load formula (equation 3.14), when expanded, gives an expression for sediment transport:

$$S = 8 . g^{\frac{1}{2}} . \Delta^{-1} . (r . s - 0 \cdot 077 d)^{1 \cdot 5}. \tag{8.31}$$

Eliminating the hydraulic radius, r, and expressing the roughness factor, f, as a power of r/d (equation 8.20):

$$S = 32 \cdot 5 \, d^{1 \cdot 5} \left\{ 0 \cdot 25 \left(\frac{q^2 s^3}{d^3} \right)^{0 \cdot 25} - 1 \right\}^{1 \cdot 5} \quad \text{in cm units } (0 \cdot 8 < r/d < 10). \tag{8.32}$$

For a given grain-size, d, this expression for sediment transport increases with discharge and slope from zero at $q . s^{1 \cdot 5} = 6 \cdot 25 d$ (fig. 8.20). For a given value of $q . s^{1 \cdot 5}$, the sediment transport first increases as grain size increases, and then falls off sharply to zero at a critical competent diameter, $d = 0 \cdot 157 \, q . s^{1 \cdot 5}$. This peak is only present when sediment discharge is expressed in terms of water discharge: it expresses the contrary effects of reducing grain size in both reducing the force needed to move material and in reducing the roughness and hence the flow depth and tractive force.

At least two possible process–response models can be derived from a bed-load model of this type. The simpler model is based on the assumption that if grain-size is controlled by lithology and weathering, then slope will be adjusted so that the prevalent grain size will be the one most efficiently carried. The total discharge is then given by the tangent

228

line in the upper graph of fig. 8.20:

$$S = 0.65\,q.s^{1.5}, \tag{8.33}$$

and the slope is controlled by the grain-size through the relationship:

$$s = 41\,d.q^{-0.67}. \tag{8.34}$$

This model has the attraction of simplicity, but is limited by its static nature, which allows for no development of the profile. To allow for development, it is necessary to go back to the continuity equation:

$$\frac{\partial S}{\partial x} + \frac{\partial y}{\partial t} = 0, \text{ or its variants.} \tag{5.1}$$

Grain size can now be considered as a function of the degree of weathering (for a given bedrock), that is of $D/(-\partial y/\partial t)$, where D is the rate of chemical lowering (chapter 9). If discharge is taken to be a known

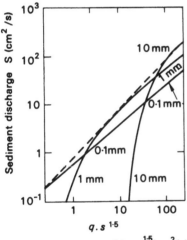
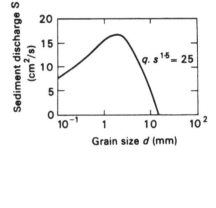

Fig. 8.20. The bed-load equation, expressing sediment transport as a function of water discharge, slope and sediment grain size. The broken line in the left hand graph is the line defining maximum sediment discharges.

function of distance from the divide, then the equation is in principle soluble as a slope development sequence. The solutions will be characterized by increasing concavity in rocks which yield coarser fragments.

Finally a correction should be made to the dimensionless bed shear ratio, θ, so that it is more accurate for thin flows and steep gradients. This ratio is between forces promoting movement and forces preventing it, so that it is a safety factor. Above it has been taken as:

$$\theta = \frac{\text{Tractive force}}{\text{Particle weight}} = \frac{\text{Tractive force} \times \tan \phi}{\text{Frictional force on particle opposing tractive force}},$$

where ϕ is the friction angle. On a slope at angle β the frictional force is no longer $W \cdot \tan \phi$, but instead:

$$W (\cos \beta \tan \phi - \sin \beta).$$

The dimensionless shear ratio, θ, should therefore be corrected to:

$$\theta = \frac{r.s}{d.\Delta} \cdot \frac{\sin \phi}{\sin (\phi - \beta)} \tag{8.35}$$

where s is the slope $\sin \beta.\cos \beta$.

If this form is inserted in the equations for suspended load and bed-load transport, it will be seen that the expected rate of transport becomes very much greater on steep slopes, and that much coarser debris may be carried. While this model has not been fully worked out, it provides a possible basis for explaining the sharp concavities and abrupt breaks in slope and grain-size which are common in semi-arid regions.

It has been seen that hydraulic equations developed for fluvial studies still remain approximately valid for the thin flows which erode hillsides, and are consistent with the available evidence on the variation of erosion rate with topography. Using models of this type, the continuity equation can be solved under particular boundary conditions, and water erosion has been shown to produce concave profiles, the degree of concavity depending on the conditions. There is some reason to suppose that coarse-grained bedrocks with little vegetation will produce the most concave profiles, with the presence of coarse debris limiting removal rates and steepening slopes. Bare fine-grained soils will yield slopes of moderate concavity, and slopes with surface flow impeded by a vegetation cover will, if moulded mainly by slope-wash, give rise to profiles which are only very slightly concave. Rainsplash erosion will in all cases help to make hill-tops convex.

CHAPTER 9

SUB-SURFACE WATER EROSION

TYPES OF PROCESS

Within the soil, water moves more slowly than over the surface. It is therefore less able to transport material physically in suspension or by rolling; but much more able to come to chemical equilibrium with the soil materials, and so remove material in solution. Even in the top-soil, throughflow velocities are no more than five metres per hour, so that only very small quantities of material can be carried through the soil pores; and the pore sizes themselves limit the size of material which is able to pass through them. Chemical removal, on the other hand, is a major form of hillslope erosion in temperate and humid tropical areas, of the same order of magnitude as all forms of mechanical erosion combined.

Sub-surface wash

Within the soil, water flow may carry large or small soil grains, clay particles, clay and other colloids, and material in ionic solution. For fine material there is a transition from suspended to dissolved transport, with a corresponding continuous transition in the transport mechanisms. Moreover some changes of state occur during transport, so that it is impracticable to distinguish rigidly between dissolved and suspended transport.

Forces acting on soil material are both physical and chemical. Since subsurface flow is mainly laminar, the main physical force is a simple viscous shear along the margins of the flow, and the magnitude of this shear is proportional to the cross-sectional area of the pores. Electro-chemical forces become increasingly important for smaller particles and dominant for colloidal and molecular particles. Weak Van der Waals' forces exist between any two molecules, and much stronger electrical forces of the polar bond type act on positively or negatively charged ions. Resistance to movement, tending to hold particles in their present positions is mainly due to cohesion, which is itself primarily a result of weak electro-chemical forces (chapter 4). In a static situation, with water in contact with a piece of clay, an equilibrium gradient will gradually develop from relatively pure clay within the material, via progressively less firmly held clay particles with larger and larger water spaces between them to relatively pure water at a distance from the particles, and this gradient will develop by means of a diffusion exchange of colloidal to sub-molecular particles between the clay and the water. In a non-static situation where water is flowing past a clay face several other mechanisms come into operation. First, the flow will produce some mixing, which will produce a sharper interface between water and solid. Second, the flow may physically shear delicate clay micro-structures and so further increase the rate of removal and reduce the rate of deposition. Lastly, the flow will transport removed debris in one direction, so that there is a steady down-flow transport in each individual removal and deposition. Small changes in the electro-chemical environment along the flow may lead to locally increased removal or deposition, and this factor appears to be very important in the formation of soil horizons.

A particle can only move in pore spaces larger than itself. Even in a deposit of uniform grain size, pores are generally smaller than the particles, so that the largest particles of a soil which form its framework are never free to move unless there are non-capillary (i.e. large) pores in the soil. Such large voids are always present at a free face, and may also be present at the interface between a fine- and a coarse-grained deposit. If a flow goes across the free face or from the fine- into the coarse-grained deposit, then large soil grains may be moved, resulting in a destruction of the soil framework at that point and the formation of a new void within the soil. Repetition of this process leads to the head-ward erosion of continuous channels within the soil mass, a phenomenon known as piping.

Piping commonly develops from steep stream banks at horizons in the soil where throughflow is concentrated. It is also common in desert sandstone bedrock (Parker, 1963) where it is accompanied by solution of chemical cements; and it may be the main means by which butte-tops are drained. Where underground cavern systems or stringers of coarse gravel along buried stream channels underlie fine-textured alluvial material or soil, piping provides one mechanism for the formation and enlargement of swallow-holes which eventually cut upwards to the ground surface. Where engineering works increase water flows, for example beneath an earth dam, piping may develop along the flow lines. In poorly consolidated materials piping may lead to collapse of surrounding materials as their support is progressively removed.

Pipes vary from a few millimetres up to several metres in diameter, and generally follow hydraulic gradients upslope, although with local deviations which may be influenced by soil structures or animal burrows. Where later collapse shows the form of a pipe network, it appears to be dendritic in form, but the evidence is clearly fragmentary. Piping is very important for sub-surface flow and debris removal, but its quantitative contribution cannot yet be estimated.

Fine-grained material which is not a part of the soil framework can be removed without collapse of the soil, but many pores will not be large enough for a particle to pass through if it has to move through soil of a similar size distribution to its own. Experiments in which soil was alternately subjected to percolation and drainage (Scharpenseel and Kerpen, 1967) have showed no measurable movement of clays in undisturbed soil profiles (grey–brown podsolic), and no movement in uniform materials of grain sizes less than 0·06 mm, when percolations totalled 50,000 mm of water. Nevertheless clay skins, forming laminated and oriented aggregates, are normal features of illuviated *B* horizons. The clays may partly be moved as physical particles, but these experiments suggest that at least some of the translocated clay is dissolved (including colloids) in the *A* horizons, and re-deposited in the *B* horizon. In this movement, the chemistry of the *B* horizon is the critical factor which encourages the re-deposition of the clays. Iron, aluminium and silicon, all of which commonly occur in soil clay minerals and are relatively insoluble, tend to move as clay colloids or by the breakdown and re-synthesis of clays (in combination with complex organic compounds derived from plant humus).

Transport in solution and as colloids

In the soil, calcium and sodium, and to a lesser extent magnesium and potassium, occur partly in chemical combination and partly as surface-adsorbed cations on colloids, and they are readily exchanged and carried away in leaching water containing anions, principally bicarbonate, sulphate and chloride. Except in areas where potential evaporation exceeds rainfall, where most of this material is re-deposited, runoff water contains an appreciable amount of dissolved or finely colloidal material consisting mainly of bases, but also including some silicon and aluminium. Because it is difficult to separate material in true ionic solution from material carried in a colloidal state (i.e. as neutral particles in the size range 0·002–5 μm which carry surface adsorbed ions, and so behave in many ways like large ions), the words 'solution' and 'dissolved' are used in this chapter to describe *both* states.

The high dissolved loads of most rivers show that solution is a very effective transport agent, sometimes the most important agent, but a part of this load may come from deep solution of bedrock where it has

233

no direct effect on the slope profile. The contribution of Si + Al derived from clay minerals within the soil is unknown. It is a small component in the *dissolved* load of streams, but re-flocculated clays cannot be separated from mechanical sources of supply (soil surface and stream banks) in the *suspended* sediment load.

Material removed in rivers involves a net lowering of the landscape, but a net transport may also occur downslope without reaching the rivers if material is removed from upslope and re-deposited lower down the hillside, in a way similar to that in which material is translocated vertically within the soil profile. Material certainly tends to be removed in solution, especially the upper horizons under conditions of leaching, and this material will move in the direction of water flow, which may be strongly or weakly downslope along the soil horizons. However, the chemical environment only changes slightly in a downslope direction, so that conditions for concentrated re-deposition are much less well developed than in the vertical direction. Most of the material leached from the upper horizons and carried downslope in throughflow may therefore go all the way to the rivers rather than be re-deposited at the base of the slope, but there is no direct evidence on this point. What seems likely is that the transfer of material from *A* to *B* horizon is accompanied by a small downslope movement. Indirect evidence which may show downslope transfer within the soil is the tendency for soils to thicken and have higher clay contents downslope. However, this thickening is often not present; and where it is, it may alternatively be explained in terms of differential weathering rates in response to variations in soil moisture and rate of surface lowering from top to bottom of a slope (see below, p. 258).

To summarize, there is definite evidence that vertical translocation of clays occurs under conditions of leaching, but it is not clear how much of the transfer takes place as clay particles, as colloids or in ionic solution. There is, however, no doubt that a large amount of material is transported and carried away by rivers in solution. Whether clays are effectively transferred and re-deposited downslope; and whether clays are effectively removed from within the soil and carried away in rivers are very open questions.

METHODS OF MEASUREMENT

Techniques for measuring sub-surface particle movement are very similar to those used for surface movement, and have therefore been discussed fully in chapter 8. The chief difficulty in measuring sub-surface movement is in obtaining a sample without cutting a free face which greatly accelerates the removal of debris; and no satisfactory solution to

234

this problem has been found. For measuring the movement of colloids and movement in solution, somewhat different methods from those in chapter 8 are generally used. Of the methods discussed there, only radio-active tracers are suitable; and then only if the tracer element consistently labels the same unit of moving material. This condition is fulfilled if movement of the tracer element itself is being followed; but if the tracer is in the form of a cation which is used to label a colloid, the label may become detached during transport.

Movement in solution is usually measured by analysing the water which carries the dissolved material. Dissolved load transport is calculated as the product of solute concentration and flow discharge, on the assumption that the dissolved material is moving at the same velocity as the water. Without going into detail about methods of chemical analysis, it is important to compare the different methods because they distinguish differently between suspended, colloidal and ionic-dissolved material. It is because standard methods do not distinguish between collodial and ionic-dissolved material that the rates of action of these two modes of transport cannot be distinguished; and that they are considered together in this book under the word 'solution'.

The separation of suspended and dissolved load is commonly made on the basis of filtration, commonly through paper which nominally allows particles of 2 μm to pass (Rainwater and Thatcher, 1960). Clearly this cut-off point is not sharp, as will be seen below when we discuss sub-surface wash in the soil, which is governed by similar factors. Thus most colloids are excluded from the suspended sediment load, and are included in the dissolved load *samples*, in which they may or may not be measured.

The simplest and most direct way of calculating total dissolved load is gravimetrically, after evaporation of the water at 110 °C (Rainwater and Thatcher, 1960), and this method is in standard use by the U.S. Geological Survey, among others. This method, provided it is applied with reasonable care to avoid loss by splattering, or gains from atmospheric dirt during evaporation, is reasonably accurate and provides a standard for comparison with indirect methods. It is also clear that it measures everything which passes through the filter paper, whatever its state or composition.

A simple indirect method for estimating total dissolved concentration is by measuring its electrical conductivity while an alternating current is passing through it (to prevent polarization). The method is very rapid and self-consistent for a single type of water, but may require calibration for comparing different streams or soil water environments. The conductivity is a measure of ionic activity, so that colloids are represented solely by their surface-active layer of ions, and the method directly measures only the ionic-dissolved material. Calibration for a particular water source is therefore mainly to allow for the colloids, on the assumption that the relationship between ions and colloids is consistent; but it is also necessary because different ions show *slightly* different curves relating conductivity to solute concentration (Jackson, 1958, p. 234).

The alternative to these methods for estimating the total quantity of dissolved material is to analyse for individual elements either by titration or by spectrographic methods. Titration is standard for calcium and magnesium carbonates (BSI, 1967; Douglas, 1968) and possible for other elements (Schwarzenbach and Faschka, 1969). If the titration is performed quickly, then only ionic-dissolved material is able to come to chemical equilibrium; whereas if it is performed slowly, then colloids too can react and so be included in the analysis. Spectrographic methods scan for characteristic spectra from each element and are available for most elements at high sensitivity (Black, 1965). Like gravimetric methods, they measure material both in true ionic-solution and in a colloidal state.

There may thus be some conflict between values for total dissolved load concentrations, and the sums of the concentrations for the constituent elements, because the former often includes the colloidal material, whereas the latter sometimes excludes it. A difference between these two figures will also arise if only the cations are considered, because only these are mainly derived originally from the parent material; whereas many of the anions, which form a large part of the *total* ionic dissolved load are derived from the atmosphere, either directly in precipitation or through the action of plants.

EVIDENCE OF RATES OF ACTION AND PROCESS MECHANISMS

Removal in suspension

Although subsurface particle movement is considered to be an important soil-forming process, experiments to measure it show very low rates of movement except in media with large pore spaces (Hallsworth, 1963; Scharpenseel and Kerpen, 1967). In experiments using material of a single textural class, the quantity of radio-active-labelled clay passing through a 50 cm soil column during the percolation of 4×10^4 mm of water decreased sharply as finer soil was used (fig. 9.1). The material which did move appeared to move as a pulse, with a similar spread for each soil grain size, but travelling a little slower relative to the water through the finer materials. This behaviour suggests that much of the material was trapped in the upper soil layers and became permanently fixed. In similar experiments with natural soils, no labelled clay passed through. It may be inferred that particle transport, even of clay sizes, is very slow except in the non-capillary pore spaces of a soil; although it does occur (Ruxton, 1958).

Civil engineering practice (USBR, 1947) follows the conclusion that soil material does not move appreciably through voids in similar soil. In the construction of filters to prevent internal scour and piping where

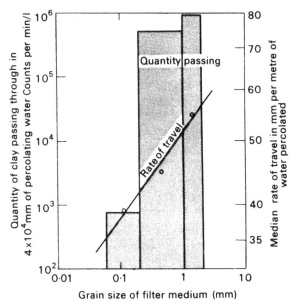

Fig. 9.1. The passage of [55]Fe-labelled montmorillonite through soils of a single textural class. Histogram shows relative quantities passing through a 50 cm soil column after the percolation of 4×10^4 mm of water. Line and open circles show the median rate of travel of the material which passes through the column (after Scharpenseel and Kerpen, 1967).

water flows out of a soil, it is considered satisfactory if the filter material has a median diameter five to fifty times greater than the material to be protected (depending on grading and grain roundness) and if grain-size curves are roughly parallel in the finer grades. Fig. 9.1 suggests that this is a very conservative requirement for preventing the movement of clays. Natural piping along stream banks and in bedrock is probably a quantitatively important process, but no measurements of its rate are available.

Chemical removal

Removal in solution is a process of the greatest importance to hillslope development, and its rate may sometimes exceed that of all mechanical

237

processes combined, as measured by fluvial sediment loads (Rapp, 1960*b*; Strakhov, 1967; Judson and Ritter, 1964). Most of the remaining part of this chapter on sub-surface water erosion is therefore concerned with the processes of removal in solution; both in ionic solution and as colloids. The literature contains few measurements which are related to the geomorphic problem of overall erosion and slope forms; and a great

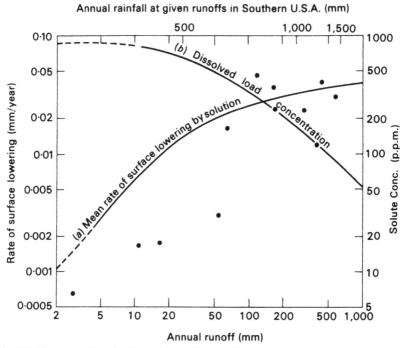

Fig. 9.2. Mean rate of surface lowering by dissolved load in streams at 170 gauging stations in U.S.A. (solid lines), and for a sample of streams in different climatic regions of U.S.A. (dots), showing the high variability to be expected; and the same data expressed in terms of dissolved load concentrations (line *b*), (lines adapted from Langbein and Dawdy, 1964; sample dots from Leopold, Wolman and Miller, 1964, table 3.10, p. 76).

many detailed discussions of soil formation and soil chemistry, which are not immediately applicable to the wider geomorphic system. The present synthesis therefore aims to combine these two types of information by selecting and simplifying the geochemical data to form models which are directly applicable to geomorphic systems, and provide tentative explanations of some observed space and time distributions of dissolved load.

Evidence of rates of removal

Figure 9.2 shows generalized relationships between annual stream runoff and dissolved load amount and concentration in the United States (Langbein and Dawdy, 1964), together with a few individual values for total load which show how great the variability is. In areas with low runoff levels, even the most soluble materials can be removed only slowly, so that they remain available almost indefinitely, and it is their saturation concentrations which decide the composition of the runoff. Material dissolved by rainfall is redeposited in the soil by evaporation, producing pedocal types of soils and high concentration runoff. In areas with high runoff levels, bases in the soil are quickly exhausted and removal is limited to less soluble material, so that total load tends to increase with runoff only very gradually. The data of fig. 9.2 are also included, for comparison, in fig. 8.17 in terms of annual precipitation, and show that chemical removal becomes more important than slope surface erosion at about 500 mm precipitation (though not more important than total mechanical erosion).

More detailed studies of solution have been concerned with single rock types or single chemical compounds. The case of limestone has received most attention because of the relative simplicity of the geochemical system with only two main solutes, calcium and magnesium bicarbonates. Even this system is complicated by interactions, but calcium carbonate usually forms 90–95 per cent of the material in solution, so that magnesium is to some extent ignored. Water from a deep circulation limestone system is usually of constant, saturated strength. Surface streams on limestone contain smaller amounts of calcium as the proportion of their drainage basins underlain by limestone declines; and show dilution reductions in response to fluctuations in discharge over a period (Smith, 1965; Pitty, 1968). These observations suggest that water is able to come to a saturated equilibrium with the calcium in the soil, but that it is not able to reach a new equilibrium as it mixes with non-saturated water in the stream channels. Basins with only part of their area in limestone seem to show a greater calcium concentration in their waters than a linear combination of equilibrium saturations for the various rock types present would suggest (Pitty, 1968). This is mainly because if only a small area of limestone is present, say 5 per cent, then a much larger proportion than 5 per cent of the slope water will flow across this limestone, and become saturated. In this way small proportions of limestone, especially in inter-bedded series, produce disproportionate amounts of calcium in the water, and the limestone is lowered unusually fast.

Cavern systems are notable features of limestone areas, but their geomorphic significance may not be very great. Pitty (1968) argues that enough calcium is removed from the Peak District of Derbyshire, England to erode the entire volume of known caves in a single year.

Further, if solution in caves were the major factor in limestone denudation, then surface lowering would be by cavern collapse, leading
to a bomb-crater type of topography dominated by collapse dolines.
Instead most removal is *close* to the surface, with perhaps 25–30 per
cent of the total erosion actually *at* the surface (Williams, 1965).
Saturated concentrations of calcium and magnesium in water, expressed
as parts per million of the carbonate, vary over a range of about 100–
300 p.p.m. (by weight). Factors controlling this rate include the calcium:
magnesium ratio in the limestone; and the concentration of CO_2 in the
percolating waters. The solubility of CO_2 decreases as temperature
increases, but the concentration of CO_2 available from biological

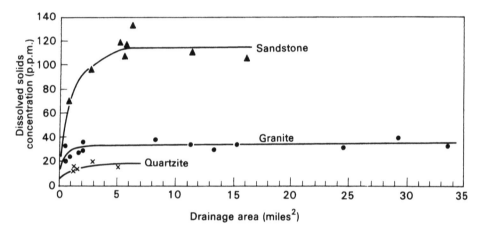

Fig. 9.3. The relationship between dissolved solids concentration and drainage basin area for
three rock-types in the Sangre de Cristo Range, New Mexico (from Miller, 1961).

processes tends to increase with temperature. These two effects counteract one another to give similar concentrations of CO_2 and hence
carbonates in humid temperate and humid tropical regions, but lower
values in the unvegetated arctic (Smith, 1969*a*). The high tropical
rainfall, however, tends to produce the highest rates of total erosion
(Smith, 1969*b*).

Miller (1961) made a study of small streams draining single rock types
in the Sangre de Cristo Mountains of New Mexico. For each rock type,
concentrations are almost constant (fig. 9.3), after an initial rise for
small basin sizes. As with the simpler limestone systems, the evidence
of fig. 9.3 suggests a saturation phenomenon, the mineral assemblage
coming into equilibrium with the water over a short period. Closer
examination of fig. 9.3 shows, however, that the lower concentrations
for small basins are an interpretation based on a single point for the

sandstones. Referring to the original data (Miller, 1961), it may be seen that this low concentration is almost entirely a deficit in calcium carbonate ($CaCO_3$) and that the sandstone contains a variable quantity of interbedded limestones, so that the critical point is of doubtful validity. Travel times of water within a basin are normally a matter of days rather than hours except where overland flow occurs, whereas

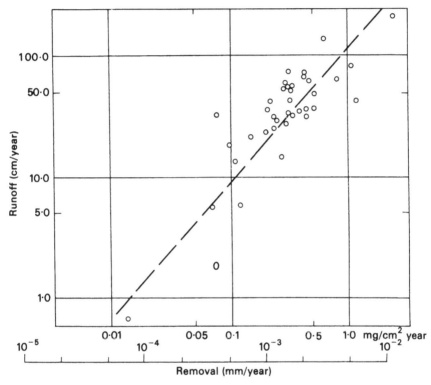

Fig. 9.4. Relation of runoff to removal of silica for drainage basins in U.S.A. Temperatures range from 0–30 °C (from Davis, 1964).

experiments with equilibration of water with soil (Bricker, Godfrey and Cleaves, 1968) suggest that equilibrium is achieved within a few hours. It may therefore be considered an open question whether an appreciable amount of water normally reaches streams in an unsaturated condition, a point which is taken up below in discussing the variation of solute concentration with discharge during floods.

A somewhat different approach is that taken by Davis (1964) who studied the concentration of silica in U.S. waters over a temperature range of 0–30 °C, and with wide variations in soil organic matter. He showed a clustering of stream silica concentrations about 15 p.p.m.,

241

independent of conditions (fig. 9.4). Because of the large number of determinations made, and the persistence of silica in rocks and soil, this value will be taken as a standard for comparison, and silica as a reference oxide for comparing with others.

A model for chemical removal

In principle the rate of removal of each element depends on the minerals of which it is compounded and the amount of each present; on their solubilities or the partition coefficients between colloid surfaces and the solution; and on the amount and constitution of water flowing through the soil and the way in which it makes contact with the soil particle surfaces. Many of the features of weathering and solution can, however, be approximately explained on a much simpler, linear model for the loss of substance of each oxide (or element). The rate of removal of material is considered proportional to (1) the saturated solubility of each oxide, which is assumed constant; (2) the quantity of the oxide present; and (3) the flow of water through the soil. Each of these proportionalities represents simplifying assumptions, which can be justified partly in terms of geochemical theory, and partly in terms of empirical results.

The assumption of a constant solubility, whatever the chemical combination in which an element is bound, is made mainly for simplicity, to avoid the consideration of phase equilibria of complex systems which are incompletely known. Its justification is that it seems to work, at least as a first approximation for most silicate minerals. Each element has a fixed coordination number in silicate lattices, which implies a similar energy level needed to remove it, and hence a similar solubility – this argument seems reasonable when applied to clays, but only shows that solubilities will be of the same order of magnitude.

The assumption that the amount removed is proportional to the amount of an element present makes little sense in terms of ionic solution unless the water is static in contact with discrete grains which differ in composition; but it has more meaning when applied to exchange at colloid surfaces, there the equilibrium concentrations of an element in the water and on the colloid surface will be related by a fixed partition coefficient, which determines the effective 'solubility' of the element.

The assumption that the amount removed is proportional to the flow of water is an assumption that concentrations reach their equilibrium values in a time which is short compared to the residence time of water in the soil, which has been partly discussed above.

In symbols, the solution model is:

$$-\frac{\partial p}{\partial t} = k \cdot \frac{\partial q_z}{\partial x} \cdot p, \tag{9.1}$$

where p is the amount of the oxide present as a proportion (by substance) of the unweathered material, t is the time elapsed, k is the solubility for the oxide, x is the distance downslope, and q_z is the water discharge through unit cross-section of the soil.

The model can be compared with empirical data in two ways: (1) by looking at the composition of stream waters and comparing them with the rocks from which they flow; or (2) by looking at the evolution of

TABLE 9.1. *An example of a 'by substance' chemical analysis. Soil developed on basalt at Kaui, Hawaii* (from Lovering, 1959)

Oxide	% by weight				% by volume or by substance			
	Soil	Subsoil	Mantle	Rock	Soil	Subsoil	Mantle	Rock
SiO_2	9·2	9·9	32·8	49·0	2·3	3·1	16·0	49·0
Al_2O_3	24·4	28·9	24·0	13·7	6·1	9·0	11·7	13·7
Fe_2O_3	35·8	35·4	21·0	13·2	9·0	11·1	10·3	13·2
CaO	0·3	0·2	3·8	7·3	0·08	0·06	1·9	7·3
MgO	0·3	0·2	2·4	13·5	0·08	0·06	1·2	13·5
K_2O	0·21	0·06	0·21	0·27	0·05	0·02	0·10	0·27
Na_2O	—	—	0·34	1·62	—	—	0·17	1·62
TiO_2	6·89	5·54	3·54	1·73	1·73	1·73	1·73	1·73
P_2O_5	0·37	0·40	0·45	0·13	0·09	0·12	0·22	0·13
H_2O	15·0	17·1	10·3	0·4	3·8	5·3	5·0	0·4
Total	92·5	97·7	98·8	100·8	23·2	30·5	48·3	100·8

Calculations on basis that TiO_2 is constant at 1·73 per cent.

rocks into sub-soils by weathering. The first of these ways can be applied directly: the second requires a solution of equation 9.1 in the form:

$$p = p_0 . e^{-k . t_*}, \tag{9.2}$$

where p_0 is the quantity of oxide present in the unweathered material, and t_* is a relative measure of time

$$= \int_0^t \frac{\partial q_z}{\partial x} . dt.$$

An experimental relationship of this form, with t_* proportional to dated real times, has been found by Ruxton (1968*b*) for the weathering of volcanic glasses in Papua. If an oxide is compared to silica as a standard, then equation 9.2 can be re-written in the form:

$$\left(\frac{p_1}{p_{01}}\right) = \left(\frac{p_2}{p_{02}}\right)^{k_1/k_2}, \tag{9.3}$$

243

where the suffices 1, 2 refer to the tested oxide and to silica respectively.

Equation 9.3 can now be applied to soil analyses, if the loss of substance can be calculated making allowances for changes in bulk. This is usually done on the assumption that titanium oxide (TiO_2) is totally insoluble in the form of ilmenite ($FeTiO_3$), so that the quantity of iimenite is unchanged. An example is shown in table 9.1 (Lovering,

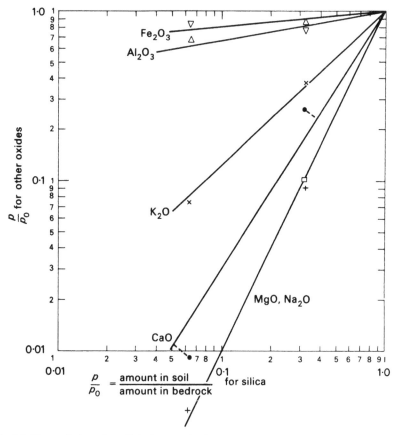

Fig. 9.5. The evolution of a soil during weathering at Kaui, Hawaii. Data from table 9.1 and Lovering, 1959.

p values refer to remaining proportions of original rock substance.

1959). In this chapter amounts have all been calculated on this basis, which is usually called a 'by volume' basis, allowing a comparison to be made between the amount of an oxide present in the soil and its amount in the unweathered material. *Initial* weathering of bedrock is assumed below (p. 262) to produce an increase in porosity equal to the loss of substance, but it is recognized that subsequent loss of *substance* during

TABLE 9.2. *Mobility of oxides, mainly in stream waters*

Oxide	Sangre de Cristo Mtns. (Miller, 1961)				Mean in U.S.A.[4]	Relative values from fig. 9.5 ($SiO_2 = 10$)	Mobility (Polynov 1937). ($SiO_2 = 10$)	p.p.m. in plants/ prop. in crust rocks	Infiltration experiments with humus[5] ($SiO_2 = 10$)
	Granite	Quartzite	Sandstone[1]	Limestones[3]					
SiO_2	14	4	5	—	25	10	10	50	10
Al_2O_3	0·1	0·6	2·8	—	—	1·7	1	} <10	3·6
TiO_2	0·3	0·3	—	—	—	0	—		—
FeO, Fe_2O_3	1·6	2·5	—	—	—	1·0	2		0·32
CaO carbonate silicates	160	950	7,400[2]	300 100	—	15	} 150	2,000	50
MgO carbonate silicates	9·0	2·0	—	—	—	20	65	} 100	13
Na_2O	45	260	15	—	—	20	120		44
K_2O	11	30	62	—	—	9	67		36
C,N	—	—	—	—	—	—	—	>10,000	—

Units are $\left(\dfrac{\text{p.p.m in water}}{\text{proportion in bedrock}}\right)$ unless otherwise stated.

[1] Approximate rock analysis only.
[2] Occurs as ½–1 per cent thin inter-bedded limestones.
[3] Smith, 1965 and Pitty, 1968.
[4] Davis, 1964.
[5] Kerpen and Scharpesneel, 1967 and fig. 9.6.

weathering causes complex changes of *bulk* as compaction occurs and clay minerals form. The basis of the analyses is therefore referred to as a 'by substance' basis, as true volume changes will also be discussed.

Figure 9.5 shows log–log plots for the soil development sequence of table 9.1, which should provide a straight line for each oxide on the basis of equation 9.3. In this and other examples fairly good straight lines are indeed found except for the surface soil (not included in fig. 9.5), where other processes besides weathering are acting. It is

TABLE 9.3. *Average dissolved quantities from one-metre columns of rock fragments and soils undergoing percolation with distilled water*

Material	Rock fragments: total losses in mg						Soils: losses in p.p.m.	
	Basalt		Trachyte		Sand		Para-brunerde	Podsol
	with humus	without humus	with humus	without humus	with humus	without humus		
CaO	1300	180	270	240	190	13	91	11
MgO	310	29	45	45	12	tr	12	1·7
K_2O	230	65	260	150	24	10	0·7	2
P_2O_5	—[1]	—	—	—	—	—	0·05	17
Na_2O	480	270	320	180	12	5	—[1]	—
SiO_2	3070	1920	1870	1550	478	222	23	23
Fe_2O_3	11	6	5	3	5	13	0·27	1·3
Al_2O_3	270	360	165	174	73	63	tr	5
organic substance	1230	0	570	0	1020	0	0	37

[1] not analysed.

Rock fragment data for rock particles undergoing continuous percolation with distilled water (pH 5·5) for one month.

Soil data for undisturbed soil columns: Parabrunerde developed on loess under forest, and podsol developed on sand under heath-forest.

Percolation consisted of 200 mm percolation followed by one month drying, and repetition of the cycle.

Table after Kerpen and Scharpenseel, 1967.

found, however, that their relative slopes and therefore solubilities differ somewhat from soil to soil, but generally remain of the same order of magnitude. The consistency of this model is also shown by the highly significant relationships found, between different measures of weathering in igneous rocks, by Ruxton (1968a). The similar orders of magnitude led Polynov (1937) to estimate 'mobilities' for each oxide indicating their relative ease of removal from the soil during weathering. The values calculated from fig. 9.5, and those of Polynov are included in table 9.2.

Studies comparing the concentration of solutes in water and in the material through which it has flowed have usually been based on field measurements, and table 9.2 includes some measures of solubility obtained by Miller (1961), from limestone studies (Smith, 1965, 1969*b*; Pitty, 1968) and from Davis' study of silica (1964). An alternative to stream studies is to allow water to infiltrate through columns of rock fragments or soil, and to compare concentrations of oxides in water and

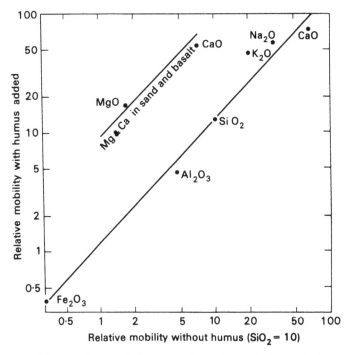

Fig. 9.6. The influence of humus in increasing the mobility of minerals during percolation through columns of rock particles. Data from table 9.3 (Kerpen and Scharpenseel, 1967).

soil. Kerpen and Scharpenseel (1967) carried out a series of experiments with and without the addition of humus to the surface and their results are summarized in table 9.3 and fig. 9.6. The relative solubilities obtained are also included in table 9.2. A final source of comparison is on the basis of plant intake, which may follow similar mobility patterns, and table 9.2 also includes a ratio between plant concentrations and rock concentrations.

All of these data can be combined to give an estimate of average solubilities for most common oxides, and some attempt can be made to separate out different forms of chemical combination for non-silicates. These values are listed in table 9.4. It is recognized that these values

247

Process: the interaction of force and resistance

vary somewhat with the geochemical environment, but it is considered (1) that they are reasonably constant within a given weathering profile (fig. 9.5); and (2) that they provide a basis for estimating overall trends for application to the problems of landform development.

Working from the solution model, weathering sequences can be deduced from any original parent material with fair accuracy, especially

TABLE 9.4. *Effective solubilities of oxides in rock and soil, expressed as parts per million by weight*

Oxide	Form	k = solubility in p.p.m. by weight	Source
SiO₂	all forms	15	1
	quartz	*c.* 4	2
	silicates	*c.* 11	2
Al₂O₃	all, pH > 3	6	3
TiO₂	all	< 0·2	3
Fe₂O₃	all, pH > 4	0·47	3
Al₂O₃ TiO₂ Fe₂O₃ FeO	all forms, grouped value	4·4	4
CaO	in carbonates (i) with organic material	130	5
	(ii) without organic material	15–50	3,5
	silicates	83	3,4
CaO MgO	in sulphates, chlorides	*c.* 250	6
MgO	in carbonates (i) with organic	43	5
	(ii) without organic	5–15	3,5
	silicates	24	3,4
Na₂O	silicates	60	3,4
K₂O	silicates	52	3,4
Na₂O K₂O	in carbonates	*c.* 400	6
	in sulphates, chlorides	*c.* 600	6

Sources for data:
[1] Davis, 1964.
[2] Miller, 1961.
[3] Kerpen and Scharpenseel, 1967, and fig. 9.6.
[4] Fig. 9.5.
[5] Smith, 1965.
[6] U.S. Geol. Survey Water Supply papers: Chemical quality of water in U.S. Rivers.

igneous parent materials. Fig. 9.7 shows the predicted changes in composition for a granite, a basalt and a limestone, from the bedrock up to the subsoil; and the loss of substance associated with this change, calculated from the solubilities in table 9.4. Composition changes follow a series of parallel curves which first move away from the 'bases'

248

corner of the diagram, and then swing round towards the 'sesquioxide' corner, as the more readily soluble oxides are each removed in turn. Fig. 9.7 also shows the predominating clay minerals formed; with kaolinite forming at an advanced stage of weathering for most rocks, but finally giving way to bauxitic materials. In practice, these clay zones

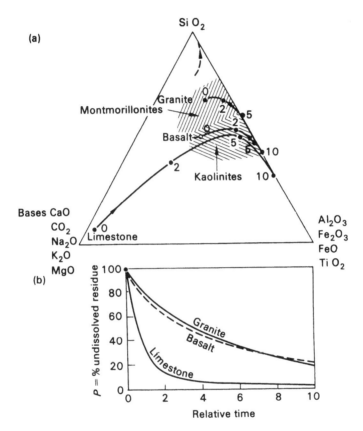

Fig. 9.7. Estimated change in composition and loss of substance during weathering for three rock types; based on solubilities in table 9.4
(*a*) Changes in composition. Figures against curves are relative ages (t_* in equation (9.2)). Broken line shows evolution of acid soils (pH < 4).
(*b*) Loss of substance during weathering.

should often move a little away from the silica corner, to allow for the presence of free quartz, commonly about half the total silica, which is not available for the formation of clay minerals. Physical breakdown of rock materials has to proceed some way before there is enough finely divided debris to form clays in large quantities, and this does not usually occur until a relative time of about 3 in the figure. This means that montmorillonites (smectites) are not the principal clays formed from *weathering* of granites, but are usually derived from rocks which have initial compositions close to the SiO_2–bases side of the triangle (that is with *low* initial aluminium and iron content).

Organic material modifies the weathering profile near the surface. Relative mobilities of elements in plants are similar to mobilities in water, so that the proportions taken from the soil are similar to those in weathering. However, there is an equilibrium between uptake and return of inorganic material, so that equal amounts are exchanged, and this interchange will tend to partition the available material according to a constant fraction, so that in the upper layers silica and alumina are reduced by the same fraction, instead of in proportion to their mobilities. This alone would cause a loss of substance, commonly 10–25 per cent, without change of composition, but the more mobile oxides can be carried into the plants from greater depths, and are returned to the surface, so that bases are somewhat concentrated in the surface layers. The organic material also combines with inorganic material in the soil to form more mobile compounds, and the increase in mobility is greatest for the most mobile oxides, and is greater for magnesium and calcium oxides in the sand and basalt (table 9.3).

If mobilities with and without humus are compared (fig. 9.6), then it can be seen that most silicates show a slight but consistent increase in mobility, averaging about 30 per cent for the least mobile and 100 per cent for the most mobile. However, the mobilities of calcium and magnesium in the sand and basalt are increased by a factor of ten times.

There are of course numerous complications to this simplified view of soil development. For example, near the silica corner in fig. 9.7, the material is almost pure sand, with an acid reaction. At low pH, first Al_2O_3 (pH $=$ 4) and then Fe_2O_3 (pH $=$ 3) sharply increase their solubility, by a factor of over 100 times. This material has the acidity to dissolve and remove the sesquioxides, and also has a coarse enough texture to allow appreciable through-wash of clay particles. These two effects reverse the ordinary evolution in fig. 9.7, and replace it by a convergence on the silica corner, shown by the broken line. Re-deposition of sesquioxides occurs at shallow depth, often in the form of a hardpan, at the level where pH increases.

Spatial variations in solution rates on a world scale

Total chemical removal from the soil profile can be estimated simply by summing equation 9.1 through the soil depth:

$$D = \sum_{\text{depth}} \left\{ \frac{\partial q_z}{\partial x} \sum_{\text{oxides}} (k \cdot p) \right\} = \frac{\partial q}{\partial x} \cdot \sum_{\text{oxides}} (k \cdot \bar{p}), \qquad (9.4)$$

where D is the total rate of chemical removal of substance, q is the total flow through the soil

$$= \sum_{\text{depth}} q_z,$$

and $\bar{p}$ is the mean oxide proportion in the profile, weighted by discharge contribution:

or $$\bar{p} = \sum \left(\frac{\partial q_z}{\partial x} \cdot p \right) \Big/ \frac{\partial q}{\partial x}.$$

For soils with a total undissolved residue greater than 70 per cent, that is, most temperate soils, the subsoil value of p is a reasonable estimate of $\bar{p}$; but for fully developed tropical soils, in which the surface residue is 50 per cent or less, then it may be preferable to use a value of $\bar{p} = 0 \cdot 70$ in equation 9.4.

Initially the quantity of water which is effective in dissolving solutes is equal to the quantity which infiltrates the soil, and this is generally a large fraction (80–100 per cent) of the rainfall. As a simple model, let us assume that all the rainfall comes to equilibrium with the soil solutes (ignoring overland flow which is generally a small proportion of total rainfall where solution is important) so that the whole system of soil water and dissolved solids is now subjected to concentration by evaporation until the soil water is finally reduced to the quantity of runoff leaving the drainage basin. If, during evaporation, the concentration of any substance in the soil water rises above the saturated concentration then the excess must be re-deposited within the soil and this re-deposition may involve some movement downslope. The quantity of each solute, $\partial q / \partial x . k . \bar{p}$ in the equation for chemical removal is therefore proportional to the product of rainfall and $\bar{p}$, or of runoff and $1 \cdot 0$, whichever is less. If the latter term is smaller, then some of the oxide must be re-deposited in the soil and this amount will be equal to the solubility, k, of the oxide multiplied by the difference between the two values [(rainfall.$\bar{p}$) and (runoff.$1 \cdot 0$)]. Fig. 9.8 shows generalized relationships between rainfall, runoff and temperature (Langbein *et al.*, 1949; Schumm, 1965) with superimposed values for rainfall.$\bar{p}$. The importance of this figure is that it shows that major soil and rock components, commonly silica and alumina, will be re-deposited in temperate areas; whereas lesser constituents, which are commonly bases, will be re-deposited only in arid areas. The re-deposition of bases complicates the model a little because the material deposited (for example, as caliche) is in a more soluble form than as a silicate in (igneous) rock. There is, therefore, a two-stage process of deposition and removal which results in net deposition only if silicates are converted to

251

Process: the interaction of force and resistance

carbonates (and other salts) faster than the carbonates can in turn be dissolved. This may also occur for silicon dioxide (SiO_2); the re-deposited silica being very much more soluble than as quartz. The relationship between silica and alumina is also complicated by the fact that they are partly, perhaps mainly, combined in clay colloids. Re-

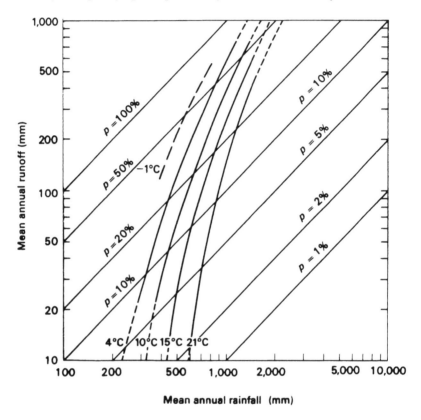

Fig. 9.8. Generalized rainfall–runoff relationships (Langbein *et al.*, 1949; Schumm, 1965) and lines of constant rainfall × labelled proportion undissolved, p. Chemical solution is proportional to runoff, or rainfall × p; whichever is the less.

deposited clays are commonly illites, with a silicon–aluminium ratio of 4, so that SiO_2 is probably much more mobile if available in higher ratios.

In fig. 9.9 the model has been applied to estimate the rates of chemical weathering of an igneous rock soil with a total of 82 per cent undissolved, 50 per cent of which is SiO_2 (25 per cent quartz and 25 per cent silicates). For each oxide, the quantity removed at low precipitations is that contained by the runoff at saturation concentration; and at higher

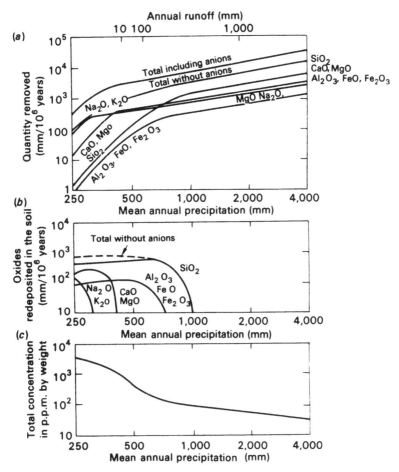

Fig. 9.9. Estimated removal and concentration of solutes from an igneous rock soil* as precipitation varies at 15° C. Values based on table 9.4.
* Assumed composition by substance: $SiO_2 = 50.0$ per cent; $Al_2O_3 = 14.3$ per cent; $Fe_2O_3 = 3.1$ per cent; $FeO = 3.7$ per cent; $CaO = 2.04$ per cent as silicate; $MgO = 2.62$ per cent; $Na_2O = 1.96$ per cent; $K_2O = 1.75$ per cent.

precipitations is that contained by the rainfall at a percentage saturation equal to the quantity of the oxide present in the soil. At low rainfalls, therefore, the total dissolved solids concentration depends on the number of oxides present, especially bases, and their summed saturation concentrations; whereas at high rainfalls, the total concentration tends

253

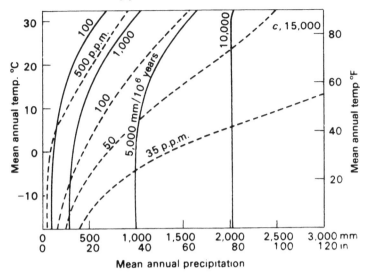

Fig. 9.10. Estimated chemical removal from an igneous rock soil (see fig. 9.9 for composition) in differing climatic conditions. Total removal (solid lines show mm/10⁶ years) calculated on basis of cations only; total concentration also includes anions (broken lines show p.p.m. by weight).

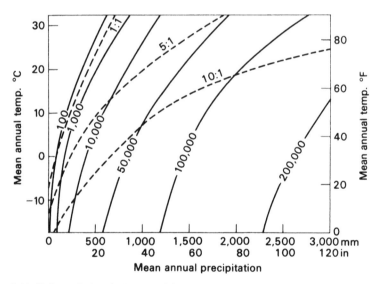

Fig. 9.11. Estimated chemical removal from a pure 100 per cent $CaCO_3$ limestone under differing climatic conditions. Total removal (solid lines show values in mm/10⁶ years) includes carbonate in rock. Total concentration is always at saturation (assumed constant at 235 p.p.m. by weight). Broken lines show values of $\dfrac{\text{limestone solution.}}{\text{igneous solution (fig. 9.10)}}$

towards a quantity-present-weighted average of their individual saturated concentrations. The estimated values of total dissolved solids concentrations shown in fig. 9.9 compare very favourably with the *measured* average concentrations in the United States (fig. 9.2). In considering total quantities of material removed in solution from a granitic soil, it is reasonable to consider only the cation total, since most of the anions, CO_3^{2-}, NO_3^-, Cl^- and SO_4^{2-}, are derived from decayed plant material, or, to a smaller extent, from rainfall. In neither case are they derived from the bedrock, but they do imply the existence of another set of controls on rates of solution, which have not been fully considered in this weathering model, but which might be important enough to limit maximum concentrations of dissolved material, especially under arid conditions.

In fig. 9.10, total chemical removal (without anions) and total chemical concentration (including anions) for an igneous rock soil of constant composition under different climates, have been computed on the basis of the runoff values in fig. 9.8 to show morpho-climatic regions based on differences in the rate of chemical solution. The graph shows that the highest rates of chemical weathering are found in areas of high precipitation, irrespective of whether they are hot or cold, and that the lowest rates are associated with hot-dry climates because in these areas all solutes are saturated. Fig. 9.10 should be compared with fig. 9.11 which gives comparable values of chemical removal for a pure limestone (100 per cent $CaCO_3$) instead of an igneous rock, though in this situation it can be seen that the limestone waters are always saturated and at a similar value.

In fig. 9.11 the calculation of total lowering includes the carbonate because this is an integral part of the rock. In the case of limestone solution maximum rates occur in extremely cold-wet conditions with the next fastest rates being found in hot-wet areas. In the real world no area is subjected to such extreme cold and wet conditions so that the theoretical maximum rates of solution do not occur, and in practice, the fastest limestone solution takes place in hot, wet areas and temperate, .wet areas. The lowest rates of limestone solution are found in the same areas as those for igneous rocks; that is, under hot-dry climates. However, what is important in determining the relief and scenery of any one area is the *ratio* of solution rates between the rocks present rather than the variations under different climates in the rate of solution of a single rock type. Fig. 9.11 therefore also shows the ratio of limestone solution to granite solution under different climates (broken lines).

Under hot-dry climates, the large number of bases in the granite produces high chemical concentrations and the granite dissolves faster than the limestone so that in arid and semi-arid areas, limestone is typically a relief-former. However, in all other climates it is the granite which is more resistant to chemical weathering and it is the granite which will tend to form the mountains rather than the plains. The greatest differences in solution rates between the two rocks are found in both hot-wet and cold-dry conditions, but in cold-dry areas any differences in rates of chemical solution between rock types tend to be masked by the dominance of mechanical removal so that it is only in the humid tropics that the much lower resistance of limestones (compared with other rocks) is expressed in the topography.

Measurements of solution rates of rocks of different composition (Corbel, 1964) indicate similar conclusions. In making the estimates of fig. 9.10 and 9.11, saturation solubility has been assumed to be independent of temperature. This appears to be true for silica (Davis, 1964) over the range 0–30 °C, and approximately true for calcium carbonate (Smith, 1965), so that even if temperature does affect the rate of solution, any resulting variation will be greatly outweighed by the influence of temperature in changing the ratio of rainfall to runoff, and by the influence of rainfall in increasing the total amount which is carried away in solution.

Variations in solution rates over time

The model in the previous section may also be applied to show how dissolved concentrations vary with stage in a stream. Part of this variation is due to direct dilution by water which has not passed through the soil, but has travelled as overland flow during a high stage; but even at relatively high stream levels, most water in temperate areas flows through the soil on its way to a stream. Furthermore, although a rainstorm displaces water into the stream, much of this water is not physically the same as the rainwater, but is water which has already been in contact with the soil for some time, and so has reached chemical equilibrium. During high flows however, there is a relatively high ratio of flow into the streams to losses from evaporation, so that the ratio of runoff to rainfall is effectively changing through time, and concentrations will similarly change through time in response to the changing conditions.

Let us consider simplified slope flow conditions in which water movement is split into three components: (1) an overland flow discharge, q_0; (2) a throughflow discharge into the stream, q_t; and (3) a flow from the slope profile into the air as evapo-transpiration, totalling q_e over the length of the slope. Then the first component will have zero concentration of solutes derived from the soil; and the throughflow will be concentrated by a factor $q_t/(q_t + q_e)$, which is analogous to the ratio of runoff to rainfall in the model of the previous section. In the same way therefore, the dissolved load of an oxide from the slope profile is given by:

$$\sum_x D = q_t.k.p.\frac{q_t+q_e}{q_t} \quad \text{or} \quad q_t.k, \tag{9.5}$$

whichever is the less. Concentrations are obtained by dividing these loads by the total flow $(q_0 + q_t)$.

Figure 9.12 has been constructed on the basis of equation 9.5 for condi-

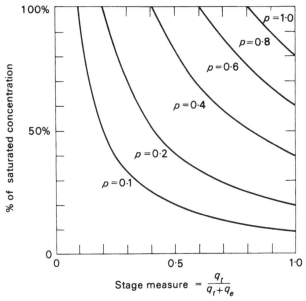

Fig. 9.12. A model for the relationship between the degree of saturation of an oxide in stream waters and the stream stage, on the assumption of no overland flow. The values of p refer to the proportion in which the oxide is present at the weathering front.

tions where there is no overland flow; and the ratio $q_t/(q_t + q_e)$ is taken as a measure of stage which takes values of zero at very low stage and of 1·0 at very high stage. The effect of an increasing overland flow discharge at higher stages is to reduce the concentrations shown at higher stages. It can be seen that rock components which are present in large proportions show very little variation in concentration with stage; but that minor components show great sensitivity, changing from saturated concentrations at low stage to very low concentrations at flood stages.

PROCESS–RESPONSE MODELS

Slope development models with chemical removal alone

Sub-surface wash of material through the soil has been described above

257

as negligible in most real soils. Coarse-textured soils (sands, gravels and perhaps decomposed bedrock in which pore sizes are large) might appear to be exceptions to this rule, but where movement does occur, the evidence suggests that a large proportion of the fine material present is more or less trapped and prevented from moving; and that the small amount that can move is carried away quickly. For these reasons, models relating rates of sub-surface movement in suspension to hillslope forms are considered to be of secondary importance, and are not included here.

By contrast, most of the material removed in solution is removed entirely from the hillslope system and carried away in the rivers. Furthermore, removal in solution is a major item in the hillslope debris

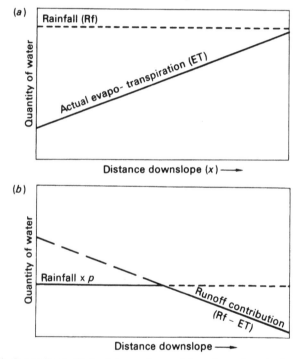

Fig. 9.13. (*a*) Generalized effect of increasing moisture downslope in increasing actual evapo-transpiration, and (*b*) runoff contribution; equal to rainfall minus evapo-transpiration, and rainfall multiplied by proportion of oxide present, *p*. The lower of these two values, shown by the solid line, is proportional to the net rate of chemical removal. Where the runoff contribution is lower, the difference between the two values is proportional to the amount of material re-deposited in the soil.

budget, especially on slopes which are not steep enough for landslides (see, for example, fig. 8.17); and may be the most important item under temperate to humid conditions, although subsidiary to mechanical removal under periglacial and semi-arid conditions. The variations in the rate of chemical removal over a hillside are therefore of great importance in interpreting the evolution of hillslope forms, although very little work has been done on this problem. All that can be done here is to show the directions in which the process models described above for removal in solution tend to lead.

On a real hillside, average soil moisture tends to increase downslope, but is not usually enough to allow evapo-transpiration to proceed at its potential maximum rate. Fig. 9.13 (a) shows schematically the way in which actual evapo-transpiration might increase with distance downslope from the divide (although there is no *a priori* reason to assume that the

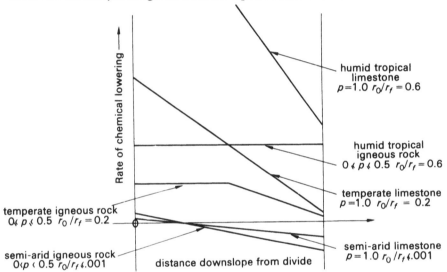

Fig. 9.14. Idealized types of variation in the rates of chemical removal downslope; calculated from models similar to those shown in fig. 9.13. (Actual differences between climates are greater than those shown in the figure.)

increase is linear as shown), in response to the increasing moisture content. In fig. 9.13(b) the contribution to downslope runoff (rainfall minus actual evapo-transpiration) is shown together with the rainfall (ignoring overland flow and interception, etc.), times a proportion p of an oxide (say, silica) in the rock. It has been argued above (p. 251) that the quantity removed in solution is proportional to the (rainfall $\times p$), or the (runoff $\times 1 \cdot 0$), whichever is less. In the example shown in fig. 9.13(b), the resulting rate of chemical removal downslope is initially constant, and then decreases downslope. The difference between these two lines is the amount of re-deposition (compare fig. 9.9(a) & (b)). If the horizontal line in fig. 9.13(b) (equal to rainfall $\times p$) and the oblique line (equal to runoff $\times 1$) are moved relative to each other as the proportion p or climate varies, then three types of variation in chemical removal down-slope can be distinguished:

(1) Rainfall. p < runoff for whole slope: chemical removal remains constant.

(2) Rainfall. p > runoff for whole slope: chemical removal decreases.

(3) The intermediate case described above: chemical removal is first constant, then decreases.

In fig. 9.14 the generalized type of variation downslope (1–3 above) is shown for single component rocks (e.g. limestones) and multi-component rocks (e.g. igneous rocks) under semi-arid, temperate and humid tropical conditions. Only in the case of temperate, multi-com-

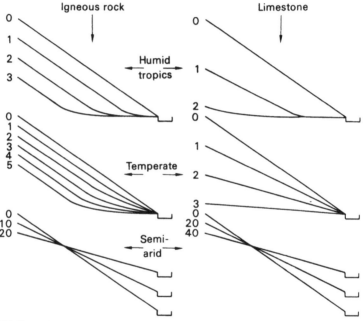

Fig. 9.15. Slope development sequences resulting from chemical solution alone. Slopes are initially of uniform slope and have fixed divides and fixed stream position with no down-cutting. The rates of downcutting are schematically related to those shown in fig. 9.14.

ponent rocks is the third, dog-legged case normal; because only in this case do several components of the rock satisfy condition (1), and several satisfy condition (2) above. For limestones, the proportion, p, of calcium carbonate is almost 1·0, so that case (2) applies whatever the climate. For igneous rocks the chemical removal pattern progresses from case (2) through case (3) to case (1) as the ratio of runoff to rainfall increases in more and more humid areas. In fig. 9.15 these schematic variations of chemical removal rate downslope have been converted to slope development sequences, under the simplest possible conditions, which are: (1) an initially straight slope, (2) divide and stream fixed in position, (3) no stream down-cutting and (4) solution alone acting on the slope.

It can be seen from fig. 9.15 that for a tropical igneous rock there is a simple parallel lowering of the profile except near to the stream level where flow stagnation limits removal, so that the steep valley sides abruptly abut broad, almost level valley floors. Perhaps the most interesting case is that of an igneous or other multi-component rock under temperate conditions where the dog-legged chemical removal curve of fig. 9.13 and 9.14 gives rise to a dog-legged slope profile, which will be converted by flow variations over time to a smooth concavity, which is associated with the re-deposition of clays. Under semi-arid conditions the strong tendency to re-deposition appears to assist the tendency to mechanical alluviation in such climates, but the amounts of re-deposition are very small compared to the rates of mechanical erosion processes.

As well as downslope variations in moisture content, there are also lateral variations in moisture content and therefore also in evaporation rates because of differences in soil thickness or the existence of small hollows and spurs. In general, moister areas will tend to have higher rates of evapo-transpiration and so lower rates of net chemical removal resulting from preferential re-deposition in these areas. In this way hollows, for example, tend to have higher soil clay contents which encourage overland flow and mechanical erosion, even though direct chemical removal may be lower in them.

Even with these very simplified models, it can be seen that rock type and climate are able to influence not only the rate of chemical removal, but also the forms of the slopes resulting from it. In temperate climates, limestones appear to be subject to a simple, straight slope recline, whereas most other rocks develop a concavity in the lower part of the slope; and neither of these forms is appreciably affected by the proximity, or otherwise, of base level. In the humid tropics, rates of chemical removal are very high, and lead to a rapid recline with retreat for limestones, and a simple parallel lowering for other rocks. Where, as is normal, the rates of lowering are limited by base level, broad erosional flat valley floors are formed, bounded by rather steep valley sides. In semi-arid climates, the effects of chemical removal are slight, and masked by the effects of mechanical removal.

Soil profile development with chemical removal alone

The weathering equation, (9.1), can be applied to the soil profile to examine the variation of removal rate with depth, and so to the general form of the profile. If equation 9.1 is summed over all the oxides, we obtain:

$$-\frac{\mathrm{d}P}{\mathrm{d}t} = q_z \cdot \sum k \cdot p, \tag{9.6}$$

where P is the total undissolved (by substance) for all constituents together. The expression $\sum k \cdot p$ is a unique function of P for any given parent material, and can be obtained as the slope of curves such as those shown in fig. 9.7(*b*). It can be seen that the value of $\sum k \cdot p$ decreases rapidly at first, and then more slowly as P is reduced (that is, as weathering proceeds). The distribution of weathering through the profile is determined much more by variations in q_z, the flow of water. In rock which is only slightly weathered ($P > 0.9$), the solution increases porosity, so that q_z increases greatly as the rock becomes more weathered. As weathering proceeds further however, it is accompanied by compaction and the formation of clays, both of which reduce the flow of water. Water flow is therefore at a maximum in the zone of incomplete weathering; that is the mantle, or arène, where the debris is fairly coarse textured with little interstitial clay. Weathering is also concentrated in this relatively narrow zone, producing a rather sharp 'weathering front'.

For the lower part of the soil profile ($P \geqslant 0.8$) the approximate form of the profile can be obtained from equation 9.6 if it is assumed that the initial loss of substance produces an equal increase in pore space. In this case, the permeability is roughly proportional to the square of the porosity (Terzaghi and Peck, 1967) and the weathering equation becomes:

$$-\frac{\mathrm{d}P}{\mathrm{d}t} \propto P \cdot (1-P)^2, \tag{9.7}$$

which can be solved in the form:

$$t = \left\{ \frac{1}{\varepsilon} + \log_e\left(\frac{1-\varepsilon}{\varepsilon}\right) - \frac{1}{1-P} - \log_e\left(\frac{P}{1-P}\right) \right\}, \tag{9.8}$$

where t is the time elapsed to weather from initial porosity ε to a state of undissolved residue P. If the bedrock is weathering at uniform rate then soil depth is directly proportional to t. The form of this solution is shown in fig. 9.16. It shows the form of the weathering front resulting from the presumed increase in porosity; and it shows the variation in total weathering rate resulting in differences of overall rock porosity which are due to local variations in joint spacing. It can be seen that a variation in overall porosity from 0.5 per cent to 5 per cent increases the depth between fresh rock and the $P = 0.9$ level by a factor of almost 20 times, so that a considerable amplitude of sub-surface bedrock relief can be attributed to such joint variations.

In thin soils it may also be seen that the low porosity of all the weathered material limits the total amount of water flowing through it, so that the total amount of dissolved material tends to be less at values of P very close to 1·0. Where the surface layers have weathered to lower values of P the total soil discharge is not limited in this way, and so tends to assume roughly constant values. This point is taken up again below, and is illustrated in fig. 9.19 below.

Soil profile development with chemical and mechanical removal: the equilibrium concept

The land surface is actually lowered by a combination of mechanical erosion, which is concentrated at the surface, and of chemical removal in

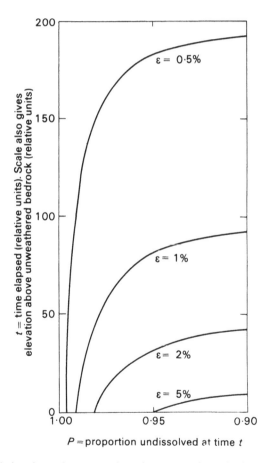

Fig. 9.16. The relative time taken to reach a given state of weathering (indicated by P), from various values of initial rock porosity, ε. The diagram can also be read as a graph of P against depth.

263

the zone of the weathering front. At a point on a hillside, the balance of factors producing a soil may be expressed as:
(Rate of bedrock weathering) − (Rate of surface lowering by mechanical erosion)

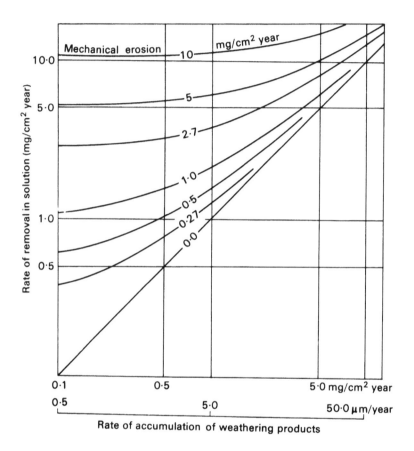

Fig. 9.17. Theoretical relation between rate of removal of material in solution, rate of mechanical removal of undissolved residue, and rate of accumulation of this residue; for a parent rock assumed 50 per cent soluble by weight; and weathering products assumed to have a density of 2·0 g/cm³ (from Davis, 1964).

= (Rate of surface lowering by chemical erosion) + (Rate of thickening of soil mantle).

In symbols
$$W - T = D + \frac{dz}{dt},$$ (9.9)

where z is the soil thickness (strictly expressed in terms of substance).
 Two simple models can be derived from the continuity equation 9.9.

264

The first assumes a sharp weathering front, at which the bedrock is converted to soil of constant undissolved residue, P. This model then allows rates of soil accumulation to be estimated. The second model examines the equilibrium which is established between mechanical and chemical removal when the soil thickness has settled down to a constant, stable value ($dz/dt = 0$).

The accumulation model eliminates W from equation 9.9 by equating:

$$D = (1 - P) . W,$$

since this amount of chemical load has to be removed to convert bedrock to soil. It follows that:

$$\frac{dz}{dt} = D . \frac{P}{1 - P} - T. \tag{9.10}$$

Davis (1964) applied this equation for the case of $P = 0.5$ to obtain the relationship shown in fig. 9.17. This model seems appropriate for extensive plains areas where mechanical removal rates are very low (Strakhov, 1967).

In the equilibrium model, simplifying assumptions about the nature of the weathering profile need not be made, but instead it is assumed that the surface and the bedrock are being lowered at equal and constant rates, so that height above bedrock in the soil is directly proportional to age. In this case, the mechanical removal carries away surface material, which is at a proportion P_s (the suffix s referring to the surface), so that

$$T = P_s . W \tag{9.11}$$

exactly, provided that the rate, T, of mechanical removal is assumed to include only materials present in the original rock, and not those introduced from the atmosphere by rain or plants. It follows that:

$$D = (1 - P_s) . W, \tag{9.12}$$

and that:

$$P_s = \frac{T}{T + D}. \tag{9.13}$$

This equation provides a simple basis for deciding whether soils are in equilibrium with slope processes; or, if equilibrium may be assumed, of estimating mechanical lowering from the state of the soil (P_s) and the rate of chemical lowering (fig. 9.18), both of which may be determined within a relatively short period of measurement. As may be expected, high rates of chemical weathering lead to very fully developed soils (low P_s), and high rates of mechanical lowering are associated with immature soils. Limestones are able to reach undissolved residues of

265

10 per cent or less, but most soils become so impermeable as a result of clay formation at P_s = 30–60 per cent that further development is exceedingly slow, and the soil may never reach equilibrium.

Fig. 9.18 includes a number of points (circles on broken line) showing typical values for mechanical and chemical erosion at rainfalls of 250–4,000 mm. The lower rainfall values refer to 4,000 km² drainage basins in the western United States and the other values are taken from a series of river basins from all over the world (Strakhov, 1967). In qualitative terms, the effects of topography can be seen clearly from fig. 9.18. At

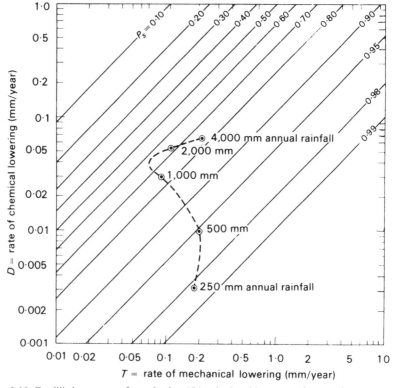

Fig. 9.18. Equilibrium stage of weathering (P_s) calculated in terms of rates of chemical and mechanical lowering. Points and broken line indicate average 4,000 km² basins at 15 °C (from Langbein and Schumm (1958), Langbein and Dawdy (1964) and Strakhov (1967)).

lower gradients and/or at increasing distance from divides (in mature valleys) rates of mechanical lowering tend to be much less, whereas rates of chemical lowering change very little. The *total* rate of weathering ($W = T + D$) at these points therefore tends to be less, but at the same time the soils become more developed (lower P_s). This will produce typical catena-like associations, progressing from immature divide and steep-slope soils to very mature bottom-land soils.

Values of mechanical and chemical removal for large river basins (Strakhov, 1967) suggest low equilibrium values of P_s which are probably never reached, so that the soil accumulation model is the more appropriate for plains areas. For northern temperate areas, like Britain, measured rates of removal again suggest low equilibrium values of P_s, but these are nowhere attained because of the short time period since the Pleistocene. In this situation, neither model is really appropriate because, although soil thickness is tending to increase, the whole weathering profile is within the 'weathering front' so that the assumption in the accumulation model of a constant value, P, is not a reasonable approximation. Equilibrium conditions are probably approached most commonly in areas of great age, especially the tropics; in areas of high total removal rates, especially periglacial and semi-arid areas; and on steep slopes, where total removal rates again tend to be high.

Chemical–mechanical removal models for slopes

In the last section the continuity equation 9.9 was applied to the problem of soil development; but it can also be applied to hillslope development because the rate of surface lowering, $-\partial y/\partial t$, is equal to $T + D$; and the rate of mechanical lowering, T, is equal to the rate of increase of debris transport downslope, $\partial S/\partial x$. Thus:

$$-\frac{\partial y}{\partial t} = \frac{\partial S}{\partial x} + D. \tag{9.14}$$

This form of the continuity equation is the most appropriate for examining slope development where chemical removal is important; and it is instructive to compare it with equation 5.1:

$$-\frac{\partial y}{\partial t} = \frac{\partial S}{\partial x} - (\mu - 1).W. \tag{5.1}$$

In equation 5.1, μ was taken to be the ratio of debris *volume* to rock *volume*; and S to be the *volumetric* mechanical transport, ignoring only organic matter. But equation 5.1 may alternatively be taken to refer to *substance* (see above, p. 246) rather than *volume*, in which case μ is the ratio of debris *substance* to rock *substance*, which must clearly be interpreted as equal to P_s, the undissolved residue (by substance) at the surface; and S is the mechanical debris transport expressed in the form of original rock *substance*. Only the latter interpretation is consistent with equation 9.14.

The more complex model of equation 9.14 will not be pursued fully here, but leads to slope profiles which are a combination of the forms due to chemical and mechanical processes. In the humid tropics, for example, where slope wash is important on landslide-stable slopes, the straight slopes of fig. 9.15 may be converted to a sequence of retreating concave slopes. In temperate regions, where creep is probably the most important process on landslide-stable slopes, convexo-concave slopes are likely to develop except on limestones where simple convexities are more likely to form. For semi-arid areas, a model of this general type

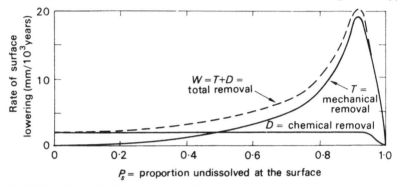

Fig. 9.19. The relationship between mechanical, chemical and total removal rates, on the assumption of the specified chemical removal rate; and of equilibrium between chemical and mechanical rates. Rates are appropriate orders of magnitude only.

has already been outlined in chapter 8 to show how the variation of grain-size near breaks in slope might be analysed.

If we make the additional assumption that the soil is in equilibrium, in the sense of the previous section, then equation 9.14 provides a suitable model for weathering-limited processes. It has been indicated above that the rate of chemical removal is roughly constant for low and moderate values of P_s, but declines at high values of P_s (only slightly weathered rock) because little water is able to enter the debris pore spaces. This relationship is shown diagrammatically in fig. 9.19; and the rate of mechanical removal, T, is deduced from it, using equation 9.13 in the form:

$$T = \left(\frac{P_s}{1 - P_s}\right) . D. \tag{9.15}$$

In a given climatic environment, decreasing P_s is equivalent to thicker soil, so that a similar peaked curve relates the total rate of surface lowering to the soil thickness (fig. 5.5a). It was argued in chapter 5 that the right hand side of the total removal curve in fig. 9.19 is unstable, small local variations in P_s producing positive feedbacks which lead to slopes being polarised either to the peak of the curve or to the unweathered condition.

Equation 9.15 can be considered to provide a more precise definition of the rate at which mechanical removal may occur during landslides in association with particular states of chemical debris weathering. If physical weathering is ignored, the values of P_s define the degree of debris breakdown, and hence angles of internal friction and moisture-holding characteristics, which together determine the stable straight

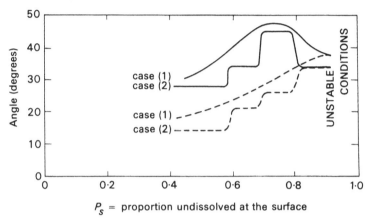

P_s = proportion undissolved at the surface

Fig. 9.20. Hypothetical variations of debris angles with the degree of chemical weathering in the absence of physical weathering.
—— angle of internal friction.
— — stable slope angle.

slope angle corresponding to each degree of weathering. In fig. 9.20 the values of P_s corresponding to the different angles of friction and stable slope angles are schematic only, although the angles are based on the data in chapters 4 and 7. Two possibilities are shown: (1) the stable angles show a stepped curve, with rather abrupt changes from one to another, or (2) the stable angles change continuously with the degree of weathering, P_s. Which of these types is followed during the weathering of a particular parent material will depend on its particular composition, but each is associated with a distinctive pattern of slope development, and provides a 'chemical' approach to the forms described in chapter 7.

At high values of P_s, the relationship between lowering rate and soil thickness has been seen to be unstable, so that material cannot accumulate before a certain degree of weathering has taken place, and it seems reasonable to associate this with a coarse talus or taluvium which is just porous enough to allow a reasonable passage of water. In other words no gradients are stable between those for in-situ weathered rock and those for talus. At lower gradients there will either be a sequence of stable gradients in case (1), or a continuous range of stable slopes in

269

each case (2); each associated with its rate of overall lowering, as shown in fig. 9.21.

In case (1) an appreciable amount of weathering can take place before a slope becomes unstable, so that the hillslopes are characterized by multi-modal slope distributions, moderate soil thicknesses and relatively large slides at long intervals. In case (2) each increment of weathering

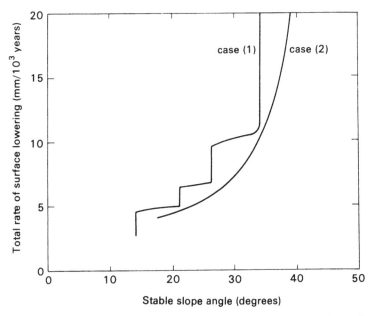

Fig. 9.21. The relationship between surface lowering rate and stable slope angle obtained by combining figs. 9.19 and 9.20. Rates shown are illustrative only. Cases (1) and (2) are described in the text.

changes the stable angle, resulting in frequent micro-slides, associated with terracette forms, and very thin soils. This dichotomy appears to correspond to observed slope types (chapter 7).

This survey of sub-surface water erosion and particularly of the dominant role in hillslope development played by chemical removal, leads to several important conclusions.

(1) The *nature* of the overall process of weathering and soil development is dependent on climate only to a minor degree and the influence of climate is mainly restricted to variations in the *rates* of the processes and the degree of their development.

(2) The rates of chemical solution are fastest in hot-wet conditions and slowest in hot-dry conditions.

(3) It is the ratios between rates of chemical removal on different rock types within one area that influences local relief, so that, in humid-

tropical and arctic areas, limestones are relatively weak rocks whereas in hot semi-arid and arid areas they are relatively resistant.

(4) On hillsides where the soil forming processes are in equilibrium with the rates of mechanical soil removal, then the most developed soils will be formed not in the humid tropics but in moist-temperate areas because the ratio of chemical to mechanical removal is greatest under moist-temperate conditions.

(5) Soils tend towards equilibrium forms which are most developed on lower gradients and at greater distances from divides, because it is in these areas of the hillside that the ratio of chemical to mechanical removal is greatest.

CHAPTER 10

SOIL CREEP

Creep in soils has been defined as any movement which is imperceptible, except by measurements over long periods of time (Sharpe, 1938). It may be caused by systematic re-working of the surface soil layers as soil moisture and temperature vary, by random movements due to organisms or micro-seisms among other causes, and by the steady application of a downhill shear stress. If movement is mainly due to the last of these agents, and produces a deformation of the soil usually without failure, then it may also be described as creep in the rheological sense, but it is important to recognize that soil creep is not necessarily a movement of this type, even though it is also true that no soil creep will occur in the absence of a downhill shear stress.

Steady movement under the direct action of low shear stresses has been called 'continuous' creep (Terzaghi, 1960), and commonly occurs both before and after major soil failures, as a part of the sequence of progressive failure which defines the strength and position of the slide surface (Terzaghi, 1962c), as has been described in chapter 7 above. Continuous creep may, however, continue for long periods and to depths of at least ten metres without leading to progressive failure (Kojan, 1967).

In a soil of uniform properties, the shear stress acts in a downslope direction, parallel to the surface, and increases linearly with increasing depth. Variations within real soils, especially in bulk density and moisture conditions, considerably modify the effective shear strength, and all that may safely be said is that effective shear strength generally increases in *some* manner with depth. The rate of shear movement, if any, will not, therefore, increase linearly with depth. Velocity profiles of movement, if it occurs, are not simply predictable (Kojan, 1967, p. 244), but movement is usually confined to a surface zone a metre or less in depth.

Where the base of the creep zone is well defined, for example corresponding to the granular zone of decomposing bedrock where soil properties change abruptly, then coarse particles may migrate towards

272

this layer of maximum shear (Terzaghi, 1953) as finer material becomes wedged behind them, ultimately interfering with one another to such an extent that an adjacent parallel shear surface is formed (Kojan, 1967, p. 234). Repetition of this process forms sheeted zones 10–50 cm thick, as have been observed by Kojan (1967) in the Berkeley Hills (Cal.). At other locations no discrete shear zone of this sort can be distinguished and continuous creep movement dies away very gradually with depth.

Continuous creep movements may cover whole hillsides and produce movement on slopes as low as 9° (Skempton and Delory, 1957). Even where no landslides occur on a slope, local variations in soil properties will lead to variations in the rate of continuous creep movement, analogous to those found in a glacier. Where these variations are

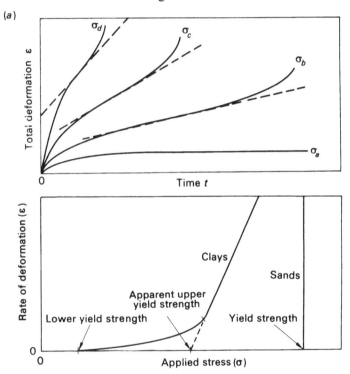

Fig. 10.1. (*a*) The form of creep curves for frozen soil at increasing stresses $\sigma_a < \sigma_b < \sigma_c < \sigma_d$ (after Vyalov *et al.*, 1969). (*b*) Schematic form of stress–rate of strain relationships for clays and sands. Values for the rate of strain (deformation) refer to the slopes of the straight-line portions in (*a*).

sufficient, relative movement may give rise to crack patterns analogous to crevasses, especially in the form of terracettes (Kojan, 1967, p. 235), or to pressure ridges. In extreme cases they may even produce tension cracks, pressure ridges and radial crack patterns which combine to form recognizable lobate features which grade into landslides. The absence of such features, however, does not mean that continuous creep is not present.

Continuous creep behaviour arises directly from the rheological properties of clays, and is absent in coarse-grained soils. If soils containing clays are deformed under constant stress, a portion of the deformation occurs at a uniform rate (fig. 10.1a) which increases with the applied stress. If this uniform *rate* of strain is plotted against the stress, a characteristic form of curve (fig. 10.1b) for clays is obtained, with a steep straight-line section for higher applied stresses. Continuation of this straight line down to the zero-strain axis gives an apparent upper yield strength which corresponds roughly with the value obtained from analysing a landslide failure. At lower stress values, the strain rate deviates from the straight line relationship, sometimes showing a lower gradient line with a definable lower yield strength, especially for thin clay suspensions (<25 per cent clay) (Houwink, 1958). In other cases, and usually for the high clay concentrations (relative to water content) found in soils, finite strain rates can be detected down to zero stress. In comparison, sands show a very simple stress–rate of strain relationship, with a single yield strength, above which deformation occurs very rapidly. Continuous creep is a flow at stresses lower than the upper yield strength and may be exceedingly slow. In soils with larger amounts of non-clay material, grain to grain contact will be more important relative to clay to clay contact and consequently continuous creep movement will be slower or completely suppressed (Kojan, 1967, p. 241).

On the molecular scale, rheological creep phenomena are thought to result from the breaking and re-forming of relatively weak electrochemical bonds at the edges of clay particles and their adsorbed water layers (Kojan, 1967, pp. 238–40), rather than a shear within the clay particles. During shear, particles will tend to rotate so that the plates of the clay molecules are parallel to the direction of shear as has been shown by Mitchell (1956), but it is clear from field measurements of creep that continued shear need not lead to progressive failure, so that a number of factors tend to maintain soil strength under conditions of low shear stress (Kojan, 1967, pp. 241–2). Remoulding and reorientation may lead to an immediate loss of strength associated with a local increase of pore-water pressure, but for movements over a long period, or over cycles of stress application and release, the reduction of clay particle size which accompanies remoulding tends to increase the effective strength, and so reduce creep rates (Mitchell, Campanella and Singh, 1968; Worrall, 1968) over a time period.

Near the soil surface, the steadily acting gravity stresses are combined with movements of the soil produced by variations in soil moisture, temperature and ice content, to produce somewhat larger rates of creep. This element of soil creep has been described as 'seasonal' creep by Terzaghi (1950) because it is triggered by seasonal variations in the soil micro-climate. On a level surface these seasonal movements will clearly produce only expansions and contractions normal to the soil surface, but in conjunction with downslope gravity stresses they may be able to accelerate the rate of movement.

Davison (1889) observed the expansion of a sloping soil mass during freezing and thawing, and proposed a model of expansion normal to the soil surface, and contraction in a direction between the normal and a vertical direction, but this model is only consistent with soil-mechanical considerations if soil strength is very large during expansion and very slight during contraction. Nevertheless some sort of zig-zag movement can be observed during freeze–thaw and also during changes in soil moisture (Kirkby, 1967), although the exact track of a particle may be more or less ∩-shaped (Kojan, 1967, pp. 237–8). During an expansion–contraction cycle three effects will tend to produce non-recoverable downslope displacements of the soil in addition to that produced by the steady application of the gravity stress:

(1) Resistance to shear will vary at different stages of the expansion–contraction cycle, so that creep movement will tend to be greater during a cycle than if the average rate were applied steadily.
(2) The direction of movement during expansion or contraction is produced as a resultant of the expansion or contraction stresses and the gravity stresses. The direction of movement will therefore always have a downslope component, even if the gravity stresses on their own are insufficient to produce any movement.
(3) Expansions and contractions parallel to the surface will also tend to produce greater downhill expansions and somewhat lesser contractions, resulting in a net downhill movement overall, but a great local variability in rate, according to whether a point is at the uphill or downhill end of a contracting block.

The gravity stress acts as the downslope component of the over-burden pressure, as in soil stability calculations; and clearly increases with depth, roughly linearly. The seasonal expansions and contractions are greatest at the surface, but die away very rapidly with depth. The seasonal creep is produced by the combined action of the expansion or contraction stresses and the gravity stresses.

Shear resulting from seasonal creep is zero at the surface, because no gravity stress acts there, and is usually very slight at depths greater than 30 cm because of the relatively slight changes of micro-climate at greater

275

depths. It is a maximum at depths of 10–20 cm for British conditions (Kirkby, 1967). With such shallow depths of operation, the rates of seasonal creep may depend strongly on the surface micro-topography and plant distribution which influence moisture content and its variation over time. Local variations in creep rate produced in these ways, and also as a result of any movement caused by lateral expansions and contraction, will lead to areal differences in rate which may be large compared to the mean rate.

Near-surface creep may also occur as a result of random displacements of particles or groups of particles relative to the soil mass as a whole (Culling, 1963). Such movements may occur in association with more systematic movements such as have been described, or independently, as a result of movements produced mainly by organisms. The formation and destruction of root-holes and animal burrows, the production of worm-casts and other burrow material at the surface, and the chemical weathering or re-deposition of material are examples of such random movements. In the presence of a concentration (porosity) gradient, these random movements will lead to a net diffusion (chapter 3, p. 59) of material from areas of low porosity to areas of high porosity, that is towards the soil surface. This movement cannot continue indefinitely, but must be balanced by forces tending to consolidate the soil, which, acting in the presence of a downslope gravity stress will lead to a net downhill displacement of debris. Little evidence is available to judge the relative importance of this type of creep, but what there is suggests that it is of less importance than other forms of seasonal creep (Kirkby, 1967). It should be noted that it is not related to creep in the rheological sense.

In periglacial areas, soil creep is replaced by solifluction, which may be considered as a rather rapid form of seasonal creep produced by annual freeze–thaw cycles in the soil. It is characteristically spasmodic in time and space, almost all the movement occurring at the times of annual freeze and annual thaw; and occurring in well-defined lobes which are able to over-ride the surface downslope (Williams, 1957). During freezing, water is drawn into the soil to produce layers of segregated ice (chapter 3, p. 58) which commonly produce heaves of 5–20 cm. Net downslope movement may be considerably greater than the downslope component of this heave (Williams, 1966), so that some sort of flow or failure must also have occurred in order to account for the amount of movement; and this is confirmed by direct measurements (Washburn, 1960).

Solifluction may occur in areas with or without permafrost, but is not widespread where mean annual temperatures are above 1 °C (Williams, 1962). In non-permafrost areas, water can be drawn into the soil from below to form ice layers during the autumn freeze. In spring, melting takes place from the top, and may lead to very wet conditions in the surface layers. Most of the solifluction movement takes place during the spring thaw. Under permafrost conditions, water cannot be supplied from the permanently frozen zone during the autumn freeze, but only from the base of the active zone. An unfrozen layer is sandwiched between the freezing surface and the permafrost, and water in this layer may be unable to drain, perhaps producing very fluid soils. Everett (1963*a*) observed that most movement occurred during the autumn freeze, but found that it was directly associated with heaves and not with flow phenomena.

Movements during the melting of segregated ice have been considered to arise in several ways:

(1) Washing of debris across the surface or from the soil as saturated soils drain. Such movements occur (Williams, 1959; Jahn, 1960), but are not a mass movement of the solifluction type.

(2) Settlement after heaving normal to the surface. This may occur, but appears to be a simplification of the mechanisms involved, and is inadequate to account for the rates of movement observed.

(3) Rheological creep or failure movements within either the wet soils or the segregated ice itself. High moisture contents will reduce safety factors and tend to encourage failure, but pore-pressure measurements are inconclusive as to whether this type of failure may occur (Williams, 1959, 1966*b*; Penner, 1967). Rheological creep movements will also be accelerated by high moisture conditions and may contribute significantly to solifluction movement but probably cannot produce by themselves the maximum short-term rates observed (10 cm in 24 hours by Williams, 1959).

(4) Lubricated flow of soil layers which have been physically separated by ice lenses. Even if soil strength is not sufficiently reduced to produce failure within a soil mass by saturation (Williams, 1959), the segregated ice produces sufficient physical discontinuities within the soil and parallel to its surface for flow to occur without soil to soil shearing. Movement under these circumstances will occur if water is thawed from an ice lens at a rate faster than it can drain into the soil pore spaces. Any soil movement which results will itself assist drainage, and the total soil movement really consists of a number of short movements along discrete slide surfaces which coincide with the upper faces of ice lenses. Movement would be most common under conditions in which the previous freezing was slow, so that large amounts of water were drawn into the ice lenses; and thawing is more rapid, so that the water cannot drain out.

Although the mechanics of solifluction have not been very fully analysed, the last two mechanisms (3 and 4) are the most likely. True lubricated flow appears to explain some solifluction features which creep movements cannot; for example (1) the low (2–5°) slope gradients on which solifluction is effective; (2) the discrete flow characteristics of solifluction as shown by the formation of lobes and overriding features; and (3) the velocity profiles of flow (Williams, 1966*a*) which typically show increasing shear rates near the surface. Creep movements should ideally show zero shear rates at the surface, corresponding to the zero downhill gravitational stress, and slides should ideally produce movements along a single slide surface, neither of which results in the velocity profile typical of solifluction movements. On the other hand lubricated flow operates under minimal stresses and is greatest where thawing is most rapid, namely near the surface. Published velocity profiles do not, however, show discontinuities of rate, as might be expected from lubricated flow at ice-lens boundaries.

On coarse-grained scree surfaces there is no rheological creep, but seasonal variations of temperature and freeze–thaw may produce some seasonal creep. Because moisture contents are low on screes, freeze–thaw can produce only small heaves, in the water held at contact points between stones and frozen from the outside so as to contain a part of this water in a casing of ice. Settlements after such heaves will always tend to produce some net downslope transfer of debris, but this will be very small unless the slope is barely stable. Temperature changes will produce small bulk expansions and contractions, but stones will generally return to their original positions if they are in point-to-point contact. In a small minority of cases there will be a shift of contact points in a downhill direction. In the less realistic case of two plane surfaces in contact, there will be movement in every temperature change cycle (Moseley, 1869). Measurements of screes in a cold environment (Rapp, 1960*b*) showed little movement in response to heavy rains by wash processes over the coarse-textured surface, but overall superficial creep rates of up to 10 cm per year (Rapp, 1960*b*, p. 175). Schumm (1964, 1967) has measured the surface creep of stones on semi-arid surfaces, mainly through freeze–thaw but partly as a result of rainsplash (chapter 8, p. 201 and fig. 8.7).

METHODS OF MEASURING SOIL CREEP

With the exception of creep produced as a diffusion process, movements involved in soil creep are mass movements; that is to say the soil moves together and neighbouring soil particles remain close together throughout the movement. In an ideal measurement, therefore, a column or other shape of soil is identified at the beginning of the measurement with reference to fixed points; and its position is re-measured periodically. In practice, the column of soil is usually identified by replacing it with different material, which immediately raises the questions of whether the introduced substance responds to movement in the same way as the original soil, whether non-representative conditions are set up at the interface between the soil and the introduced material and to what extent the soil has been disturbed in introducing the measurement material.

Other major problems arise in the selection of fixed points, which must be in bedrock either at the base of a measurement profile or at an outcrop elsewhere on the slope. In practice it may be impossible to reach sound bedrock at the base of a measurement profile without serious disturbance of the measurement itself; and measurement from more distant reference points introduces greater surveying errors which may invalidate the measurement. A solution commonly adopted is to assume that creep movement dies away with depth, so that the lowest part of a measured profile, if it shows no measurable shear, is unmoving. The ten-metre deep movement profiles found by Kojan (1967) must cast serious doubts on this assumption, but the method remains valid for measuring relative movement.

Soil creep has been considered (Kirkby, 1967) to show great variations in rate within a small area, corresponding to the large variations found in other hillslope variables (Carson, 1967a), and it may therefore be uneconomic to make measurements of great precision at sufficient points to analyse the local variation. Measurements therefore tend to use simple materials with relatively large numbers of replications, and to continue the measurements for relatively long periods to offset any resulting loss of sensitivity. As an extension of this principle, it may be advantageous to install relatively inexpensive materials at a large number of sites, and combine these installations with a single relatively expensive and sophisticated device which is able to record the movement at each installation in turn.

The simplest kind of measurement which has been attempted is that used by Davison (1889) in which small markers are placed on the soil surface and their movement measured from or relative to a fixed point, usually by optical means. This method has been used in the field by Washburn (1960) in combination with a theodolite. Each surface marker is simple, but the theodolite platform must be very carefully constructed to ensure its stability. If sub-surface markers are used, optical means can no longer be used to measure the movement, but Everett (1963 *a* and *b*) has used a buried plate 5–15 cm in diameter as a marker, and connected this by a thin rod to a potentiometer transducer capable of

279

measuring movements of less than 0·05 mm. Selby (1968) has similarly buried 60 mm aluminium cones with wires attached: movements are measured relative to fixed points, using the wires. As for surface markers, a reference point must be secured to sound bedrock and protected from movement of the surrounding soil.

Where shear rates are more or less uniform, a simple vertical stake will show the movement through a horizontal translation of the top of the stake and a downhill rotation of the whole stake. Where the shear is not at uniform rate, is not in material of uniform strength throughout the depth of the stake, or is not simultaneous at different levels, there is considerable doubt about what is being measured, and comparative tests (Kirkby, 1967, p. 370) suggest that the stake rotations may be 20 per cent or less of the average shear rate. Expansion and contraction movements of the soil perpendicular to the surface will also cause rotation (downhill during expansion) of stakes inserted *vertically*. Stakes inserted perpendicular to the surface will not show this tilt, but may show false tilts due to rotation under their own weight, so that vertical insertion is usually preferred.

Leopold and others (Leopold, Wolman and Miller, 1964, p. 352; Leopold, Emmett and Myrick, 1966, p. 228) have installed mass-movement lines consisting of a series of rods or tubes, 6–12 mm in diameter and 20–25 cm long, at 1·5 m spacings along a line of sight established between two fixed points. Movements were recorded as deviations from the line, observed with a theodolite or transit. Iron rods (1–2 m long) and bolts in large trees were used as fixed points. Young (1960) has used a similar method, using rock outcrops upslope from the stakes as fixed points. In these measurements tilting of the stakes was measurable, and was taken into account. Kirkby (1967) has inserted T-shaped stakes with a screw to level the cross-piece of the T with reference to a (removable) spirit level bubble. In this method great sensitivity (10 second of arc) can be achieved which allows the tilts to be recorded for individual soil expansions and contractions, but the inherent measurement errors and great likelihood of disturbance with a cross-piece above the ground make an instrument of this type very suspect for recording net movement over a period of months or years.

A single measurement of creep over a period can be made by inserting marker material in the soil with a minimum of disturbance, and measuring the movement after a period by digging a hole at the side of the markers and observing the profile of movement, either with respect to fixed points, or, for relative movement by reference to a plumb-line arbitrarily located. The latter method can be considered to give absolute movements if and only if the lowest markers have not moved. Most investigators have made small (1–5 cm) augur holes and filled them with marker material to depths of up to two metres. There are many possible tracer materials; for example Everett (1963*b*) has used cylinders of modelling clay, frozen in dry ice to assist insertions and Rudberg (1958) used 2–3 cm sections of dowel rod, inserted down a tube which was afterwards withdrawn. Methods of this type have the advantage of extreme simplicity, a minimum of disturbance during introduction, an expectation that the tracer material will behave very similarly to the surrounding soil, and a direct visual impact in that one can see the profile of movement and whether it has discontinuities or other special features. An additional advantage of these methods is that they allow some measurements to be made of creep resulting from diffusion processes. If, for example, a mixture of glass beads in clay is inserted, then diffusion would result in a mixing of the original cylinder with the surrounding soil, and a correspondingly greater mixing for grain sizes for which the diffusion is most rapid (presumably fine grades). The great disadvantage of these methods are that they destroy the site so that no repeat measurement can be made. This means that no correlations can be made between short-term changes in, for example, soil moisture or soil ice, and the corresponding soil movements.

The last objection is partly met in the 'Young Pit' (Young, 1960), for which a pit is dug with a measurement face which is vertical and runs up-slope–down-slope in direction. Horizontal pins, 1–5 mm in diameter and 5–15 cm long are inserted into this face in one or more vertical lines, the relative positions of which are measured with respect to a plumb line or stakes secured to bedrock in the base of the hole. The measurement face is covered with paper, and the hole carefully re-filled. Re-excavation of the hole allows a measurement of movement to be made, and this can be repeated a number of times. It is clear, however, that the initial installation and each subsequent measurement involves a considerable disturbance of the soil and the water flow within it, even if the actual soil in which the pins are inserted has not been disturbed directly. It is also possible that pins of this sort may allow the soil to flow past them to some extent.

Flexible inserts can be instrumented to give a continuous record of movement provided that the soil moves as a mass rather than by diffusion, and provided that the soil moves *with* the insert rather than *past* it. This method has been used with various insert materials, for example flexible metal strips (Williams, 1959), simple plastic tubing (Williams, 1966*a*) and flexible ducting tubing consisting of a thin-walled plastic tube built around a wire helix which keeps the cross-section constant (Hutchinson, 1967).

The last of these insert materials is the most satisfactory, because it is the most flexible and the only one which is able to extend and contract along its length. Instruments may be fixed permanently to the insert, giving high reproducibility but at high cost; or else mounted on a carriage which slides within the tube, lowering both cost and reproducibility. They may measure deviation from a central position directly, or, more commonly the inclination of the tube (1st differential) or its curvature (2nd differential). Measurement of the higher differentials is simpler technically, but makes more stringent demands about the continuity and smoothness of the velocity profile and the shape of the tube.

To measure deviation directly, the tube must be wide enough to allow a central datum stake to be inserted, so that a carriage sliding on it can measure the distance to the wall of the tube. Without a datum stake, measurements can be referred to an arbitrary vertical to give relative movements. Inclination is usually measured with a sliding inclinometer, consisting of a leaf-spring with a weight on the end: as the inclination increases, the leaf spring bends under the weight, and its change of curvature is registered by an electrical strain gauge which is calibrated directly in inclinations (Hutchinson, 1967; Kojan, 1967). Alternatively, inclination can be recorded by a pendulum moving in a potentiometer transducer (Brunsden, 1968, oral communication). Small inclinations might also be observed optically, using an inclinometer consisting of a small reflecting mercury bath. Curvature of an insert can be recorded sensitively using electrical strain gauges, mounted directly on the insert (Williams, 1959) or on a sliding carriage (Williams, 1962).

These methods of measuring creep using a flexible tube insert in combination with strain-gauge carriage-mounted measuring devices come close to the ideal in many ways. The measurement is sensitive; it can be repeated at frequent intervals without risk of disturbance; the inserts are relatively inexpensive and only the single probe is relatively costly. Doubts only arise in ensuring that the soil and the inserted tube move together as one, but this problem probably becomes important only where the profile shows definite discontinuities of rate.

QUALITATIVE EVIDENCE OF SOIL CREEP

Displaced natural and artificial materials are widely held to indicate soil creep. Sharpe, for example, (1938, pp. 22–3) lists moved joint blocks; trees with curved trunks concave upslope; downslope bending and drag of bedded rock, weathered veins, etc.; displaced posts, poles and monuments; broken or displaced retaining walls and foundations; turf rolls downslope from creeping boulders; a stone line at the approximate base of the creeping soil; and roads and railroads moved out of alignment. Clearly most of these phenomena indicate at least local movement, but there must be some doubt how many indicate continuous or seasonal soil creep on a general scale; or conversely whether creep processes may be active despite the absence of such evidence.

Many of these supposed creep indicators involve the displacement of relatively large objects which are resting on the soil surface or buried to shallow depths, so that accelerated soil movement around features like poles, surface boulders or roads may indicate only that the additional stresses imposed by the feature were able to move the soil, and not that the soil was already undergoing creep. Nevertheless where a *small* additional load has been able to produce movement, it is reasonable to infer that the soil is in a state of incipient movement, or is already moving. Movement, if present, need not be a seasonal creep movement, but is more likely to be a continuous creep.

The curvature of trees, the downslope drag of bedded rock, and the occurrence of stone lines are phenomena with more complex interpretations, which are considered below.

If trees were uniformly geotropic and were sheared at uniform rates throughout their lives, then tree-trunks would show a constant curvature throughout their length and tree roots would show an equal and opposite curvature, so that trunk and roots would form an S-shape. Parizek and Woodruff (1957*a*) compared this model with the curvatures observed in actual trees, which usually have a single sharp curvature near the base (if at all); and their roots tend to be extended slightly more downslope than up. They also found examples of neighbouring trees curving in opposite directions. Since curvatures are produced when the growing point is at their level, observed curvatures refer to incidents in the first few years of the tree's life, which were not repeated. Possible explanations include a relative instability of the tree in its youth while its root system is not well developed; the destruction of terminal buds at ground level (although this is not very common at other heights from the ground); and a tendency towards phototropism which is strongest in young and therefore lightweight trees, which encourages them to grow somewhat away from the vertical on any hillslope. This effect will be most marked on trees along the edges of woodlands, where the light comes very strongly from one side. Curvature of trees cannot therefore be accepted as evidence of soil movement. Small shrubs however do show an S-shaped form due to opposite curvatures of stem and roots during growth, and this effect may be observed in semi-arid (Schumm, 1964) and arctic (Williams, 1966*a*; Hopkins and Sigafoos, 1952) locations. There are, however, examples of sites which show no soil movement (Williams, 1966*a*, p. 196) but which show deformation of woody shrubs by creep within surface snow accumulations (Andrews, 1961), so that even evidence from shrubs must be accepted with some reservations.

Downslope bending and drag of bedded rock and weathered veins clearly indicate some movement, but it is again dangerous to deduce process from it. At some locations the direction of drag varies within a short distance from downslope to upslope: at other locations the depth of dragged features appears to be too great for the penetration of contemporary movement. In such situations contortions and solifluction movement produced under periglacial climate are commonly proposed as the formative processes, but the diagnosis is again uncertain. Downslope bending is most clearly developed in thinly-bedded and steeply-dipping strata, and many examples (e.g. Stearn, 1935; Strahler, 1951, p. 319) show the strata with a single, rather abrupt, bend instead of a

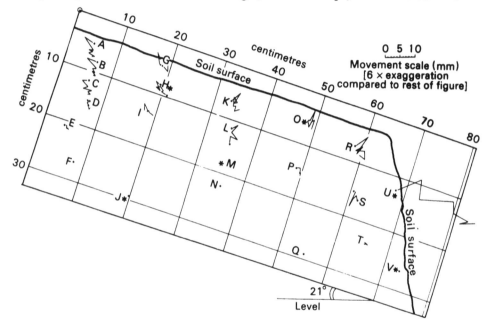

Fig. 10.2. Movement of wires in an inclined soil-box, over three cycles of alternate wetting and drying (from Kirkby, 1963, fig. 43, p. 152).

progressively increasing rate of displacement as might be expected from creep velocity profiles. In such a case it seems possible that frost and moisture changes acting within the more weathered bedding planes have operated to heave the beds apart by tipping them over like a pack of cards, and movements of this sort are more allied to cambering than to soil creep as it is normally understood. In other cases the drag increases rapidly away from undisturbed rock, and soil creep may well be the most satisfactory explanation.

Stone lines in soil may form during continuous creep in zones of most rapid shear movement (above, p. 272), leading to the formation of a sheeted zone with a concentration of coarse particles. Where the creep does not have a basal zone of rapid shear, relative movement will tend to work coarse particles to the surface (Corte, 1966 has demonstrated this for freeze–thaw action). Surface concentration may also arise as a lag deposit from wind or water erosion at the surface, or, in a channel. Ruhe (1959) and Parizek and Woodruff (1957*b*) have argued that the stratigraphic relations of many buried stone lines shows that they are former surface deposits, since buried by colluvial material. Stone lines

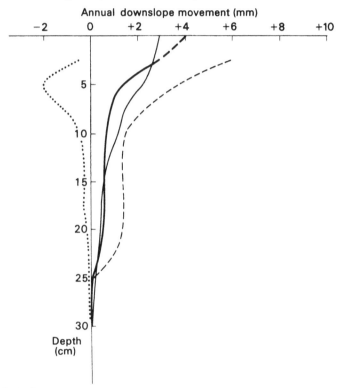

Fig. 10.3. Soil creep velocity profiles from a deep stony soil near Baltimore, Md. Gradient = 17°.

⋰⋱Rudberg test pillars ⎫
⟍⟍Flexible tube   ⎬ mean of 5 values each.
⟍⟍Young Pit   ⎭
⟍Above combined; 15 values in all.
Each set of values was measured over 2½–3 years.

cannot therefore be taken as sure indicators of creep movement, unless they occur in association with other signs of shear; and not in a depositional setting.

RATES OF SOIL CREEP

Seasonal and continuous creep

Most measurements of soil creep have concentrated on the seasonal creep component at relatively shallow depths which is associated with frost or moisture heave. Ward (1953) has shown that vertical heaving of the ground is associated with changes of moisture content, and is most marked in clay soils and in the neighbourhood of trees where soil moisture loss is greatest. Davison (1889), Taber (1930), Penner (1963*b*)

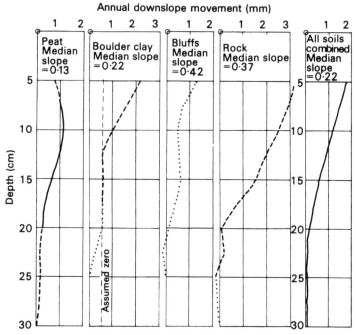

Fig. 10.4. Soil creep velocity profiles from Young Pit measurements in south-west Scotland, for four types of soil, and the four together.
Relative reliability: { ···. value based on 6–10 measurements
ᐭ value based on 11–20 measurements
ᐳ value based on more than 20 measurements
(from Kirkby, 1963, fig. 73, p. 250).

and others have shown that the freezing of soil water can produce large vertical heaves, and that these are greatest for silty soils in which inter-stitial ice lenses can more than double the soil thickness during freezing.

Observation of soil blocks under laboratory conditions (Young, 1958; Kirkby, 1967) show similar moisture expansion movements, but over a number of wetting and drying cycles there is a net downslope movement, expansions and contractions each tending to be slightly downhill of the preceding movements (fig. 10.2). Under field conditions,

286

this correlation between expansion and downhill creep is difficult to observe directly, but may be inferred from T-peg (above, p. 280) data, which show large short period tilts and returns and a smaller long-term component of increasing tilt. The short-term tilts correlate significantly (95 per cent level) with an antecedent soil moisture index; and the long-term component is significantly downhill (95 per cent level) (Kirkby, 1967). T-peg data are, however, unreliable for measuring net rates of creep movement.

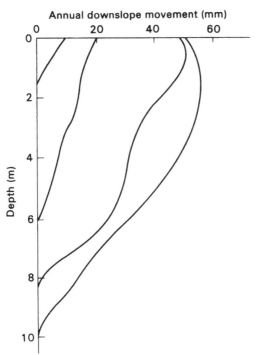

Fig. 10.5. Representative soil creep velocity profiles from the Berkeley Hills area, California (from Kojan, 1967). Note different horizontal and vertical scales.

Velocity profiles of seasonal soil creep have been obtained using Young Pits, flexible tubes and Rudberg test pillars. A comparison of the three types of method (fig. 10.3) for a 17° slope near Baltimore, Maryland, shows comparable profiles for flexible tubes and Young Pits, but surprisingly different values for the Rudberg test pillars. The large variation between neighbouring measurements and the small number of replicates (5) does not allow the methods to be distinguished significantly and even the mean velocity profile for all methods together is not critically defined. Fig. 10.4 shows velocity profiles of broadly similar form, obtained for different soil types in south-west Scotland (Kirkby, 1967) for a larger number of Young Pit measurements (36 in all). Except perhaps in the top 5 cm of soil, the profile in both figs. 10.3 and 10.4 might be considered as a constant shear of the top 15–25 cm of the soil.

The orders of magnitude of soil movements are very similar in these two figures, but are smaller by a factor of over 100 than the measurements obtained by Kojan (1967) for slopes in the Berkeley Hills, California (fig. 10.5), where horizontal displacements are ten times greater, and depths of creeping soil extend to ten metres or more. It is clear that these measurements refer to continuous, and not seasonal creep, because of their great depths and the fact that some velocity profiles show indications of progressive failure, but it should be stated that very few measurements of seasonal creep extend deep enough to compare with Kojan's results, and that the rates of shear strain are comparable in figs. 10.3, 10.4 and 10.5 (3–6 mm per m depth/year). It may therefore be necessary to treat many creep measurements with scepticism. However, creep rates do appear to become very small at the base of the profiles in figs. 10.3 and 10.4, even though no absolute bench marks were used. In measurements where bedrock bench marks have been used (Young, 1958; Everett, 1963b; Leopold, Emmett and Myrick, 1966) surface creep rates are generally not more than 5–10 mm/year, so that Kojan's measurements seem to refer to unusually mobile soil conditions.

Table 10.1 summarizes average rates of soil creep from a wide range

TABLE 10.1. *Measured rates of soil creep*

Area	Mean slope	Rate in cm³ per cm year	Source
Pennines, N. England	26°	1·5	Young, 1958
Neotoma valley, Ohio	20°	6	Everett, 1963(b)
S.W. Scotland	17°	2·1	Kirkby, 1963
S. England	10°	30?	Kirkby, 1963
Sangre de Cristo Mtns, New Mexico	45°	8	Leopold, Emmett and Myrick, 1966
Berkeley Hills, California	19°	650	Kojan, 1967
Baltimore, Maryland	17°	1·3	Kirkby (unpublished)
S. Alps, N.Z.	?	3·2	Owens, 1969
W. Colorado	10°	8 mm/year linear	Schumm, 1964 for movement of
	20°	18 mm/year linear	25–50 mm surface markers by
	30°	60 mm/year linear	creep and wash

of areas, usually for a range of slope gradients with means in the 15–25° range. Values of 1–6 cm³ per cm/year appear to be normal except where

continuous creep extends to some depth. Schumm's data for the movement of surface markers (1964) has been included, although it is a particle rather than a mass movement, and probably includes some movement due to wash processes (chapter 8).

With such large variations in creep measurements, whether from
random or systematic causes, little correlation has been obtained
between creep rates and topographic factors, although the data of fig.
10.4 do suggest that creep rate might increase linearly with the sine of

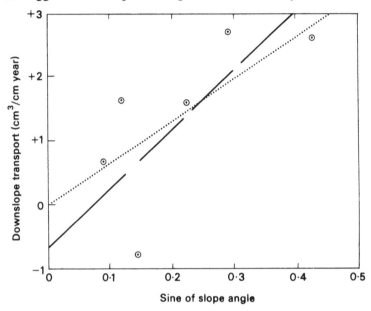

Fig. 10.6. Variation of soil creep rate with slope gradient. Each point represents the median
slope and rate for 5–7 sites, measured over 1–2 years in a Young Pit. Data for Galloway,
Scotland from Kirkby (1963). The correlation is not significant.

the slope angle, as theoretical considerations suggest. The relationship
(fig. 10.6) is not, however, significant with the amount of data available.
Correlations with distance from the divide show no apparent trend.
Moisture changes have been identified as the main agent causing
measured creep in some cases (Kirkby, 1967; Everett, 1963b), though
frost cycling is recognized as an important subsidiary cause. In other
sets of measurements (Owens, 1969), frost cycles have been identified as
the major agent of causation. Burrowing by animals, followed by re-
filling of holes from the upslope side; and growth and decay of roots,
again with filling from the upslope side, may both cause small amounts
of creep movement, and table 10.2 lists possible rates, each calculated
for ideal conditions. It may be seen that these organic causes are

TABLE 10.2. *Estimated maximum rates of soil creep from organic causes*

Agent	Estimated rate in cm^3/cm year
Wedging by grass roots and refilling	0·003
Worms: (a) forming of burrows and refilling	0·15
(b) Uneven distribution of surface casts	0·25
Rabbits: forming of burrows and refilling	0·10

Data from Dittmer (1938, p. 655);
      Darwin (1882), pp. 262–73).

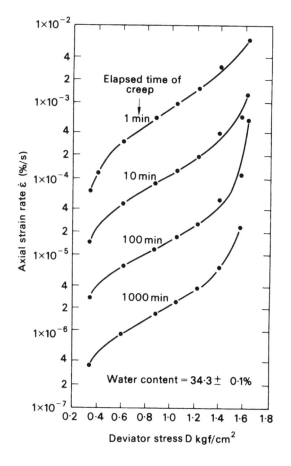

Fig. 10.7. Variation of axial strain rate with deviator stress for undrained creep of re-moulded, saturated illite (from Mitchell, Campanella and Singh, 1968, fig. 11, p. 245).

secondary in importance to systematic moisture and frost cycling in producing soil creep.

A more basic approach to soil creep measurement is in triaxial compression tests carried out at very low deviator stress levels – well below the nominal yield value. Mitchell, Campanella and Singh (1968) have investigated the creep rate as a function of deviator stress and time elapsed (fig. 10.7), and have interpreted the results in terms of rate pro-

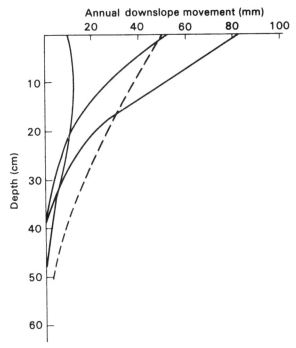

Fig. 10.8. Velocity profiles for soil movement during solifluction.
Rudberg (1962); Kärkevagge, Sweden.
Williams (1966*a*); Schefferville, P.Q., Canada.

cess theory (below, p. 296). As might be expected, movement rates increase with the deviator stress, but they are also shown to decrease with time, for periods of up to 1,000 minutes in these experiments. The observed strain rates are, however, high compared to field measurements of creep movement, and extrapolation of the time dependence shown in fig. 10.7 suggests that field rates correspond to times elapsed of about a year (5×10^5 minutes on a 10° slope). This seems reasonable, but the comparison is partly vitiated because it compares *linear* strain rates under *undrained* conditions in the laboratory with field values of *shear* strain rates, under naturally *drained* conditions.

291

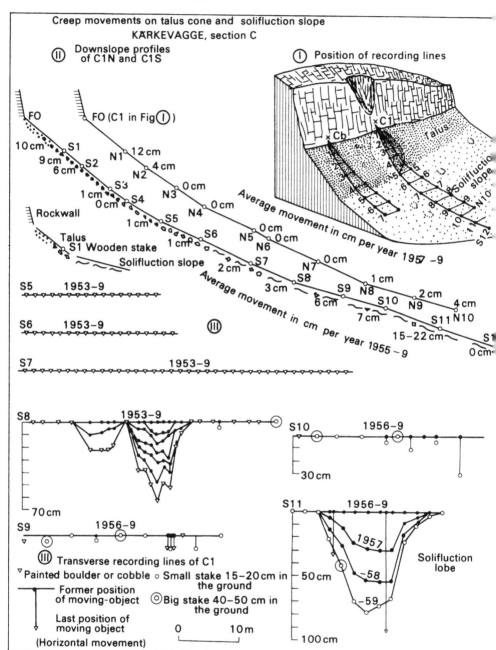

Fig. 10.9. Creep and solifluction movements in Kärkevagge, north Sweden. Sketch shows positions of downslope and transverse movement lines (from Rapp, 1960, *Geografiska Annaler*, vol. 42, nos. 2–3, fig. 65, p. 173).

Solifluction

Individual velocity profiles of solifluction movement (fig. 10.8) show movements which are as large as those for rapid continuous creep (fig. 10.5), but which only extend to depths of 40–50 cm. Rates of shear are therefore very much higher (100 mm per m/year) than those occurring in soil creep. Since movements usually appear to occur only during periods of annual freeze and thaw (Williams, 1966*a*; Everett, 1963*a*), higher rates of mean annual shear probably result from high rates of instantaneous shear during the actual movement, suggesting that movement occurs at a rate more characteristic of flow or creep in *water films* than of creep in *soils*.

The areal pattern of solifluction shows that movement is concentrated in lobes of rapid movement, with intervening areas of negligible

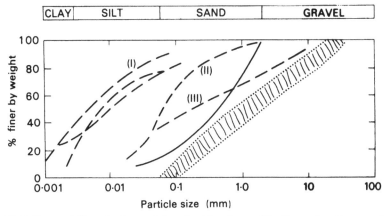

Fig. 10.10. Susceptibility of soil materials to frost heaving, solifluction and sorting associated with patterned ground.

Range of materials susceptible to size-sorting by frost in periglacial deposits (Corte, 1963*b*).

Coarse limit of frost-heaving morainic soils (Taber, 1929, 1930; Williams 1957).

Examples of solifluction soils

   (i) Three soils near Schefferville, P.Q., Canada (Williams, 1966*a*).

   (ii) Frost-heaving morainic soil (Williams, 1957).

   (iii) Solifluction lobe, Kärkevagge, Sweden (Rapp, 1960*b*, p. 180).

movement (fig. 10.9). For Kärkevagge, Sweden, Rudberg(1962) measured a rate of 125 cm^3 per cm year for an active site, but Rapp (1960*b*, p. 182) considers that the average rate of movement is about 50 cm^3 per cm year, when allowance is made for the relatively static areas on a solifluction hillside. Williams (1966*a*) has recorded similar rates of mass movement, which are large compared with rates of seasonal soil creep but small compared with rates of wash, which is also active in periglacial environments.

The same forces which occur during annual freeze and thaw of soil water and cause solifluction, also produce other features. Corte (1963*a*) has shown that horizontal layers of coarse material on fine silt will be sorted by freeze–thaw into domed, sorted polygons. On a level surface no net mass transport occurs, but on a slope the polygons become elongated because of the asymmetry of the dome. When the general ground slope becomes as steep as the gradients formed on the dome, the polygons will develop into stripes. Movement due to frost heaving will therefore always have a downslope component so that in each freeze–thaw cycle some soil particles will not only be sorted laterally according to their size, but will also move a short distance down the hill. Sorting, and hence downhill movement during sorting is probably a once-for-all movement, and its rate has not been measured.

Susceptibility to sorting of this type, and also to frost heaving with the formation of segregated ice, is largely dependent on the grain size distribution of the soil. Fig. 10.10 shows empirical curves obtained for the two processes by Taber (1929, 1930), Williams (1957, 1966*a*), Rapp (1960) and Corte (1963*b*). It appears that fine-grained soils are susceptible to frost heave with segregated ice and to solifluction; that somewhat coarser soils are susceptible to size-sorting and the formation of polygons and stripes; and that the coarsest debris is unaffected by frost heave. The three types of soil overlap little, if at all, so that solifluction movement and movement in sorted stripes are mutually exclusive processes.

MECHANISMS OF CREEP PROCESSES

Seasonal creep is produced by expansion and contraction cycles which are mainly at right angles to the soil surface, and occur mainly in response to soil water freezing or soil moisture changes. In the simplest model (Davison, 1889) of perpendicular expansion and vertical contraction (above, p. 274) the total downslope displacement, measured in a horizontal direction, at depth z in the soil is given by:

$$C(z) = k \cdot \sin \beta \cdot \int_{z}^{\infty} M(z) \cdot dz, \tag{10.1}$$

where $C(z)$ is the horizontal creep movement at depth z, k is the soil expansion per unit moisture change (assumed constant), β is the slope gradient angle, and $M(z)$ is the *accumulated* moisture change in the soil at depth z. If the moisture change, $M(z)$ is assumed to fall away exponentially with depth, then the resulting velocity profile will also be exponential, of the form:

$$C(z) \propto M \sin \beta \cdot e^{-b \cdot z}, \tag{10.2}$$

where M is the accumulated moisture change at the surface, and b is a constant. A profile of this form is shown in fig. 10.11a. This simple model ignores the forces tending to produce soil movement.

A more realistic model assumes that expansion and contraction take place in the direction of maximum applied stress, a minimum work condition; and this direction is produced as a resultant between the expansion or contraction forces and the deviator gravity stresses, which increase with depth. To a first approximation, this model leads to a shear rate which is proportional to the deviator stress and the accumulated moisture change (Kirkby, 1967, p. 362). That is:

$$C(z) \propto k . \sin \beta . \cos \beta . \int_z^\infty z . M(z) . dz, \tag{10.3}$$

with the same notation as before. If moisture change is again assumed to decrease exponentially with depth, then:

$$C(z) \propto M . \sin \beta . \cos \beta . \frac{1+bz}{b^2} . e^{-bz}, \tag{10.4}$$

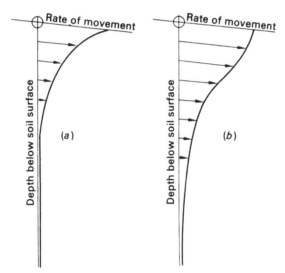

Fig. 10.11. Idealized velocity profiles for soil creep.
(a) $C(z) \propto e^{-bz}$, an exponentially decreasing velocity with depth, as in the Davison (1889) model.
(b) $C(z) \propto (1+bz)e^{-bz}$, a velocity profile with zero rate of shear movement at the surface.

a type of velocity profile which is illustrated in fig. 10.11b. Comparing it with the previous model, it can be seen to differ most importantly near the surface, where it predicts a zero rate of shear, corresponding to zero gravity deviator stress.

Random diffusion of soil particles, as suggested by Culling (1963) will produce a diffusion of particles outwards; that is towards the soil surface in response to lower bulk densities near the surface. In equilibrium, this upward and outward movement will be balanced by consolidation which occurs under gravity and cohesive forces, in a direction between the perpendicular to the surface and the vertical. This case may be treated as a special case of the preceding model, and produces a similar type of velocity profile. However, it will differ in detail because diffusion scatters particles which were initially close together in the soil, so that the extent of diffusive creep may be detected by this type of effect.

Rate-process theory (Glasstone, Laidler and Eyring, 1941) may be used to predict rates of continuous creep, and it may be argued (above, p. 274) that seasonal creep is really continuous creep wholly or partly accelerated in the presence of seasonal factors which reduce soil strength. Under given soil strength conditions, the rate of shear strain is given by:

$$\frac{dC(z)}{dz} \propto - \sinh{(A.z\sin\beta)}.\cos\beta, \tag{10.5}$$

for constant A (Mitchell, Campanella and Singh, 1968). This type of formulation is, as yet, of little practical value in assessing rates of seasonal creep, because the relationship between moisture content and the constants of proportionality has not been investigated. It does show, however, the effect of deviator stress. For low values of deviator stress, near the surface, the hyperbolic sine can be approximated by:

$$\sinh{(A.z.\sin\beta)} \approx A.z.\sin\beta, \tag{10.6}$$

which is equivalent to the form in which deviator stress appears in the previous model. At greater depths, the approximation:

$$\sinh{(A.z.\sin\beta)} \approx \tfrac{1}{2}.e^{A.z.\sin\beta} \tag{10.7}$$

is more appropriate, and fits the data shown in fig. 10.7. Only rather careful experiments will distinguish the form of velocity profile obtained from this theory from that obtained above (fig. 10.11*b*), but the dependence upon moisture change will be very different for rate-process theory.

Very similar process mechanisms must be invoked to explain the velocity profiles obtained during solifluction movement. In both cases, all theories predict a strong dependence on slope gradient; approximately a first power law for moderate gradients. The only factor which may appreciably alter this relationship in the field is the presence of strong correlations between slope gradient and soil strength or moisture parameters. If for example steeper slopes have stronger soils, then the increase of creep rate with gradient will be less than the simple law predicts. Downslope variations in creep rates are not a simple result of moisture flow patterns, as they are in the case of wash processes, but may arise if either the accumulated moisture change or the soil strength alters in a consistent fashion downslope. Although moisture *content* increases steadily downslope on many hillsides, there is no evidence that moisture *change* has a corresponding trend. It therefore seems reasonable to assume that distance downslope has at most a secondary effect on the rate of soil creep, and that it is essentially independent of distance from the divide.

The only non-creep mechanism which can occur in solifluction is the process of lubricated movement of overlying soil as ice lenses melt so that a film of water forms which can act as a shear zone (above p. 277). The formation of an ice lens physically separates layers of the soil above and below it. While the ice remains frozen, the ice–soil contacts have a strength which is probably at least as great as that of the unfrozen soil without ice lenses; but on thawing, the ice–soil contact has only the strength of a water film, which is negligible. Soil strength may thus be lost over the whole area of an ice lens, leading to local failure.

PROCESS-RESPONSE MODELS FOR CREEP

If the rate of soil creep is assumed proportional to the slope gradient, and slope gradient is taken as the tangent of the gradient angle, an approximation which is valid for moderate slopes ($<20°$), then the slope development equations take their simplest possible form. Development is clearly transport-limited, since soil creep is only present if there is an adequate soil cover, so that the equation, in its two-dimensional form is, from equations 5.3 and 5.4, for $f(x) = K$ and $n = 1$:

$$\frac{\partial^2 y}{\partial x^2} = \frac{1}{K} \cdot \frac{\partial y}{\partial t}, \tag{10.8}$$

where y is elevation, x is horizontal distance from the divide, t is time elapsed and K is a constant of the process rate. For the boundary conditions of a fixed divide at $x = 0$, expressed by $\partial y/\partial x = 0$ at $x = 0$; and a slope base fixed in vertical position but varying in elevation according to:

$$y = \phi(t) \quad \text{at} \quad x = l;$$

and with the initial slope form specified by:

$$y = f(x) \quad \text{at} \quad t = 0;$$

then the complete series solution has been worked out for the heat conduction case by Carslaw and Jaeger (1959, p. 104), and quoted in the geomorphic context by Culling (1963, p. 146). It is:

$$y = \frac{2}{l}\sum_{0}^{\infty}\left[e^{-K(2n+1)^2\pi^2 t/4l^2} \cdot \cos\frac{(2n+1).\pi.x}{2l}\left\{(-1)^n.\frac{K(2n+1).\pi}{2l}.\right.\right.$$

$$\left.\left. \int_0^t \phi(\lambda).e^{K(2n+1)^2.\pi^2.\lambda/4l^2}\cdot d\lambda + \int_0^l f(x')\cdot\cos\frac{(2n+1).\pi.x'}{2l}\cdot dx'\right\}\right] \quad (10.9)$$

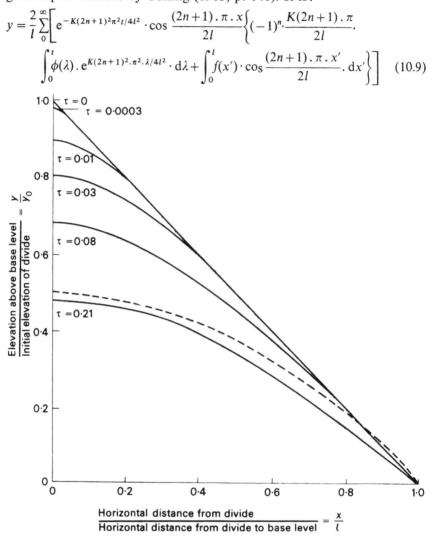

Fig. 10.12. Dimensionless graph (solid lines) showing slope development by creep of an initially straight slope with a fixed divide (at $x = 0$) and a fixed base level (at $x = 1$). $\tau = $ relative time elapsed $= Kt/l^2$.

Broken line shows a steady state profile in equilibrium with slope-base downcutting at constant rate.

Two examples of this rather complicated formula will show roughly the forms to be expected. For an initially straight slope, and a base level which does not vary in elevation ($\phi(t) = 0$), then the slope profile develops as shown in fig. 10.12. A convexity grows from the divide and extends farther and farther downslope as time passes. After very long times ($t \to \infty$), the form of the slope tends towards a simpler form of the equation:

$$y \propto e^{-K\pi^2 t/4l^2} \cdot \cos\left(\frac{\pi x}{2l}\right), \tag{10.10}$$

and this is the characteristic form, as defined in chapter 5.

A second simple.example is the profile obtained during downcutting of the slope base at constant rate, A. After a long period ($t \to \infty$), the equilibrium profile attained is:

$$y = \frac{A}{2K}(l^2 - x^2) - At. \tag{10.11}$$

This form is shown by the broken line in fig. 10.12, relative to a moving origin at the basal point, for the case $A = Ky_0$. The slope angles do not decline with time as with the fixed base elevation case above. The form is seen to be somewhat more convex at the slope base, as may be expected in the case of downcutting. If the slope base aggrades, then a concavity will develop at the base, but the equilibrium form becomes meaningless because the slope base ultimately becomes higher than the divide.

Culling (1963) applies other solutions of the heat conduction equation (Carslaw and Jaeger, 1959) to the cases of a laterally migrating slope base, to landforms with radial symmetry, and to landforms where either an upper or a lower surface extends indefinitely. For example, if two such surfaces are initially separated by a vertical step, then the solution takes the form of a normal curve integral function. A convexity works back into the upper plateau and a concavity spreads out across the lower surface.

Measured rates of soil creep are usually so slow that creep processes can rarely be considered dominant in shaping hillslope profiles. Soil creep generally removes less material than chemical solution on gentle slopes, and less than mass movements on steep slopes. Solifluction operates under periglacial conditions at a higher rate than soil creep, so that it may be more significant as a slope development process in periglacial areas than creep is in temperate areas. However, it is probably also true that soil wash processes are even more effective than solifluction in a periglacial environment, and operate on similar materials and at similar gradients. Near divides in all environments soil creep has been widely considered to be the most effective process, because it is one of the few which do not increase in rate with distance from the divide. Gilbert (1909) has argued that this property of soil creep is responsible for the convexity of divides. He argues, in effect, that if the divide is to be lowered, then increasing volumes of material must pass points at increasing distances from the divide. If the dominant process increases *only* with increasing gradient (and not with *distance* from the divide), then these increasing rates of transport require increasing gradients for their removal. Gradient therefore increases away from the divide, which is thus convex in profile. What Gilbert did not state was that this argument applies not only to creep but to solifluction and to rainsplash (chapter 8) as well, and these processes may equally be responsible for convex divides in periglacial and semi-arid regions respectively. Gilbert's argument is put in mathematical form in appendix C (case 1), where it may be seen that divide convexities in fact form under a still wider range of processes, although not as a result of surface wash.

PART THREE: FORM: COMPARISON OF REAL FORMS WITH PROCESS-RESPONSE MODELS

CHAPTER 11

HUMID TEMPERATE AREAS

In part two we examined the mechanisms of the major processes operating on hillslopes, and, in each case, suggested possible models of hillslope evolution through time assuming that one particular process was dominant. The form of actual hillslope profiles in nature is not the product of a single process, however, but the result of many of these processes acting upon the slope. The relative importance of the different processes varies according to the climatic and lithologic environment and with the stage of development of the landscape. In this chapter we shall examine the typical slope form of humid temperate areas, the so-called 'normal' climate of W. M. Davis. We shall tackle this under three main headings: the typical slope profile; the probable sequence of slope development through time; and the influence of paleoclimatic conditions on the form of present slopes.

THE TYPICAL SLOPE PROFILE

Undoubtedly the slope profile most characteristic of humid temperate areas, at least where the slope is cut in relatively uniform rock, consists of a convex upper part and a lower concave slope (fig. 11.1a) as noted by many of the early workers. There are, however, very many slope profiles which contain a distinct straight segment between the upper convexity and the basal concavity. This was apparently not recognized by Davis; 'in the scheme for the normal cycle there is no room for straight erosional slopes' (Dury, 1959, p. 64). The accumulation of field evidence in the last two decades (Strahler, 1950; Young, 1958; Carson, 1967b) indicates quite conclusively that straight hillside slopes do exist. Some slopes, in addition, especially those in relatively youthful valleys, also contain steep cliff faces of bare rock above the lower slope components.

301

Considerable attention has been given in Britain to definitions of the various components of hillslope profiles and to the development of a particular terminology. This was originated primarily by Savigear (1956) and elaborated upon by Young (1964) more recently. The usefulness of these contributions is not immediately apparent and, in our opinion, the classification is unnecessarily complex. We prefer to recognize the four main hillslope components suggested by Wood (1942) and King (1953); cliff face (free face), straight segment (constant slope), upper convexity (waxing slope) and basal concavity (waning slope). Each of these is discussed in turn below.

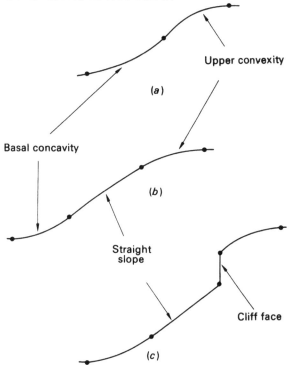

Fig. 11.1. Typical slope profiles in humid temperate areas.

The cliff face

In areas of uniform lithology, the existence of a cliff face in a slope profile is probably very short-lived. Steep bare rock walls may be produced by rapid down-cutting into strong rock, but, as noted in chapter 6, as soon as active undercutting ceases, the cliff face will retreat, and possibly flatten also, under weathering-limited processes. The result of these processes is the production of a talus slope (fig. 6.12) which eventually extends upslope and masks the initial rock wall. It is true that if some mechanism existed whereby fallen talus could be removed as quickly as it accumulates, then the cliff face would not become obliterated, but although this probably occurs in cold and arid areas (chapters 12 and 13), this is uncommon in humid areas. The upslope extension of the talus slope is thus the major mechanism by which the initial cliff face rapidly disappears from the slope profile.

The continued persistence of cliff faces on slope profiles in humid-temperate areas probably depends upon mixed lithology, with a strong cap rock overlying weaker strata. This point has been echoed by many workers in semi-arid areas (Everard, 1963; Sparrow, 1966) and is perhaps even more important in humid areas. Such rock walls as the grit-stone edges of the Pennines, where Millstone Grit caps much weaker shales, are typical of this type of cliff face.

The actual mechanism by which a cap rock will maintain a cliff face is relatively simple and arises from the tendency of the weaker underlying rock to weather and retreat at a faster rate than the cap rock itself. As a consequence the cap rock is continually undercut and its face maintained at a steep angle. There is insufficient time for the cap rock to retreat relative to the underlying material and develop its own talus slope. Once undercutting of the cliff face halts, a talus slope will accumulate and eventually extend upslope obliterating the cliff.

The straight segment

Sparks (1960, p. 63) notes that 'the straight slope section appears to be more characteristic of areas of high relief, while in lowlands the slopes are essentially convex above and concave below'. There are a number of possible explanations for this. The most probable reason is that, although straight slopes were produced in lowland areas during an initial period of downcutting, they were inevitably small and, as soon as active stream erosion ceased, they were quickly obliterated from the slope profile through transport-limited processes. In the deeper valleys of upland areas, the same processes would need a much longer period of time to eliminate the longer straight segments. It is possible, however, that in lowland areas stream down-cutting is so slow that transport-limited processes, particularly soil creep, are able to keep pace with stream incision (fig. 10.12) and prevent the slope being steepened to the angle of limiting stability of the debris. In these cases, straight slope sections probably never emerge, at *any* stage of landscape development; slope profiles are initially convex, changing eventually into convexo-concave forms. Where straight slopes do occur they usually indicate an early stage of landscape development under weathering-limited processes and relatively rapid stream down-cutting. Transport-limited processes act to obliterate any straight segment on a slope profile rather than maintain one.

It has already been noted that Davis did not seem to recognize the existence of straight hillsides. Actually it is very difficult to follow Davis' treatment of the development of valley-sides because he never explicitly refers to the *geometry* of the slope profiles. At an early stage in the cycle of erosion he does refer to the talus slopes produced by cliff retreat as 'those evenly-slanting waste-covered mountain sides which have been reduced to a slope at the angle of repose' (Davis, 1954, p. 267), but the impression from his sketches of slope development (Davis, 1930) is that these straight talus slopes are very rapidly converted to convexo-concave profiles.

One of the first specific references to straight components of the slope profile was by Lawson (1932) in an interesting but rather confusing paper. He emphasized that straight slopes are characteristic only of the mature stage of landscape development, and after this are fairly rapidly eliminated from the slope profile by extension of the upper convexity and basal concavity. He speculated that during its existence the angle of a straight segment on a given hillside would remain unchanged, and corresponds to an equilibrium slope where soil removal by rainwash is just balanced by soil formation due to weathering. Lawson was extremely vague, however, as to the mechanism responsible for the straight slope in the first place. At one stage, as noted by Sparks (1960, p. 69), he even suggested that the straight slope marks a transition area where processes tending to produce the upper convexity alternate with processes responsible for the basal slope, an idea apparently accepted by Baulig (1950) also.

The much-quoted article by Wood (1942) unfortunately does little to help clarify the picture. The initial part of Wood's slope sequence corresponds to the Davisian cycle with the emergence of a straight talus slope (constant slope) beneath a retreating rockwall. The angle of this constant slope depends on the character of the mantle and, as the mantle becomes more weathered, so the angle of the constant slope diminishes. This is an interesting idea, but the exact nature of this dependence between angle and the mantle character is left undiscussed by Wood. The idea that the slope flattens over time resembles Davis' ideas, except that Davis leaves the nature of the slope geometry obscure. In addition, however, Wood argued that a certain amount of retreat of the constant slope, due to the processes of soil wash, stripping material from the slope, would take place, producing at the base of the straight slope a gentle waning slope which is concave in profile. The upslope extension of the basal concavity, along with the downslope encroachment of the upper convex part, eventually results in the complete elimination of the straight segment from the profile.

These ideas of Wood appear to form much of the basis for King's (1953) famous canons of landscape evolution. King adopts the four hillslope components of Wood, namely upper waxing slope, free face, constant slope and waning slope, but assumes a much narrower point of view than the earlier worker. It was King's contention that straight slopes remain at the same angle throughout the time in which it under-goes retreat until it is eventually consumed by the other components of the slope profile. In other words, King advocated the principle of paral-lel retreat of straight slopes which is commonly, although mistakenly (Simons, 1962), associated with Walther Penck. The concept of retreat is derived from Wood's earlier article, although it must be emphasized that Wood did not specify *parallel* retreat, but allowed flattening of the straight slope at the same time. A corollary of King's argument is that the extent of retreat of the straight slope, before its eventual obliteration, and thus the extent of the basal waning slope, is much greater than envisaged by Wood.

It is a great tragedy that King, basing his ideas on careful observations in semi-arid areas, primarily southern Africa (King, 1951), attempted to plant these ideas, unmodified, on humid temperate landscapes. The model developed by King may well be applicable in many semi-arid environments (chapter 13) but it depends upon particular conditions which are not evident in temperate areas. The underlying feature of his model is that soil wash is powerful enough to remove soil debris from the constant slope as soon as it is produced by weathering. As a result the mantle remains basically coarse talus and does not weather in the manner envisaged by Wood, so that the straight slope exists at an un-changed angle during its retreat. Now, although this may be valid for the intense precipitation and almost bare talus slopes of semi-arid areas, it is unlikely in humid temperate areas where, as noted earlier, the combina-tion of much less intense rainstorms and a dense vegetation covering on slopes renders soil wash a relatively ineffective process.

The concept of parallel retreat of a straight talus slope, as developed by King, is not supported by field evidence in humid temperate areas. It has been stressed by Carson (1969b) that in these areas scree or talus slopes *do* undergo a sequence of progressive weathering, and that this process is not aborted at the talus stage as King claims. As a result, straight slopes develop at gentler angles according to the extent of weathering of the waste mantle. These more recent ideas support, in a restricted sense, the ideas of Wood. The evidence presented in chapter 7 suggests, however, that the relationship between stable angle and mantle character is not a continuous one as conceived by Wood, but comprises three distinct mantle types (talus, taluvium and residual soil) rather than a continuous spectrum. Superficially there is some resemblance between the model outlined in chapter 7 (p. 181) and the ideas of King, since both involve parallel retreat of the initial talus slope; in the former one, however, the retreat of the talus slope produces a new straight slope, whereas in the latter a concave basal slope emerges.

Clearly, a great deal more work needs to be done in connection with straight hillslopes. We should emphasize that most of the classical models discussed above have assumed landscape development in an area of strong, well-jointed rock, and it is only under such conditions that the sequence of cliff face, talus slope, basal slope is valid. The more recent ideas on straight slopes discussed above and in chapter 7 are, however, sufficiently general that they can be adapted to a wide range of lithologic settings.

The upper convexity

Attempts to explain the upper convexity of hillslopes, so characteristic of humid areas, have usually invoked the mechanisms of unconcentrated soil wash or soil creep.

One of the first discussions was by Gilbert (1877) in his classic work on the geology of the Henry Mountains, Utah. Gilbert argued that typical hillslope profiles resemble the long profile of streams, concave upwards, and at the divides where two profiles meet, the crests are usually very sharp. This law of divides apparently did not agree with observations in many badland areas where convex summits were often found. Davis (1892), commenting on Gilbert's work, attributed these badland convexities to the process of soil creep. Subsequently, in another classic paper, Gilbert (1909) recognized that convex summits are not peculiar to badlands and elaborated on soil creep as a mechanism for them.

Gilbert's ideas have been neatly summarized by Sparks (1960, pp. 66–7) but are briefly noted here also. The basis of this theory is (1) the *observation* that, apparently, along convexo-concave slopes, the thickness of the waste mantle is fairly constant, and (2) the *assumption* that the amount of waste passing any point on a hillside is proportional to the distance of that point from the crest. In order to reconcile these two points it follows immediately that the velocity of the creeping waste mantle must increase downslope so that a continuous increase in the amount of waste moving downslope can occur. Now, the major force in determining the variations in the velocity of creep on a given hillside was thought to be the component of gravity tangential to the slope. The necessary increase in the velocity of soil creep downslope therefore demands that the slope angle continuously increases downslope also. In this way a convex profile is needed to satisfy the initial bases.

Actually Gilbert's idea is essentially the same as the process–response model suggested in chapter 10, based on the belief that soil creep is a distance-independent process, that is, $f(x) = 1$ and $n > 0$ in equation 5.4. Under these conditions, along a slope of constant gradient, the amount of waste creeping downslope would be the same at all points along the slope, so that no *net* loss of material takes place except at the divide which, accordingly, becomes rounded. Gradually the convexity is extended downslope, as indicated in fig. 10.12, until a fully convex profile results.

The critical point is whether sufficient field evidence exists to verify that soil creep is indeed a distance-independent process. Theroretically, even along a slope of constant gradient, it is possible that the downslope discharge of debris by soil creep could increase with distance from the divide due to (*a*) an increase in the velocity of creep, or (*b*) an increase in the thickness of the moving mantle. Insufficient evidence exists to comment on the relationship between velocity and distance from divide on straight slopes, but there is no obvious mechanism which might be offered for a possible downslope increase. The evidence relating to the thickness of the creeping mantle is also rather equivocal. Young (1958) noted that some straight slopes in the southern Pennines showed an increase in the thickness of the soil mantle downslope and he offered this as a means by which effective ground loss [*sic*] could occur at all points along a straight slope, even if the velocity of creep was constant downslope. Subsequent work (Carson, 1967*b*) has, however, indicated that a systematic increase of soil thickness downslope on straight slopes is probably an exception to the general pattern. Moreover, if it is assumed (as evidence seems to indicate) that seasonal soil creep is restricted to a very narrow depth below the ground surface, then the thickness of the creeping mantle could be constant even though the total soil depth is a function of position. So far, therefore, little evidence exists to contradict the belief that soil creep is essentially a distance-independent process.

The hilltop convexity of humid temperate areas is thus almost certainly a product of soil creep. Moreover, in view of the relative importance of soil creep *vis-à-vis* soil wash in humid areas, as stressed in part two, it is not surprising that well-rounded summits are so characteristic of these areas. In contrast, in semi-arid areas, soil creep is subordinate to soil wash, and, as a result, convex divides are poorly developed, as noted by Gilbert.

A number of renowned geomorphologists have from time to time believed that the upper convexity was the product of unconcentrated soil wash rather than soil creep. One of the first was Fenneman (1908). He argued that soil wash on the upper parts of a hillslope took place in a rather different way to soil removal on the major part of the slope. Near the divide, he suggested that rainwater flows downslope in tiny rills or as a sheet of 'unconcentrated wash' which was incapable of corrasion [*sic*] of soil material; as the depth of flow increases downslope, a point is reached when gullying takes place and channel development begins. These ideas are, of course, very similar to those developed by Horton (1945). Now, Fenneman argued that while gullying on the lower slope would tend to produce a concave slope profile, unconcentrated wash on the upper slope would result in a convex slope.

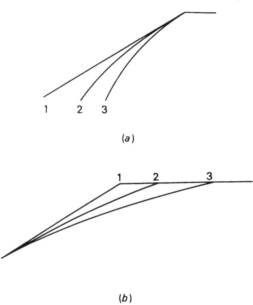

Fig. 11.2. Theoretical development of convex slope profiles according to (*a*) Fenneman (1908), and (*b*) Lawson (1932).

According to Fenneman, unconcentrated wash, although incapable of actual corrasion of debris, nevertheless moved a certain amount of soil, picking up soil particles released by weathering. He thought, in fact, that at every point on the upper part of the slope, unconcentrated wash would be loaded to capacity with these soil particles. The carrying capacity of flowing water increases as the depth of flow increases, and since the depth of unconcentrated wash increases downslope, it was not unreasonable to assume that the load also increased downslope. Fenneman attempted to deduce from this that the net amount of material removed from each point also increased downslope, thus steepening the slope and producing a convex profile. This idea, as shown below, is unfounded.

Slightly different ideas also based on the process of unconcentrated wash were offered by Lawson (1932). He argued, in contrast to Fenneman, that the 'removal of soil by rainwash proceeds by regularly *diminishing* increments from the upper to the lower part of the curve' (p. 706). The assumed effects on slope geometry are shown in fig. 11.2b and compared with the ideas of Fenneman (fig. 11.2a). It is unfortunate that neither Fenneman nor Lawson attempted to treat this problem mathematically. Both arguments as they were presented were incomplete and, in fact, illogical. The development of a convex curve from an initially straight slope is *not* dependent upon the rate of increase (or decrease) of net surface lowering with the distance downslope. A linear increase in net surface lowering due to soil wash downslope will start to produce a steepening slope over time, and a linear decrease in net surface lowering due to soil wash with distance from the divide will start to produce a flattening of the slope over time, but the slope will initially remain straight. Whether or not a curve is produced depends not only on variations in the rate of surface lowering, but also on the conditions at the divide and basal removal point. This problem is analysed in appendix C, where it is shown that for *transport-limited* removal without appreciable solution, the conditions for concavity and convexity may be related to the values of the distance exponent m in the slope transport equation 5.4. It is shown that the upper portion of the profile will become convex if $m < 1$, and concave if $m > 1$; and that the base of the profile (for a fixed basal removal point) will become concave for all $m > 0$.

Data relating to values of m and n for *unconcentrated* soil wash are unfortunately scarce. General data for soil wash given in chapter 8 indicate that a concavity, rather than a convexity, results from soil erosion by overland flow. Similar data for rainsplash, however, indicate that under this process a convexity should develop. Inasmuch as rainsplash is an important 'upslope wash process', Fenneman, therefore, may be credited with offering a potential mechanism for convex summits, notwithstanding certain flaws in his actual reasoning. Whether or not rainsplash actually does contribute, significantly, to the development of rounded summits in humid temperate areas is a debatable point. The relative inefficacy of rainsplash on vegetated surfaces leads us to believe that, in humid areas, the contribution of rainsplash erosion is subordinate to that of soil creep.

So far we have assumed that the upslope convexity is essentially a superficial form and independent of the geologic nature of the underlying rock. In certain areas, however, cambering of the underlying rock strata will also impose a shallow convexity on the ground surface. This mechanism is illustrated in fig. 11.3 which depicts the typical situation in Jurassic areas of central England (Hollingworth, Taylor and Kelloway, 1944) where beds of fairly resistant sandstones and limestones overlie much weaker clays and shales. The tremendous stress on the clays due to the overburden results in a plastic deformation of these weak beds and a 'squeezing out' of the clay into the lower part of the valley. The result is to bend the contact between the two strata downwards near the valley slopes and thus produce a convex bedding plane which is reflected in the ground surface.

Slopes in humid temperate areas which do *not* show summit convexities are rare and only occur under special conditions. We have already noted that where the retreat of a straight slope is slow, it will be eventually obliterated by the downward encroachment upon it of the upper convex slope. Occasionally the reverse of this happens. In areas

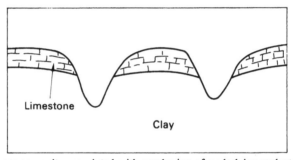

Fig. 11.3. Upper convexity associated with cambering of underlying rock strata (based on Sparks, 1960).

where the rate of retreat of the straight slope is very rapid there may be very little chance for a convexity to be maintained. This is one reason why rounded hilltops are poorly developed in semi-arid areas; the retreat of the free face and talus slope is so fast, due to the intensity of soil wash and related processes, that there is insufficient time for the convexity to develop. The same thing can also happen, although rarely, in humid areas. An instance is illustrated in fig. 11.4. This depicts the north wall of the Valley of the Rocks, near Lynton, North Devon. The valley runs parallel to the coast and the outside slope of the north wall is rapidly being undercut by marine action. As a result, retreat of the cliff face and the long constant slope takes place at a rate which prevents the development of a summit convexity.

The basal concavity

Although all slopes in humid temperate areas do not possess basal concavities, this feature is an extremely common component of slope profiles, and we must take exception to Strahler's (1950) statement that concave-up slopes are merely illusions from viewing profiles which are not truly orthogonal to the contour.

Application of empirical data for soil wash, from chapter 8, to the continuity equation ($m > 1$, $n < 2$ in equation 5.6) provides a simple explanation for concave profiles in humid temperate areas. However, implicit in equation 5.6 is the assumption that the prevailing slope processes are entirely transport-limited; this is a questionable assumption in many cases, especially where overland flow is weak and the shear strength of the surface soil is large. It is therefore necessary also to consider the weathering-limited case.

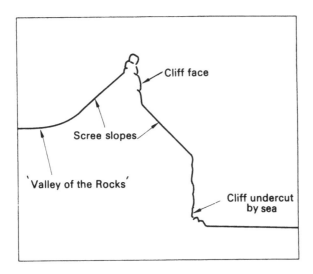

Fig. 11.4. The north wall of the Valley of the Rocks, near Lynton, North Devon.

In discussing unconcentrated surface wash, in chapter 8, it was pointed out that the erodibility of a hillside surface is strongly influenced by the condition of the surface skin of material. After a long period of drought, for instance, the top few millimetres of soil are usually extremely friable, easily broken up by rainsplash and rapidly washed away. The newly exposed soil underneath this skin, usually more moist, is commonly much more resistant to erosion. Accordingly, after the easily eroded material is quickly washed away, a rapid dilution in sediment concentration (fig. 11.5) often takes place. It is not until the next inter-storm period, after drying out of the surface skin, that the ground surface again becomes readily erodible under thin overland flows. Extrapolation of this process through geologic time (assuming concomi-

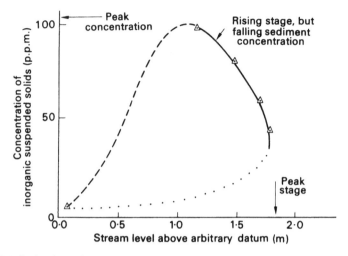

Fig. 11.5. Dilution in sediment load of a small stream during flood conditions (Data for Turlow Brook, Derbyshire, 31 December, 1965).

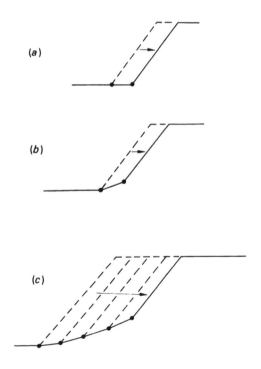

Fig. 11.6. Theoretical development of a basal concavity by retreat of initial straight slope.

312

tant lowering of the rock-soil interface) would produce parallel retreat of the hillside slope.

Under conditions in which *weathering-limited* soil wash produces retreat of the hillside, a new slope element must emerge at the base of the retreating profile. This new small element is unlikely to be horizontal (fig. 11.6*a*) because material being transported off the slope above would accumulate there. The new slope element must, therefore, be at least steep enough to enable transport of material over it from the retreating main slope. Moreover, it is unlikely that the angle of this new element could be much greater than this threshold gradient because, if it were, erosion by surface water flowing from the straight slope would probably strip away any excess material. In this way we may conceive an equilibrium slope inclined at that angle which just permits transportation of material from upslope, but not steep enough to initiate erosion on it. The sketch in fig. 11.6*b* represents this new element after an infinitesimally small amount of retreat of the straight slope. As the straight slope continues to retreat it will continue to add new slope elements at its base which together form a basal slope. In the case where each element possesses the same angle of slope, this basal slope would be straight. This is, however, a rather unlikely situation. The angle of the new slope element depends upon, *inter alia*, the calibre of the surface soil and the discharge of the surface water which, at times of heavy rains, flows over it. As the calibre of the soil particles increases or the discharge of the surface flow decreases, the angle of slope needed to transport a given quantity of sediment from the straight slope above must increase. Now, the amount of water flowing over the slope must increase continuously with distance from the divide, or, alternatively, the discharge of the surface water flow gradually decreases upslope from the initial position of the base of the straight slope. This means that, as the straight slope retreats further from its initial position, the angle of the new slope elements must gradually increase; in this way a concave basal slope (fig. 11.6*c*) is formed.

Clearly this model, which is very similar to Davis' ideas on the development of the graded stream long profile, hinges on a number of very dubious assumptions. One of these is the idea that surface soil wash is a sufficiently important process in humid temperate areas that it is capable of producing enough retreat of the straight slope so that a large basal component is produced. Another is the belief that, even if soil wash occurs at a reasonable rate, it produces *retreat* of the straight slope and not simply a decline in slope angle hinged about the slope base. And, lastly, even accepting the idea that soil wash will produce sufficient retreat of the straight slope to produce a basal component, it has not yet been demonstrated that the decrease in discharge between the lower and upper limits of the basal slope is large enough to produce a noticeable increase in the gradient of this slope.

With regard to the first of these assumptions, it is true that surface soil wash operates at a minimal rate in humid temperate areas (chapter 8) relative to soil creep. Nevertheless, soil wash does occur in most storms and, given a long enough period of time, it will produce removal of material from the slope. The assumption that surface and sub-surface wash are capable of producing slope recession, whether parallel or central recession, is difficult to assess in the light of presently available evidence. Schumm (1965 *a* and *b*) has shown that on straight badland slopes in New Jersey and South Dakota, soil wash appears to produce parallel retreat, but whether or not these results are valid for grass-covered slopes is uncertain.

Although, in both *weathering-* and *transport-limiting* cases, theory indicates that soil wash should produce concave slope profiles, whether the actual basal concavities of humid temperate areas are due to contemporary soil wash is still uncertain in view of the extremely slow rates of soil wash found in these areas today. If the ideas of chapter 9 are valid (figs. 9.13 to 9.15), solution in multi-component rocks in humid temperate areas offers an equally plausible mechanism for basal slope development, and, in view of the high rates of solute-removal in these areas, this process may be expected to be at least as important as mechanical soil wash in producing basal slope elements. Measurements by Young (1963*a*), on soil mantled slopes at 25–30° in the Pennines, suggested that a *combination* of surface wash, sub-surface wash and chemical removal of solutes results, under present conditions, in a rate of slope retreat horizontally in excess of 5 m per 10,000 years. This is sufficiently rapid to produce a reasonable size basal concavity even in the time that has elapsed since the last glacial period. Erosion rates during late-glacial times were possibly much higher than today.

Irrespective of the validity of the ideas just discussed, it is very probable that other mechanisms exist which produce concavities at the base of slopes in humid areas. In areas of extremely high drainage density, such as shales and clays, a mechanism not unlike the lateral planation theory of pediments (Paige, 1921; Blackwelder, 1931; Johnson, 1932) may produce a concavity at the base of valley side slopes. Surface streams occupying very broad, shallow depressions, in the order of 1–2 m deep, are common on the low-angle side slopes of clay areas. Often these small seepage depressions are dry and indistinct since there is no channel form; in periods of wet weather, however, surface flow is seen to concentrate in them. In the upslope parts of the side slope they are usually separated by broad interfluves so that the profile of the small stream is independent of the main slope profile. On the lower part of the slope, there is a tendency for many of these shallow depressions to coalesce, so that the profile of the hillslope now becomes determined by the long profile of these micro-valleys. In this way the shallow concave profile of the seepage depression is enforced on the lower part of the hillslope. This idea was also suggested by Davis (1930) and is shown in fig. 15.2*b*, taken from Davis' paper. Smooth concave lower slopes are characteristic of many areas of high permeability also, provided that valleys have become dry. In these cases it is probable that slope processes, particularly soil creep, have gradually resulted in deposition of material in the small valley bottoms, producing a superficial concavity that is not reflected in the profile of the underlying bedrock.

Not all slopes possess lower concave slope profiles, but those slopes which are entirely convex or convexo-straight usually occur in valleys where downcutting is still active or has only recently stopped. Observations in the Verdugo Hills and in other areas studied by Strahler (1950) suggest that this is probably the explanation of the straight basal slopes in these areas and the basis of Strahler's previously noted comments.

THE PATTERN OF SLOPE DEVELOPMENT THROUGH TIME

The notion that the landscape evolves through time in a systematic manner was the magnificent contribution by William Morris Davis to the study of geomorphology. It was noted earlier (chapter 1) that strong criticism has recently been levelled against this idea by Chorley (1962) who has emphasized that not *all* elements of the landscape change in form during the cycle of erosion. In an extremely imaginative paper, Hack (1960) also challenged the Davisian idea, aiming his comments particularly at the form of hillside slopes. He asserted that the form of many hillsides is unchanged through time, even though erosion is taking place, and that such slopes should be regarded as existing in a time-independent state of dynamic equilibrium. Subsequently Schumm and Lichty (1965) attempted to resolve this conflict of opinion by suggesting that slopes might exist in a state of dynamic equilibrium over a very short time scale, but, nevertheless, during the span of geologic time, undergo a progressive change in shape.

In part one we emphasized that, with reference to slope profiles, this book accepts the basic evolutionary system of W. M. Davis. Stream incision produces slopes which are continually undercut in the early stages of the development of a landscape and may, temporarily, exist in a steady state; eventually, however, rapid undercutting ceases, denudation of the hillside slope becomes independent of stream activity, and the slope undergoes a progressive change in geometry as the landscape is lowered.

The major difficulty in testing the idea that slopes do evolve over time is that it is almost impossible to actually observe the effect of time, except for very short time spans, on the profile of hillsides. Hutchinson (1965), with the aid of historical maps and current field survey, has examined the development of coastal slopes in London Clay over a period of centuries, but it is difficult to expand this time span to the dimensions of geologic time. It is therefore tempting to circumvent this difficulty by allocating present-day slope profiles to some particular historical sequence and, in essence, substitute space for time. Studies by Savigear (1952), Palmer (1956), and Carter and Chorley (1961) adopted this procedure.

The study by Savigear is especially interesting. It was undertaken along the coast of South Wales in an area where 120–150 m high cliffs of Old Red Sandstone have been undercut by the sea. In certain places,

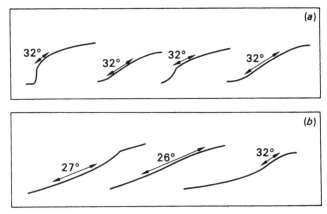

Fig. 11.7. Slope profiles in Old Red Sandstone, South Wales: (*a*) undercut by sea; (*b*) protected from marine undercutting by spit (based on Savigear, 1954).

spits have developed protecting the base of the slopes from further marine activity. This distinction between slopes exposed to marine erosion and those protected from it by the spit is in some ways similar to the change from stream downcutting to the development of a flood-plain between stream and slope. The important point in Savigear's study is that the spit examined, developed progressively from west to east so that slopes at the eastern end of the spit have only recently been protected from undercutting, whereas those at the western end have been sheltered for a much longer period of time. There is thus, theoretically, a time sequence from east to west behind the spit. The detailed changes in form over time do not conform to any clear-cut pattern of development, although a comparison of the most recently sheltered slopes with the oldest slopes in the sequence (fig. 11.7) does suggest that,

316

over time, a fairly steep straight slope is replaced by another at a lower angle. Savigear concluded that the protection of the slope base from marine undercutting produces a gradual flattening of the slope profile.

Similar conclusions were reached by Strahler (1950) in a study of hillslopes in the Verdugo Hills, California. Strahler compared slopes which were still actively undercut by streams with those where a floodplain had developed offering protection to the slope base. The maximum angle of slope was determined on a large number of slope profiles in both types of situation, and the frequency distributions of the two sets of

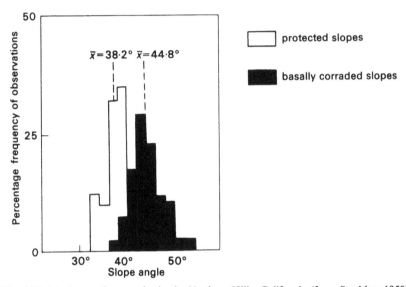

Fig. 11.8. Maximum slope angles in the Verdugo Hills, California (from Strahler, 1950).

angles are shown in fig. 11.8. The difference is statistically significant and geomorphically impressive. Strahler concluded that 'slopes which have been protected from recent basal cutting have significantly lower angles. The statistical analysis does not explain this difference, but in view of the fact that conditions of climate, vegetation, soils, bedrock and tectonic history seem to be essentially the same for both samples, it is concluded that the one group of slopes declined in the angle during the period when only sheet-runoff, creep and other mass-wasting processes operated on the slope' (Strahler, 1950, p. 813).

The results of the studies of Savigear and Strahler thus tend to support the Davisian idea that slopes flatten over geologic time. As Strahler puts it: 'some suspicion, at least, is thus cast upon the Penck assumption of parallel retreat, while some support is given Davis' scheme of declining slopes'.

317

It is possible that the results of these two pioneer pieces of research have been very misleading. In the first place, very little attention was directed to the processes operating on the hillslopes. Secondly, the notion of flattening is based solely on the observation that slope angles in areas of undercutting are greater than angles on slopes which are protected from basal corrasion; as shown in fig. 11.9, this difference could be due to slope retreat just as easily as slope flattening. And, thirdly, it is still very difficult to justify expanding the short time scales relevant to these studies to the much longer spans of geologic time. Neither study offers any support to the idea that slope flattening is a *continuing* process.

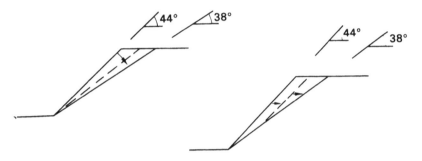

Fig. 11.9. Alternative modes of change from steeper threshold slope to gentler threshold slope.

In the Verdugo Hills, for instance, the slopes are mantled by fairly thin veneers of coarse, cohesionless waste. Now the angle of repose, or, more accurately, the angle of internal friction, of this type of material depends very much on the density of packing of the material (fig. 4.7) as emphasized in chapter 4. Tests by Skempton and Bishop (1950) long ago demonstrated that coarse, angular, sandy material could stand at 43–44° in a dense state of packing, but only at angles of 33–37° in a loose state. Inspection of slopes in the Verdugo Hills area emphasizes this difference between the two sets of slopes. Slopes at 38° are mantled by debris in a very loose state of packing, standing, in fact, at the angle of repose of the material in its loosest state. Steeper slopes at 43–44° are mantled by similar material, but in a much denser state of packing, presumably because, with continual undercutting at the base of the slope, most of the mantle is essentially decomposed rock not yet affected by the loosening processes of weathering. It is possible then that the change from 44° to 38° is simply the replacement of one stability angle by another, and, most important, there is no continual process operating which would result in indefinite flattening of the slope profile.

Similarly, it is possible that the decline in slope angle noticed in

Savigear's study is simply the replacement of one threshold slope by another. The Old Red Sandstone of South Wales is very similar to the gritstone rock of Exmoor and the southern Pennines where it has been shown (Carson and Petley, 1970) that most of the steep talus slopes at about 33° correspond to the angle of repose of this material, and that gentler slopes at about 26° are probably newer stability slopes related to a more advanced stage of weathering of the mantle. It seems possible, therefore, that the studies of Savigear and Strahler, rather than supporting the Davisian ideas of slope flattening, relate more to the general model of straight slope development under instability processes outlined in the last section of chapter 7.

The opportunity to substitute space for time in studying the evolution of slope forms, as in the article by Savigear, is rare, and, notwithstanding the potential contribution of this type of work, we believe that the most fruitful approach to the study of hillslope development is through the study of slope processes and the construction of process–response models based on these processes. It is continually frustrating to find that so few of the classic studies of slope forms really appraised, in any critical way, the processes supposedly responsible for them.

The development of models based on particular processes of denudation has been dealt with at length in part two of this book. Although no purpose would be served in repeating the contents here, it is, perhaps, pertinent to emphasize the main points arising from those chapters as they relate to the development of slope profiles in humid temperate areas.

(1) The incision of streams into the surface of a landmass will, usually, tend to produce straight hillside slopes inclined at a particular angle of stability which depends upon the type of material in the mantle. In strong rock, vertical cliff faces may be formed if stream incision is rapid; but, if downcutting is slow, weathering of the rock surface may result in a talus slope, maintained at the angle of repose of the talus. This second situation corresponds with the $c = 0$ case of equation 6.16. In areas of unconsolidated material, slopes will be gentler but, particular angles of straight slope will still prevail during the process of stream downcutting. If stream incision is very slow, relative to the weathering of the rock mass, soil creep may be able to keep pace with downcutting, and an equilibrium convex profile (fig. 10.12) will result. This may be the mechanism responsible for convex slopes in the uppermost headwaters of most drainage systems.

(2) Once downcutting by streams ceases to be rapid, slopes begin to develop independently of stream action. Three groups of processes now dominate on the slope and the resulting pattern of slope development depends very much on which group of processes assumes most impor-

tance. These three groups of processes are: weathering and mass failure; surface, sub-surface soil wash and solution; and soil creep.

(3) On slopes where weathering is relatively slow, there will be little change in the mantle character for some time, and the various transport-limited processes will act, in ways described previously, to obliterate the straight slope by downward extension of the upslope convexity and, to a smaller extent, by the upslope extension of the basal concavity. This is probably the pattern of development of many slopes in areas of weak sandstones. Sandstones weather relatively rapidly to produce a sandy, cohesionless soil mantle, and, in many areas where streams are cutting down into such sandstone masses, e.g. the Bunter Sandstone areas of central England, straight slopes at the angle of stability of the sandy mantle may form. Once downcutting ends, weathering may continue to thicken the surface soil mantle but probably does little to change the character of the mantle, because sandy quartz particles are fairly resistant to further alteration. The mantle remains stable at this angle of slope and the dominantly straight slope is then converted to a convexo-concave profile by the transport-limited processes. A similar, fairly simple pattern, probably takes place in most clay areas, although, as noted in chapter 7, p. 178, a second straight slope may develop after downcutting has slowed, prior to conversion of the straight slope to a sigmoid profile.

(4) It is perhaps unusual, however, that the first straight slope produced by stream incision into a landmass will remain stable throughout the subsequent history of the slope. The mantle on the hillside slope will undergo progressive changes due to weathering, which, through altering the cohesion, internal friction or permeability, and thus pore pressures, in the mantle, will inevitably affect the stability of the slope. The number of phases of instability, and thus the number of threshold slopes which a hillside will experience, will depend on the history of disintegration of the waste mantle. One sequence seems to be especially common. It is restricted to strong, well-jointed rocks, but this description includes a wide variety of rocks. The first stable, debris-mantled slope developed in such areas is usually a talus (scree) slope at the angle of repose of the debris and usually in the range 32–38°. Subsequent weathering alters the frictional and permeability properties of the mantle so that it becomes unstable. Such mantles are commonly only stable at angles near to 25–28°. Ultimately a completely weathered soil mantle will be produced and this, in turn, will only be stable at a particular slope angle, depending on the type of soil. Only at this stage, assuming that no further marked changes occur in the soil mechanics properties of the mantle, can the slope be considered completely stable. The numerous reports of slopes at about 32–38° and at 25–28°, listed in chapter 7,

indicate that this sequence is probably very common in humid temperate areas.

(5) Once the soil mantle becomes so weathered that little further change takes place and the slope adjusts to the angle of stability of this type of soil, the processes of soil creep, soil wash and solution begin to emerge as important. The effect of these processes is to develop a convex profile at the summit of the slope and a concavity at the slope base. Gradually these new components encroach on the stable straight slope intermediate between them, and eventually convert the profile to a convexo-concave, or sigmoid, form, although, in upland areas where straight slopes are very long, this may take a very long time.

(6) Upon attainment of a sigmoid profile, little further change in form is likely. The slope is continually lowered but the basic convexo-concave form is maintained. There is probably no threshold slope angle for any of these transport-limited processes, although as the slope becomes very subdued the efficacy of the processes becomes very small and the rate of development of the slope profile is very slow. In this way, a landscape resembling the Davisian peneplain is produced, marking the ultimate landform of humid temperate areas.

THE MODIFICATION OF HUMID TEMPERATE SLOPES UNDER
PALEOCLIMATIC CONDITIONS

On a geologic time scale, the climate of an area is continually changing and, in the recent past, geologically speaking, much of the area which today experiences a humid temperate climate experienced radically different conditions. During the course of the multiple glaciations of the Pleistocene period, for instance, many lowland areas were modified by advancing ice sheets so much that the landscape geometry is inexplicable in terms of processes described here, and can only be understood with reference to the mechanics of the movement of ice masses. Similarly in highland areas the advance of valley glaciers sometimes produced so much modification of valley cross-sections that it is more appropriate to think of valley-sides as components of the boundary of a glacial channel than in terms of hillside slope profiles. Unfortunately it is not always easy to separate the effect of glacial scour from normal hillslope processes on the moulding of slope profiles in upland areas subject to Pleistocene glaciation.

It is even more difficult to disentangle past and present processes operating on hillslopes in areas which, during Pleistocene times, were immediately beyond the ice sheet limits. The development of slopes under periglacial conditions is dealt with in chapter 12, and, in the remainder of this current chapter, we shall only briefly comment on the effect of periglacial processes on slope profiles in areas currently experiencing humid temperate climates. The suspicion that many slopes in temperate latitudes owe much of their form to processes operating during periglacial times stems from two main points.

The first of these is that evidence suggests that certain slopes in temperate latitudes have undergone virtually no alteration of form since periglacial times. Waters and Johnson (1958) identified deposits on limestone slopes in the Pennines which, on the basis of mollusc shells contained within them, clearly date back to a long cold spell. Williams (1964) has described stone stripe patterns on hillslopes in the Breckland of eastern England which also presumably date back to periglacial times; and Hutchinson (1967) interpreted slope mantles on London Clay slopes as deposits produced by solifluction. Similar reports have been made elsewhere in Europe (Judson, 1949; Dylikowa, 1964; Starkel, 1967), North America (Smith, 1949*a* and *b*; Denny, 1956) and New Zealand (Cotton and Te Punga, 1955*a* and *b*) among many. The important aspect is that, assuming deposits which date from periglacial times do exist high upslope on many valley-sides, then little denudation of these particular hillslopes can have taken place since then.

This in itself is no basis for assuming that the *form* of such slopes was moulded by periglacial processes during the last cold phase. It is quite possible that the basic geometry of these slopes was determined under humid temperate, or other, conditions prior to the Pleistocene glaciation. The second point which must be considered, therefore, is the claim by a number of workers that under periglacial conditions associated with the last glacial period, a marked acceleration of slope processes took place, profoundly altering the form of slopes in profile, and resulting in accumulations of debris, variously termed as *head, combe rock*, and other names, at the base of the slopes. Among many, this claim was made by Palmer (1956) in northern England, and, especially, by Waters (1964) working on Dartmoor. Now, it is true that slope processes appear to operate more rapidly in periglacial conditions than in humid temperate environments, but it is still very difficult to substantiate the idea of massive modification of slope profiles during the last cold phase. A number of very important points are often overlooked.

In the first place it is not always easy to identify particular deposits at the base of a slope as 'periglacial'; the criteria used in this task are often very vague and it is possible that many such slope deposits predate periglacial times. Secondly, observations by a number of workers noted in chapter 7 (Skempton, 1964; Carson, 1967*b*) indicate that many hillside slopes stand at angles which are apparently entirely explicable in terms of the type of material mantling the slope. This does not support the idea that active solifluction during the last cold phase produced large-scale denudation of slopes and a flattening of profiles. And, thirdly, even allowing widespread solifluction, it is difficult to understand why any appreciable modification of the *form* of slopes in profile should take place. Most slopes are cut in strong rock and mantled by a residual soil layer. In order to modify the profile in any substantial way, a great deal of weathering of the underlying rock mass would be necessary to supplement downslope sludging of the waste mantle. A limited amount of downslope movement of the waste mantle through solifluction would be quite sufficient to produce bare rock hillslope summits and basal accumulations of debris (fig. 11.10) without any marked alteration of slope profile geometry.

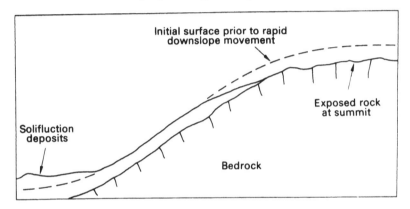

Fig. 11.10. Possible effects of Pleistocene mass-movements on a typical slope profile. (*N.B.* Thickness of waste mantle is highly exaggerated for clarity.)

Clearly nothing definitive can be said on the role of periglacial processes in moulding the profiles of slopes in areas now belonging to a humid temperate climatic pattern. Any advance in understanding the part played by past periglacial processes in these areas will only be made by more detailed research on the effects of periglacial processes currently operating in the high-altitude and high-latitude parts of the earth's surface today. Some of this work is described in the ensuing chapter. We would like to emphasize, however, that, with the evidence currently available, we believe that the effect of periglacial processes on slope profile geometry, as distinct from the character of the debris mantle, was not as great as others believe.

CHAPTER 12

THE PERIGLACIAL LANDSCAPE

The problem of disentangling the effects of past and present processes on the development of hillside slopes is not peculiar to humid temperate areas, and must be recognized in each of the four climatic environments discussed in part three of this book. In the previous chapter we emphasized that much more information about denudation in present-day cold areas than is currently available is needed before any valid appraisal of the contribution of past periglacial processes in moulding slopeforms in temperate areas can be made. It is particularly unfortunate then that much of the present-day periglacial area was covered by continental ice sheets in Pleistocene times, so that, ironically, there exists in these areas the similar problem of separating the effects of past glacial and contemporary periglacial processes on the form of slope profiles.

In much of the existing periglacial area it is, in fact, pointless attempting to explain the major geometry of the landscape in terms of present-day processes. Much of the area has been left bare by glacial removal of the former waste mantle and much of the relief of these areas probably results from differential scour by moving ice sheets. In upland areas too, a large part of the landscape geometry has probably been inherited from the Pleistocene period. It is especially difficult, for instance, to assess the relative importance of glacial *vis-à-vis* periglacial processes in the production of U-shaped troughs, so typical of high-latitude upland areas. It is imperative that more research is directed at those few cold areas which escaped major glacial modification during the Pleistocene, so that an unambiguous effort to relate slope development to periglacial processes may be made.

At this stage it is perhaps useful to define the *periglacial area*, as it is used here. The term was originally introduced by Lozinski (1912) to refer to areas lying near to the margins of Pleistocene ice, but, today, the term is used to convey a wide variety of meanings. We shall follow Embleton and King (1968, p. 448) and use the term to describe the zone peripheral to glacial ice, whether this is Pleistocene or present-day. As it stands, the term is very vague, but it does serve a useful purpose. A large part of these high-latitude and high-altitude areas, but not the entire zone, is underlain by permafrost. The largest areas of the peri-

324

glacial zone are, of course, in the Northern Hemisphere and correspond, *to some extent*, with the tundra areas of Asia and North America.
Smaller alpine areas much farther south should also be considered part
of the present-day periglacial landscape. It is important to emphasize
that, within the zone, climate, particularly moisture conditions, varies
widely. Vegetation is also very variable and may range from a complete
mat of mosses and lichens through to bare rock desert; this marked
contrast in conditions must exert an appreciable influence on the rate
and type of denudational processes operative. Notwithstanding this,
much of the landscape owes its distinctive character primarily to two
features, a sparse vegetation mantle and extensive freeze–thaw activity;
in projecting a 'type periglacial area' for this chapter, these are the two
main points which we shall emphasize.

There appears to be, *prima facie*, no shortage of research on slope
development in cold areas. The articles by Antevs (1932), Deevey (1949)
and Bird (1959) all provide long lists of references, and Hopkins and
Wahrhaftig (1959) have produced an extensive annotated bibliography
of English-language papers intended to relate specifically to the evolution of slopes under periglacial conditions. A cursory glance at these
lists of articles unfortunately reveals that very few are directly related to
the problem of slope development under cold conditions. The majority
are concerned with either the identification of fossil periglacial features
in temperate areas or the origin of special micro-relief features peculiar to
cold areas. The more recent texts by Bird (1967) and Embleton and
King (1968) include useful summaries of previous work, although they
do not treat the problem of slope development directly. Indeed, with the
exception of a brief essay in Birot's (1960) *Le Cycle d'Erosion Sous Les
Differents Climats*, there is little published material relating to the
course of slope development in cold areas. In this chapter, we shall be
concerned only with the major features of landscape development in
periglacial areas, and we shall ignore the distinctive embroidery of
patterned ground and related features peculiar to cold areas.

One of the first real attempts to put the denudation of cold areas on
a systematic basis was made by Peltier (1950) through the development
of a periglacial cycle of erosion. The major changes involved in this
sequence are depicted in fig. 12.1. The most important process throughout
the cycle is the gradual downslope mass-movement of mixed rock rubble
and soil which, in this book is termed solifluction, but which was labelled
by Peltier, *congeliturbation*, following Bryan (1946). In the initial
stages of landscape development, the downslope movement of congeliturbate results in exposure of bedrock on the upper slopes. This
occurs because under these conditions weathering is relatively slow and
movement of congeliturbate away from the upper slopes occurs more

rapidly than conversion of the rock to a new congeliturbate mantle. On the more gentle upland surfaces, frost shattered rock (congelifractate) accumulates in waste sheets often referred to as felsenmeer. On the steeper upper slopes the same process of frost shattering produces rockfall from the exposed rock walls. As a result, the rock walls retreat into the mountain mass and, at the base of the cliffs, talus accumulates, slowly merging into the smoothly concave congeliturbate-mantled lower slopes. The continual retreat of the frost-riven cliffs eventually

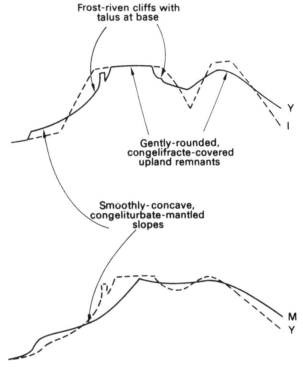

Frost-riven cliffs with talus at base

Gently-rounded, congelifracte-covered upland remnants

Smoothly-concave, congeliturbate-mantled slopes

Fig. 12.1. Slope development in the periglacial cycle of erosion according to Peltier (based on Peltier, 1950; reproduced by permission from the *Annals* of the Association of American Geographers, 40, 1950).
I : initial surface; *Y* : youth; *M* : maturity.

results in their disappearance from the landscape, and the stage of maturity is first reached at the time when all bedrock exposures have been eliminated from the slopes and congeliturbate-mantled slopes dominate. After this stage, the continued operation of solifluction results in further flattening of the slopes without marked change in geometry from the stage of maturity shown in fig. 12.1.

The Peltier scheme suffers from a number of drawbacks. The most serious is that, like the Davisian sequence, it is too vague. There is no precise statement of the changes in landscape geometry which take place during the course of the cycle, nor is there detailed treatment of the processes responsible for these changes. This is perhaps not too surprising for the state of our knowledge about landscape development in cold areas is little better than our understanding of it in humid temperate areas at the time Davis produced his cycle. A synthesis such as a geographic cycle should only be contemplated after a great deal of analysis of landscape form and process. In addition there is perhaps a special problem in cold areas; it is doubtful, in view of the continual fluctuations of the climatic belts, if the full course of slope development has ever taken place. The periglacial cycle of erosion is therefore probably doomed to be a model which can never fully be tested.

A large part of the early fieldwork on slopes in cold areas was concerned with the asymmetric character of many valley-side cross-sections. Shostakovitch (1927), for instance, noted that in areas of

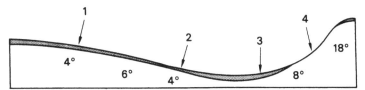

Fig. 12.2. Asymmetric slope profiles in the Chiltern Hills (based on Ollier and Thomasson, 1957).
1 : flinty clay mantle (0·7 m thick) overlying chalk; 2 : veneer of flinty loam over 0·3–0·7 m flinty clay above 0·3 m of chalky head overlying chalk; 3 : about 0·7 m of chalky head over chalk; 4 : bare chalk.

Siberia where the ground was permanently frozen at depth, south-facing slopes were usually steeper than north-facing slopes. Several workers (Smith, 1949b; Ollier and Thomasson, 1957) have suggested since that similar asymmetry in areas beyond the permafrost boundary may reflect a former extension of a cold climate. The dangers in making this type of interpretation are, however, very great, and the general difficulty in attributing present day forms to past processes is clearly illustrated in Ollier and Thomasson's work. This study was undertaken on the chalk slopes of the Chiltern Hills where slope asymmetry is very well developed, occurring on west–east sections (fig. 12.2) as well as north–south ones. The authors found it very difficult to pinpoint the probable processes responsible for it, and, in addition, even assuming that the appropriate processes were known, admitted that they were hesitant to argue that solifluction processes would result in any one of slope retreat, steepening or decline. Clearly, a great deal more work is needed in areas where present day slope geometry is the product of processes observable at the present time.

Such a study was undertaken by Currey (1964) in the Ogoturuk Creek area of north-west Alaska inside the Arctic Circle. It is interesting that his observations contradict the pattern considered normal for cold areas by previous workers. The Ogoturuk Creek basin is underlain by various mudstones, siltstones and sandstones of Jurassic and Cretaceous age; the structural pattern is very complicated, but no evidence was found which might indicate structural control of valley asymmetry. Vegetation is sparse. Trees are absent, and although a tundra vegetation mat exists on the lower parts of south-facing slopes, most of the area is bare. The ground is permanently frozen at depth, although seasonal thaw of the ground surface, varying between a few inches and a few feet, occurs from April to November. A summary of Currey's observations, based on an examination of 232 sections using 1 : 12,000 aerial photographs and 1 : 63,000 topographic maps, supported by field survey, indicates (table 12.1) a distinct tendency for north-facing slopes

TABLE 12.1. *Asymmetric character of 76 valley cross-sections in the Ogoturuk Creek area* (from Currey, 1964)

Azimuth of stream flow	Left bank steeper	No asymmetry	Right bank steeper
1–180°[a]	17[c]	32	51
181–360°[b]	56	36	8

[a] right bank faces north
[b] left bank faces north
[c] percentage figure

to be steeper than south-facing ones. Sections showing no asymmetry are usually those transverse to north–south aligned valleys. A typical valley cross-section, emphasizing the asymmetric development of the surface mantle, is shown in fig. 12.3.

The north-facing slope is steep, commonly above 30° over much of its length, mantled by a thin veneer of scree-like debris derived from the underlying bedrock, and often unstable. The upper part of the south-facing slopes is similarly strewn with frost riven debris, but it occupies a smaller part of the total profile and rapidly grades into a shallow concavity mantled by radically different material. The surface of the lower slope is covered by a thin mat of tundra vegetation, and, underneath, occurs colluvium. This colluvium is described as subangular cobbles, not unlike the upper slope debris, but now set in a silt matrix. The colluvial mantle is permanently frozen except for parts just beneath the tundra mat which thaw out during the summer. The basal slope resembles a solifluction terrace described by many other workers in cold areas (e.g. Williams, 1957), and presumably results from sludging of material downslope.

Currey attributes the asymmetry to the contrasting thermal environments of the two slopes and the resulting difference in the morphologic activity of the associated mantles. Insolation is greater on the south-facing slopes, weathering is more intense there, and the mantle of coarse rubble derived from the underlying bedrock is more rapidly broken down. The production of an appreciable silt matrix results in the mantle becoming unstable and it moves downslope as a solifluction sheet. Vegetation develops on this more weathered material and further assists in moisture retention. On north-facing slopes, in contrast, the input of thermal energy is naturally much less, weathering proceeds much more slowly, and the slope remains mantled by coarse debris. The debris appears to stand close to its angle of repose. This contrast

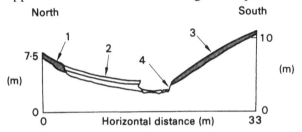

Fig. 12.3. Asymmetric slope profiles in the Ogoturuk Creek area, Alaska (based on Currey, 1964).
1 : frost-riven bedrock rubble at angle 25–30°; 2 : shallow concave slope flattening to about 5°, mantled by thin layer of tundra vegetation mat over 1–5 m of colluvial material; 3 : thin veneer of bedrock rubble at 30–35°; 4 : bedrock exposure.

in morphologic activity is only the initial cause of slope asymmetry. The colluvial mantle moving down the south-facing slope pushes against the stream channel, and apparently results in a lateral shift of the stream towards the toe of the north-facing slope. This undercutting of the north-facing slope thus assists in maintaining a steep slope and a mantle which is barely altered by weathering. 'Much as it is the ultimate cause of valley asymmetry in temperate regions, asymmetric lateral corrasion by streams that have been displaced through asymmetric deployment of slope materials, appears to be the process by which most asymmetric valleys in the Ogoturuk area have been ultimately shaped' (Currey, 1964, p. 96). These ideas seem to be supported by the work of Malaurie (1952) in Greenland, and by Hopkins and Taber (1962) in Alaska. The importance of Currey's work is, however, not simply in the development of a model of asymmetric valley development, but in its discussion of the various processes operating on slopes in cold areas. The impression gained from this study is that slopes not subject to rapid weathering and solifluction closely resemble straight, rubble-mantled slopes in temperate areas, and that the most distinctive shaping of slopes in cold areas is the development of a shallow concave basal slope which results through solifluction processes.

The extent of slope development, in terms of time, indicated in the Ogoturuk area is unfortunately distinctly limited. Although Currey argued that solifluction produced a low-angle concavity at the slope base, he was unable to comment on its effect on the main part of the slope. It is doubtful whether solifluction is capable of producing appreciable corrasion of the underlying bedrock, but continuous downslope removal of debris and associated weathering of the underlying rock must affect the main slope in some way. If we regard solifluction as simply an accelerated form of soil creep then, for reasons outlined in chapter 10, the most probable effect is a rounding of the top part of the slope, and an eventual extension of the convexity to meet with the concavity at the base. The development of slopes in cold areas would then be comparable to the sequence in humid areas, although occurring at a more rapid rate. This is, however, merely conjecture. The observation that the bedrock surface underneath the colluvial mantle is also concave might be taken to support the idea that retreat of the main slope has taken place, although it is more probable that it relates to the lateral migration of the stream away from the slope base.

A fairly lengthy discussion of Currey's work on slope asymmetry in the Ogoturuk area has been given above because it is one of the few attempts to construct a process–response model for periglacial slopes that is readily accessible. Notwithstanding Currey's contribution, however, a large part of the course of slope development in cold areas cannot be answered by observations in the Ogoturuk area. Extensive work on slope asymmetry is available in the Russian literature, but, unfortunately, there are few English translations. Recently Kennedy (1969) and French (1970) have both discussed the topic of slope asymmetry in cold climates, but, so far, the most useful contributions to the *general* problem of slope development in cold areas have been provided by Rapp (1960*a* and *b*) and Jahn (1960). The basic ideas of these workers, derived from detailed observations of present-day processes in, respectively, northern Sweden and Spitsbergen, are summarized below.

The work of Rapp was confined to a small valley, Kärkevagge, just south of the Narvik–Tornetrask axis, within the Arctic Circle of Sweden. The valley, a former glacial trough, is cut in predominantly mica-schist material, and the slope forms (fig. 12.4) are typical of northern Scandinavia. Mean monthly temperatures range from $-11\,^{\circ}\mathrm{C}$ to $+11\,^{\circ}\mathrm{C}$ and precipitation averages 800 mm per year. Snow, which accumulates especially on the east-facing slopes, melts in June in most years. The side slopes are usually unvegetated except the lower parts of west-facing slopes. Rapp's work in this area during the spring and summer months of 1952–60 ranks as the first real attempt to assess quantitatively the rates of different processes in a cold area.

The elements of the typical slope profile, the rockwall, talus slope, and lower taluvial slope, are illustrated in fig 12·4. The rockwalls appear to be retreating fairly rapidly. Scars on these walls, where rock-falls have taken place, appear to be fairly evenly distributed and show no the preferential location on the upper face which might be expected if

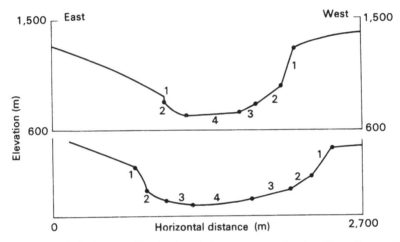

Fig. 12.4. Typical slope profiles in the Kärkevagge area, Sweden (from Rapp, 1960, *Geografiska Annaler*, **42**, no. 2–3).
1 : rockwall in mica-schist; 2 : talus slope; 3 : vegetated lower slope; 4 : giant boulders in valley bottom.

walls were declining in angle. Rapp concludes therefore that the walls are undergoing essentially parallel retreat. No mechanism is offered to account for the angles of the rock walls and, presumably, they may well relate to former glacial erosion in the trough. The profile below the cliff face is predominantly concave with a change from dry talus to more colluvial material. Talus, from rockfall, might be expected to accumulate and eventually extend upslope masking the cliff face, as in humid temperate areas, but, in fact, snow avalanches appear to be very effective in pushing newly-accumulated talus downslope, and continually re-exposing the rockwall base. This appears to be particularly important on east-facing slopes. Westerly winds result in the accumulation of snow in the lee of east-facing walls, and spring avalanches tend to predominate on these slopes rather than on those on the opposite side of the valley. This may account, therefore, for the distinct asymmetry in the profiles shown in fig. 12.4.

The steep, dry talus slopes gradually merge into gentler, wetter, grass-covered taluvial mantles lower down the profile, especially on the east-facing slopes. It is worth remarking that, except for the presence of the free face, these slopes closely resemble the south-facing slopes in the Ogoturuk area noted previously. These lower slopes are the location of the major slides, earthflows and solifluction lobes. This instability is presumably associated with the increased content of soil material in the mantle, reduced permeability, wetter conditions and reduced shear strength of the mantle. The upper talus slopes stand at about 35°. The majority of the slides occur on elements of the profile between 30° and 25°, and downslope movement of material usually comes to a halt by the

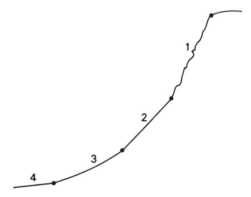

Fig. 12.5. Typical slope profile in Spitsbergen (based on Jahn, 1960).
1 : bare rock wall; 2 : talus slope; 3 : solifluction terraces; 4 : wash slope.

time the slope profile has been reduced to about 20°. Unfortunately Rapp made no conventional stability analyses of these slides, such as have been made by Williams (1957), but the description he provides closely resembles the sequence of slope development noted in the case of strong rocks in humid temperate areas. The transition between the dry talus slopes and the more weathered taluvial-mantled slope at about 25° is more gradual in these cold areas, but a definite similarity exists.

Solifluction lobes rarely extend as far as the morainic ridges in the centre of the valley floor. Sometimes between the downslope end of the solifluction slope and the valley floor proper, a very gentle mud-covered wash slope, patterned by braided furrows, exists, but Rapp believes that on the whole slope wash from the lower slopes is of very minor importance, and, presumably, regards the retreat of the lower slopes to be an extremely slow process.

In the context of slope development, the work of Jahn (1960) is

332

perhaps more directly useful than Rapp's since Jahn makes a specific attempt to use observations on slope processes to understand the course of slope evolution. The typical Spitsbergen slope is depicted in fig. 12.5 and comprises the following components: (1) rock wall, (2) talus slope, (3) solifluction terraces, and (4) wash slope. The weathering rock walls usually stand at angles greater than 40° and are retreating through the process of rockfall which supplies the debris for the talus slope. The talus slope and solifluction terraces sometimes merge gradually into each other through an intermediate moist taluvial slope. Sometimes this transition is so gradual that the upper talus slope and lower solifluction lobes combine to produce a steep concave profile between the rockwall base and the wash slope. Usually, however, it appears that the change from the talus zone to the solifluction zone is marked by a distinct break of slope. The talus cones and scree slopes stand at the angle of repose of the coarse material, which, depending on the nature of the debris, may be anywhere between 30° and 40°. The solifluction terraces may extend up to 25° on the upslope side and decrease to 10° on the downslope side, although some rock streams may extend much further with angles as low as 3°. The wash slope comprises very low-angled colluvial cones at the base of the solifluction slopes produced by soil wash from the steeper slopes above. These basal slopes are particularly common in those parts of Spitsbergen where the talus material weathers rapidly into sandy–silty particles; without the protection of a vegetation cover, these small particles are rapidly washed off the main slope by rain and meltwater.

Jahn's description of a typical slope profile agrees with Dylik's (1958) observations in the southern part of Spitsbergen. The same sequence of zones was also reported by Rozycki (1957), although he attempted to link each zone with a particular altitudinal climatic zone, an idea which does not appear to be supported by more recent evidence. The slope profile depicted in fig. 12.5 is also similar to those described by Rapp in Kärkevagge. The major difference between the Spitsbergen and Kärkevagge profiles is the development of a substantial wash slope at the base of the former slopes. Jahn disagrees with Rapp that soil wash from the main slope is negligible in cold areas, and his observations are supported by the earlier opinions of Mortensen (1930) and Dege (1941) working in the same area.

The development of a talus slope from rockfall acting on a steep, bare rock face is usually thought to correspond to the emergence of a straight slope at the angle of repose of the debris. This is, indeed, the assumption most commonly made in mathematical models of talus slope development (chapter 6), although, increasingly, field evidence (Rapp, 1960*a*; King, 1956; Young, 1956) indicates that many talus slopes are, in whole or part, concave and not straight. So far there has been no satis-factory explanation. It is worth noting, however, that the same concavity has been simulated in laboratory experiments of talus slope develop-ment. In unpublished experiments by Kirkby in which cobbles are dropped individually on an inclined cobble-mantled plane, the distri-bution of distances travelled downslope approximates to a negative exponential pattern. Actually the distance between the top of the talus slope and the place where the cobble comes to rest (s) is a direct function of the height of the fall (h) from the cliff face. Using the dimensionless number $m = s/h$, Kirkby's data indicate that the cumulative probability distribution of m values is approximately given by

$$p = e^{-nm} \qquad (12.1)$$

where n is an exponent varying with angle of slope and other variables. The important point is that, assuming this pattern is constant during the entire rockfall process, a profile will be built up in which the relative depths of material are identical to the values of the frequency distribu-tion and a concave talus slope results.

It is doubtful whether laboratory results using a model in which cobbles drop onto an existing slope mantled by similar cobbles bears any relation to conditions in arctic areas where much of the rockfall accumulates on a fairly deep snow cover. It seems much more likely that the existence of a soft snow mantle would reduce the amount of bounc-ing and rolling of boulders after the initial fall and thus radically alter the distribution of travel distances.

An alternative model to explain the concavity of arctic and alpine talus slopes has been offered by Caine (1969) based on observations on slopes cut into a Jurassic sandstone in the Southern Alps, New Zealand. It is based on two processes: the direct free fall of loose rock from a cliff face and accumulation at the rockwall base; and slush avalanching from gullies in the rockwall. The latter process is probably the more impor-tant of the two since it not only results in a redistribution of existing material on the snow from rockfall, but also introduces a much larger amount of debris onto the talus slope from the gullies. The pattern of debris accumulation on the snow cover, that is, the volume of debris as a function of distance downslope, was noted by field survey. Accumula-tion amounts derived from rockfall were found to vary *inversely* with distance from the head of the talus slope in an approximately linear manner, diminishing to nothing on the 25° segment at the talus slope base. Mean particle size showed no consistent variation with distance downslope. Subsequent slush avalanching produced the opposite pattern with debris amounts showing a *direct* linear relationship with distance downslope. On the basis of these observations, Caine developed a simple simulation model combining the two processes, and claims that a concavity is produced at the base of the talus slope; actually the assumed concavity is only weakly developed. Moreover, it is rather difficult to understand why a combination of two linear distribution processes, as against exponential patterns, should produce a concave profile at all. Nonetheless, the approach used by Caine is an extremely attractive one.

The nature of the developing talus slope clearly depends intimately on the pattern of distribution of debris downslope. This point was readily recognized by Rapp (1960) who paid a great deal of attention to the source of rockfalls on the cliff walls and the resultant effects on the patterns of debris accumulation. He emphasized that rockfalls seemed to belong to two distinct groups. Many originated high up on the rock wall and usually produced a rapid movement of debris all the way to the talus slope base, while those falls from smaller heights resulted in most accumulation near the top of the talus slope. Assuming that the height of the retreating rockwall is gradually reduced over time, most accumulation would be expected to occur near the rockwall base. This would continually raise the talus slope towards the angle of limiting stability for the debris, eventually resulting in large-scale mass movement and internal readjustment. Unfortunately little is known about this readjusting process, but it may be very important in understanding the origin of concave talus slope profiles.

Undoubtedly there are many other processes working on the talus slope that we have not considered. It was noted above that solifluction movement is rare on dry talus slopes since it depends on the presence of an appreciable matrix of fine debris among the talus. In fact, however, it appears that a flow type movement of coarse talus can take place provided that the interstices are filled with ice. Such lobes of moving ice-filled rock material, usually occupying valley floors rather than side slopes, are often referred to as rock glaciers (Wahrhaftig and Cox, 1959; Caine and Jennings, 1968). The mechanics of this movement are not clearly understood, but the development of small-scale rock glaciers on talus slopes may be one mechanism by which internal redistribution of debris on the slope is accomplished and the development of a concavity produced.

The slope profiles described by Currey, Rapp and Jahn are, in many ways, rather similar: they all, to varying extents, belong to a well-defined idealized sequence of rockwall, talus slope, taluvial-solifluction slope, and basal slope. In many cold areas, the typical slope profile is radically different from this. It comprises, throughout its length, a repetitive sequence of alternating rock outcrops and benches (fig. 12.6) which are described *via* many different terms. The flat benches between the rock outcrops are usually termed *altiplanation terraces*; the outcrops themselves are called *tors*.

The term altiplanation was first used by Eakin (1916) in Alaska, and subsequently applied by Jorré (1933) in the Urals, and Guilcher (1950), Te Punga (1956) and Waters (1963) in south-western England. The exact origin of altiplanation terraces is still debated; they appear to be independent of the structure and the denudation chronology of the

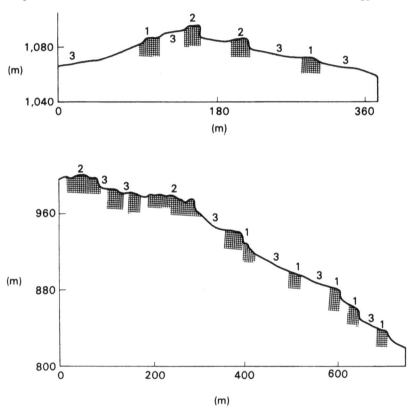

Fig. 12.6. Slope profile with tors and altiplanation terraces in the Hrubý Jesenik Mountains, Czechoslovakia (from Czudek, 1964).
1 : frost-riven cliffs; 2 : tors; 3 : altiplanation terraces.

areas in which they are found. Although they often occur with tors, they do constitute a distinctive landform in their own right. In the Urals, Jorré (1933) described them as follows: a succession of low-angled terraces (0–3°) usually several hundred metres wide alternating with scarp slopes (20–50°) which may take the form of a rockwall or a talus slope. The thinness of the soil mantle (about 3 m) indicated that these features are definitely rock-cut and not superficial phenomena. Guilcher

(1950) suggested that such terraces might have originated as nivation hollows; subsequent frost and meltwater action would produce retreat of the scarp, lowering of the tread and a general enlargement into a fairly wide terrace. A great deal of uncertainty still exists, however, and no satisfactory explanation of why altiplanation terraces are found in some places, and not in others, exists. Demek (1964) and Czudek (1964), working in Bohemia and Moravia respectively, both describe tors on altiplanated slopes and regard them as features of accelerated altiplanation.

Throughout this chapter we have not, so far, made any comment on the sequence of slope development, but merely described the typical profiles, as reported in the few studies that have been made in cold areas. The question of whether rockwalls retreat indefinitely until they meet a retreating rockwall on the opposite side of a mountain mass, or whether they disappear after a relatively small amount of retreat due to the upslope extension of the talus slope, is very difficult to answer with the meagre evidence available. Peltier (1950) believed that rock faces were quickly obliterated, but processes such as slush avalanching are clearly capable of continual removal of debris from the upper talus slope and may permit much further retreat of rock walls than in comparable situations in temperate areas. The incidence of rock faces in cold upland areas is certainly greater than in areas below the timber line, but it is very possible that this is simply a heritage of oversteepening during glacial times.

We should also remember that most of the studies of slope development in cold areas, Currey's (1964) excepted, have been biased towards high mountain terrain. Outside these areas, assuming the Ogoturuk Creek basin is representative of other areas of closely-jointed rock in high latitudes, straight, debris-mantled slopes are probably typical. Very little is known about the development of these slopes over time. According to the Peltier cycle, solifluction should result in a systematic replacement of the debris-mantled slope by the lower, concave, congeliturbate-mantled slope, but there is little evidence for this. Indeed, assuming that the effects of solifluction are comparable to those of soil creep, it would be more logical to speculate that the straight slope disappears through an extension of *both* an upper convexity and a lower concave slope, as in humid temperate areas. A great deal more work is obviously needed before any generalization can sensibly be attempted.

CHAPTER 13

SEMI-ARID AND ARID LANDSCAPES

A limited vegetation cover is the most striking characteristic of deserts, and critically influences their landforms. This chapter is mainly about the slopes of temperate semi-arid areas, for which the south-west United States is taken as the prototype. Extremely arid areas are considered as a special case of the semi-arid model in which water erosion has been reduced to a minimum. Comparisons are also made between temperate and tropical semi-arid areas: the latter are considered more fully in chapter 14 with the other tropical areas, and show interesting contrasts with temperate deserts as can be seen from comparing African and Australian desert landforms with those of Arizona.

Semi-arid landforms are very dramatic in appearance because the slope profiles tend to be more angular than in other areas. Summit plateaus are bordered by steep cliffs, which in turn stand above talus-like straight slopes at 25–35°. At the base of these debris slopes there is a narrow zone of very marked concavity in which slopes change from more than 20° to less than 5°. This zone is so narrow that it is commonly referred to as a break in slope which separates the debris slope above from the low-angle desert plains below, which are commonly very extensive in comparison with the mountain mass. Where the level of the plains has changed from time to time, a series of terrace-like features is formed and the characteristic angularity of the desert landscape preserves the form of the separate terraces as clearly as a flight of steps.

Deserts are defined above in terms of limited vegetation cover, but it is useful to express this definition in climatic terms as well, with the proviso that the climatic definition is considered to be secondary. Under natural conditions, semi-arid landforms tend to grade into more humid forms at mean annual rainfalls ranging from 200 mm at 0 °C, to 750 mm at 30 °C, because the higher temperatures require higher evapo-transpiration for the maintenance of a marginal cover. In southern Arizona, the classic area for the description of pediments in America, mean annual temperatures are 20–22 °C; and mean annual precipitations range from 80 mm (Yuma) to 300 mm (Tucson). Rainfall is seasonal with a peak of low intensity rains in December and January; and a peak of high intensity convective storms in July to September.

The critical role of a vegetation cover is shown most graphically by the forms produced in humid-temperate areas where the vegetation is artificially absent. Schumm's (1956*a*) study of a sterile tip-heap at Perth Amboy, New Jersey, describes badland slopes and gullies in the steeper parts and broad alluvial flats below, which appear typical of more arid areas. The extent to which vegetation is needed to maintain soil structure and infiltration rates has been described in chapter 8. With low infiltration rates, overland flow is frequent and therefore highly erosive; and throughflow is limited in amount so that chemical removal and weathering are slight (chapter 9): as a result soils tend to be thin and stony. The absence of a full vegetation cover is in these ways a critical determinant of the assemblage of processes which are effective in moulding the landscape of deserts.

Because a sparse vegetation cover is common to desert and periglacial areas, they share some of the process and landform characteristics which result from the limited vegetation cover. However, semi-arid landscapes differ from periglacial landscapes in important ways. First, the combination of aridity and the thin vegetation and soils means that the surface materials are generally rather dry, so that mass movements, including movements of a solifluction type are comparatively rare in semi-arid conditions (except at high altitudes). Second, semi-arid areas do not have such widespread or intense frost as do periglacial areas. The important feature that they share, a lack of vegetation, may be expected to introduce similarities between semi-arid and periglacial landforms, with wash slopes, free faces and talus slopes more common in both than in temperate areas. The greater importance of mass movements and of frost action in periglacial areas will, however, tend to reduce and obliterate the steep-slope elements relatively sooner than in semi-arid areas, where these elements are correspondingly more persistent features of the landscape.

THE TYPICAL SLOPE PROFILE

Forms can conveniently be analysed in terms of an idealized slope profile, with the four components used by Wood (1941) and described in chapter 11. These components are (fig. 13.1) an upper convexity (waxing slope), cliff face (free face), straight segment (constant slope) and basal concavity (waning slope). These components are commonly all present in arid and semi-arid profiles, although examples of almost any partial combination can be found. Either the cliff face or the straight slope may be absent, but it is unusual for both of these steep slopes components to be missing, whereas this is more common in temperate areas.

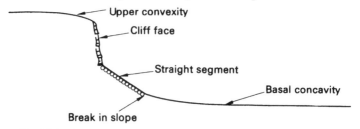

Fig. 13.1. Elements of an idealized slope in semi-arid and arid areas.

Characteristic features of semi-arid and arid landscapes are wide valleys and plains between the valley sides and isolated residual outliers (buttes) which have one or both steep slope components. In most areas the large valley widths are probably a function mainly of great age (Lustig, 1969, for the Basin and Range province) of the valleys, but the *persistence* of steep slopes is highly characteristic of the geomorphic processes acting in semi-arid areas, and requires closer examination. The relatively sharp break in slope between the steep slope components and the basal concavity also appears to be characteristic of arid and semi-arid lands, and is therefore presumed to depend on the specific combination of processes which operate in these areas, and in no other.

The upper convexity

The summit divide in an arid landscape may be bounded by straight slope segments (fig. 13.2a); or else be an almost flat butte top bounded by cliff faces. In either case Gilbert's (1909) argument which has been

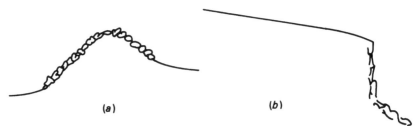

Fig. 13.2. Types of upper convexity in arid and semi-arid lands. (*a*) Convexity at a divide between two straight segments. (*b*) Convexity at the edge of a plateau above a cliff face or straight segment.

described in chapter 10 (p. 299) indicates that they will become convex under the action of processes which depend on gradient alone, rather than distance from the divide. Under semi-arid conditions the most important process of this kind appears to be rainsplash (chapter 8).

Arid topography is characteristically stepped, with a succession of low-gradient surfaces separated by steep slope segments above and below. Narrow convexities are common in this sort of landscape just above each cliff or straight segment, as is shown in fig. 13.2*b*. If the steep slope and the flat are in equilibrium, it can be seen that the rate of surface lowering must increase sharply as the steep slope is approached. The rate of transport must also increase rapidly over a short distance; and, since distance from the divide changes only comparatively slightly (in proportion), the increase in transport rate must be taken up mainly by an increase in surface gradient. This argument differs from that of Gilbert for the convexity of divides because it makes no assumptions about the type of process acting. The convexity can therefore be produced by any process, including surface soil wash which is likely to be the dominant one in a semi-arid area.

Rainsplash operating on the convex slopes moves material at a rate comparable to that of soil creep, but mass movement processes remove debris much more rapidly from straight slope segments (in all climates) so that convexities adjacent to steep slopes are relatively narrow. Furthermore, because removal on stable slopes by soil wash is more rapid than the corresponding humid temperate processes, semi-arid summit convexities above stable slopes are also narrower than the convexities of temperate lands.

Thus convexities, though present under arid conditions, are not such important features of the landscape as they commonly are in temperate areas.

The cliff face

Removal from cliff faces is clearly *weathering-limited*, and occurs in a number of ways which have been described in chapter 6. Large falls may take place through slab failure along major joints or other lines of weakness; or through rock avalanches in rocks with many small joints. The detached material usually begins to move as a block but disintegrates to some extent at the base of the cliff if the debris ever moves fast. Closely-jointed cliff faces also suffer from small rockfalls of more or less individual joint blocks. Lastly, individual grains and crystals may be detached directly by processes of grain release, which are most effective in sandstones with soluble cement.

Initial rates of movement in slab-failure (Schumm and Chorley, 1964), and rates of grain release (Schumm and Chorley, 1966) are closely related to the amount of precipitation, but also show increases in periods of freeze–thaw (figs. 6.5 and 6.10). Moisture is able to act directly, by applying tension forces to loosen grains, but its long-term role in chemical weathering, including the solution of cements from

341

sedimentary rocks, is more important. It also plays a part in freeze–thaw action, though freeze–thaw is not common to all desert locations, and in salt crystallization, but these are less important aspects of moisture action. Evidence from the Colorado Plateau on freeze–thaw action and rates of rock breakdown indicates that semi-arid rates of cliff retreat are probably very much slower than periglacial rates, for similar rock types.

The immediate fate of fallen material on landing at the base of a cliff differs widely, but in many instances the impact itself reduces the debris to a rubble which is dominated by fine material. This appears especially common for sandstones with soluble cement from which large recent falls may produce relatively small accumulations of talus-sized debris at the base of the cliff (Schumm and Chorley, 1966). On the other hand, igneous rocks commonly yield joint-block material, either directly from rock fall or else in the breakdown of debris originating in a rock avalanche, and in these cases, relatively little fine material is produced as a result of the fall. Sandstones and coarse-grained igneous rocks also produce some material directly as fines by grain release, a little at a time, and this material produces no observable accumulation of debris at the base of the cliff face.

The straight segment

Typically, the straight segment is similar in appearance to a talus slope but differs in that the boulders on it usually form only a veneer on a slope which otherwise consists of bedrock. Gradients on the straight segment are commonly somewhat less than the angle of repose although this property is also common in temperate talus slopes; and this observation led Melton (1965) to doubt whether the straight segment is a 'boulder-controlled slope' at the angle of repose of the 'average-sized joint fragment' as defined by Bryan (1923, p. 43).

The fact that semi-arid and arid debris slopes have only a veneer of boulders shows that they are not slopes of accumulation, like typical temperate talus slopes, but that they are instead an equilibrium transportation slope on which equal quantities of material are being supplied (from the cliff face and from weathering of the rock underlying the slope) and removed (downslope, often after breakdown of the rock fragments to sand sizes). The debris slope is therefore equivalent to an extreme case (of total removal) in the Lehmann (1933) scree model, described in chapter 6 (p. 144), for $c = 0$. The predicted form from this model is a straight talus slope at the angle of repose, with a thin veneer of scree overlying a bedrock slope.

Actual debris slopes, as we have seen, only approximate to this model, so that we must examine what other, non-talus processes are operating on the debris slope. These are the processes of debris breakdown; and second, the processes of debris transport, which are themselves modified by the breakdown of some of the original coarse debris. The process of debris breakdown is one of grain release produced by frost and chemical

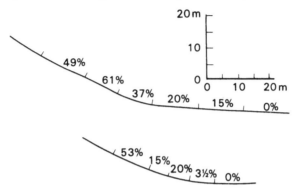

Fig. 13.3. Profiles through the base of the straight slope and break in slope region. Sacaton Mountains, south-east Arizona. No vertical exaggeration. Figures show percentages coarser than 25 mm in each section of the profile. Data from Kirkby and Kirkby (unpublished).

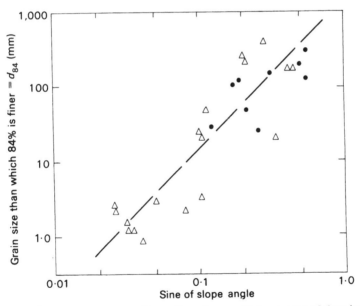

Fig. 13.4. The increase of grain size with gradient above and below breaks of slope in south-east Arizona.

△ Sacaton Mountains (granite)
● Picacho Peak (basalt)

Data from Kirkby and Kirkby (unpublished).

343

weathering acting over the surface of boulders and producing an increasing admixture of fine debris downslope which settles under and around the boulders. In addition, breakdown occurs on relatively stable slopes through the formation of desert varnish which hardens the

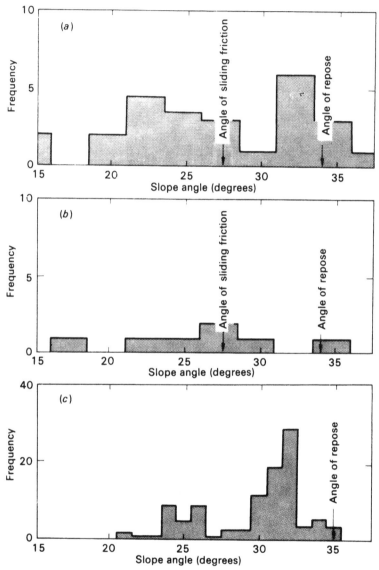

Fig. 13.5. Gradients of debris slopes in south-west U.S.A.
(a) South Arizona, granites, after Melton (1965).
(b) South Arizona, volcanics, after Melton (1965).
(c) New Mexico, mainly volcanics, after Koons (1955).

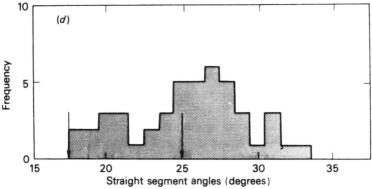

(*d*) Laramie Mountains, Wyoming, granites after Carson (1971*a*).
Arrows in (*d*) show calculated stable angles from direct shear tests for the relevant soil
and the assumption of debris saturation in extreme storms.

exterior of the boulder at the expense of weakening the material inside
so that once it is moved and the outer crust broken through, the whole
boulder is liable to disintegrate.

As well as talus processes, wash processes are potentially very active
on desert slopes. Fine material is rapidly removed from steep slopes as
fast as it is made available through weathering with the result that
undermining of coarse boulders will accompany removal of the fines. In
this way slope-wash is able directly to increase the rate of movement of
the coarse material, and is increasingly effective as the proportion of
fines rises downslope and on lower gradients (chapter 8). Furthermore,
at gradients approaching the angle of repose, debris requires relatively
small hydraulic forces to move it. Thus there is a complete gradation
from pure talus slopes at the angle of repose to pure wash slopes at
low gradients; on the steepest slopes talus movement under gravity
occurs; on slightly less steep slopes, very small hydraulic forces can
move even coarse debris directly, and the presence of a little fine material
will assist indirect movement by undermining; and on lower slopes,
hydraulic forces are more important than gravity forces, but can rarely
move coarse material except indirectly, with the greater amount of
fine material allowing movement by undermining.

The hillslope form produced by the combined action of gravity and
wash processes has the following characteristic: it is slightly concave in
profile (fig. 13.3) with coarser debris on steeper slopes (fig. 13.4), and
very little debris coarser than 25 mm diameter is able to move on gradients of less than 5–10°. This composite model helps to explain why the
debris slope should be at gradients which are usually a few degrees
less than the angle of repose and at the same time, it seems reasonable
to describe the slope as being 'boulder controlled'. The dominance of
wash processes may explain why the base of the straight slope is in

345

fact concave, and why on the lower slopes, material becomes finer downslope (the reverse grading to that normal on screes formed by accumulation), but it does not fully explain why the maximum slope angles are somewhat less than the angle of repose. Koons (1955), Melton (1965) and Carson (1971a) have measured maximum angles and find that they exhibit three distinct modes at approximately 20°, 25° and 32° (fig. 13.5). In addition Koons finds that bare rock slopes have a modal gradient of 34°, and an upper limit of 38°, somewhat higher than for debris slopes. Melton interprets the two modes for debris slopes as limited by the angle of repose (34°) for the higher mode, corresponding to conditions in which sliding will not begin; and limited by the angle of sliding friction (28°) for the lower mode, corresponding to conditions under which moving material will come to rest. Carson (1971a) analyses the separate modes in terms of differences in the degree of breakdown of the mantle, and so in angles of internal friction, on the assumption that the debris can become saturated in extreme storms (for the 20° and 27° modes in fig. 13.5d). This type of analysis is the same as that used successfully for temperate straight slopes (chapter 7; Carson and Petley, 1970).

The lower concavity and break in slope

In terms of area, the lower concavity is the dominant landform in semi-arid lands. All observers agree that desert plains are concave, and that they are usually surfaced with fine-grained material. In other respects they differ greatly. Some consist of thick alluvial accumulations (baja-das) while others have bedrock close to their surface for long distances (classic pediments). Some have braided channel patterns with channels only a few centimetres deep on their surface: others have dendritic networks incised tens of metres into a surface which stands above them as a very well-defined residual surface. Some desert plains are very simple in form, whereas others have several terrace levels, some of which are no longer attached to the mountain front, but are separated from them by reversed scarps. Because only the surface forms of desert plains are all alike, and because transport processes are active close to the surface, the discussion below concentrates mainly on the develop-ment of the surface form, rather than on the sub-surface composition of the plain. Two cases are distinguished: (1) areas of the plain which are unaffected by flow from streams leaving the mountain area, and (2) areas of the plain on which streamflow is common. This distinction is most meaningful in the upper part of the plains, closest to the steep slopes.

(1) Where no streams issue from the steep slopes, wash processes are dominant. A profile from the base of the steep slopes out onto the plain

will be a profile developed by wash, modified by gravity where the slopes are steeper. As with all wash profiles, it would tend to become concave (chapter 8) even if grain size were constant along the profile. However, the processes of debris breakdown and selective transportation produce a progressive decrease in grain-size downslope, and this tends to accentuate the concavity of the profile. All these arguments apply equally to the profiles of rivers in humid areas, but three factors tend to sharpen the concavity at the base of the debris slope in a desert and thus produce a sharp break in slope which is not parallelled in river profiles:

(i) The importance of gravity at gradients close to the angle of rest accentuates the rate of decrease of grain-size with gradient, and hence with distance downslope. At high gradients, small differences in slope produce large differences in transport rate, and the slope tends towards the talus form; that is, a straight profile, with parallel retreat. At low gradients, the influence of gravity is at its least, so that the form tends towards a simple wash concavity with slope decline. At intermediate gradients, in the 5–20° range, gravity is still effective in reducing grain-size downslope, and the wash processes acting on this difference in grain-size produce the most marked concavity. This effect is less apparent in temperate rivers because of their much lower gradients, so that in them hydraulic processes usually predominate throughout.

(ii) Desert plains sometimes begin as alluvial fans deposited by streams issuing from the mountain front, so that, between the mountain front and the fan a sharp break in slope forms, as a depositional contact. The Tucson basin, for example, contains over 1,000 m thickness of alluvial fill. If aggradation then ceases, water moving across this contact will encounter a sharp break in debris size between the straight slope boulders and the finer material in the fan, and wash processes will indefinitely *preserve* a sharp concavity as a result. In this case, slope processes are merely maintaining a break in slope, and not producing it.

(iii) Some rocks, notably coarse-grained igneous rocks and sandstones with soluble cements, break down to produce a bimodal grain-size distribution, with one mode in the gravel to boulder class and one in the sand class. Where these rocks form the steep slopes, the grain-size change at the break in slope is further accentuated, so that such rock types have particularly abrupt breaks in slope. Such features also occur in humid rivers (Yatsu, 1955), but they are generally less dramatic in appearance.

(2) At a location where the surface is shaped by rivers flowing out of the mountains, the river regime is critical to the development of the landform. Two patterns appear to be common, a superficial braided net and an incised dendritic net. Both types can be found at the same time within an area such as south-east Arizona; and both types can often be shown

to have operated at different times at one site. Typically, braided streams are associated with aggradation and the deposition of alluvial fans; and dendritic networks are associated with erosion concentrated along stream courses, which necessarily become incised below the general plain level. However, the typical associations may be reversed with braids occurring under conditions of equilibrium (Leopold and Wolman, 1957) or even of downcutting (Fahnestock, 1963); and non-braided, usually meandering, streams associated with aggradation. Nevertheless the bulk of stratigraphic evidence suggests that the two predominant channel network forms, braided and dendritic, are usually associated with aggradation and erosion respectively.

Widespread aggradation leads to the formation of alluvial fans, usually at low gradients (< 5°). These fans generally have their greatest extent along the direction in which the stream leaves the mountains, but distributaries are able to deposit a little material even at large angles of deviation from this direction, so that the whole of a plain can become covered with fan material and it will everywhere be deposited up against the steep mountain slopes, producing sharp depositional breaks in slope.

Localized stream erosion leads to the incision of pre-existing plain surfaces which will ultimately tend to be reduced by slope processes operating towards the incised rivers. The steep slopes produced by the incision are subject to the same processes as the original mountain slopes, however, so that the same characteristic forms are developed at a lower level. The plain therefore develops into a two-level landform, with relatively smooth surfaces at the upper, original level and at the lower level of the incised rivers; and the two levels are separated by steep, often straight slopes.

At the base of the main mountain slopes, runoff from the steep slopes tends to produce higher moisture contents and consequently deeper chemical weathering, particularly in tropical climates. During incision this zone of weathered material is commonly eroded by streams which then flow along the original break in slope (Twidale, 1967; Clayton, 1954), and some authors (Johnson, 1932a and b; Clayton, 1954) have held that streams in this position actually form the break in slope or the appearance of a break in slope. It seems more reasonable to say that such streams can only destroy a simple break in slope, if one ever existed; and it seems clear from numerous observations that simple breaks in slope, without marginal streams, do exist under some circumstances. Marginal streams have been held to produce a break in slope not only by downcutting, but even more importantly by lateral cutting which steepens the mountain slope (Johnson, 1932a and b; Rahn, 1966). Although marginal streams can cause significant steepening of the break

348

in slope (Rahn, 1966, p. 221), they are relatively rare in south-east Arizona where simple breaks in slope are common and where mountain slopes without marginal streams at their bases do not show any evidence of former basal undercutting. Lustig (1969) argues that streams flowing out of the mountains can only deviate in angle enough to flow along the mountain front for a very small proportion of the time. The balance of evidence therefore leads to a rejection of marginal streams as a main cause of breaks in slope.

There is no doubt that both aggradational desert plains and incised desert plains exist, and in between the two lies the classic pediment which is an exact equilibrium form. In the classic pediment, supply and removal of debris are balanced, so that the surface is neither incised by erosion nor progressively buried in alluvium, and bedrock is always at a relatively shallow depth below the surface. Such forms have been described (Tuan, 1959, for south Arizona), but the evidence is necessarily fragmentary because of limited sub-surface exploration. If the depositional and erosional types of plain can exist, it is clearly possible for the equilibrium form to exist too, although the field evidence indicates that it is a minority landform and not the typical desert plain form. If these spatial frequencies may be taken as an indication of temporal frequencies at one site, the implication is that desert plains are not generally in equilibrium, but are instead undergoing alternations of cut and fill.

THE EFFECT OF CLIMATIC CHANGE

Alternations of cut and fill are very common in semi-arid lands throughout the world. Indeed they are so common that they might be considered as a normal part of semi-arid landscape development, especially of the lower-concavity plains. The sensitivity of landscapes to small differences in vegetation cover is shown both by data for total sediment production (Langbein and Schumm, 1958) and also by data for drainage density (Melton, 1957). Under either fully arid or humid conditions, vegetation cover is unaffected by small changes in external conditions, but in semi-arid conditions even very small changes in rainfall, human water use or grazing patterns may produce considerable changes in vegetation cover, to which the landscape responds.

Under relatively sparse vegetation, sediment production is high and equilibrium drainage density is high (fig. 16.20). A thinning of the vegetation cover therefore produces a tendency to erosion and the formation of new drainage channels; that is, a period of incision, which will tend to continue until a new equilibrium is attained. Under relatively dense vegetation, sediment production is low and equilibrium drainage density is low. A change to denser vegetation therefore tends

349

to prevent sediment from being removed and causes pre-existing channels to fill in; in other words, a period of aggradation. Archaeological
evidence from south-western U.S.A. supports this view that sparse
vegetation periods produce cut, and dense vegetation periods fill (Hack,
1942).

Some changes from fill to cut are relatively well documented, especially
the period of erosion which began between 1880 and 1910 in the south-
western U.S.A. (Thornthwaite *et al.*, 1942), and this evidence shows that
very small changes in the vegetation cover are sufficient to change the
situation from one of fill to one of cut. In the south-western U.S.A.
however, it is not clear whether the crucial factor was the introduction of
grazing animals or small changes in the intensity of summer rains
(Leopold, 1951; Thornthwaite *et al.*, 1942). In the absence of cultural
factors, increases in low intensity rains, especially in the growing season,
will favour denser vegetation, whereas increases in higher intensity rains
may have no significant effect on vegetation density, but will encourage
higher erosion rates, so that there is no simple relationship even between
rainfall and cut or fill. On the whole, however, periods of decreasing
aridity tend to be associated with aggradation, and periods of increasing
aridity with downcutting.

In areas where cut and fill alternate frequently, the amplitude of each
is likely to be relatively small because of the limited time available. Each
time the rivers erode, they do so into a consequent surface of their own
alluvial deposits, and are therefore likely to cut down in slightly different positions in each episode of erosion. The underlying bedrock will
therefore gradually be cut down at all points to form an undulating
bedrock surface at a depth of a few metres below the plain surface. In
this way a surface not dissimilar from the classic pediment in overall
form might be produced not as an equilibrium form but by the repeated
alternations of cut and fill about a mean position.

EQUILIBRIUM MODELS AND THE BALANCE BETWEEN PROCESSES

Small fluctuations in the environment, initiated by changes in the
climate, vegetation cover or local river bed levels, have a dispropor-
tionately large effection on semi-arid landscapes. In the previous section
the effect of this alternation has been considered, with its influence on
landforms, but it is also fruitful to take a different point of view. The
fluctuations can alternatively be taken to be about a mean position
which represents an overall dynamic equilibrium, and this equilibrium
may be considered as the primary focus of interest. Denny (1967) proposes that alluvial fans and pediments are mainly formed as an equilibrium process, rather than primarily in response to fluctuations in

vegetation cover or base levels. He shows that a given area of highland can produce only enough coarse debris to cover a similar area of desert plain with alluvial fans (fig. 13.6). It follows that if the desert plain is much larger than the area of mountains supplying it, then the remaining area of plain must be subjected to erosion, in some places trenching earlier fans and in other places cutting down to bedrock. Extensive

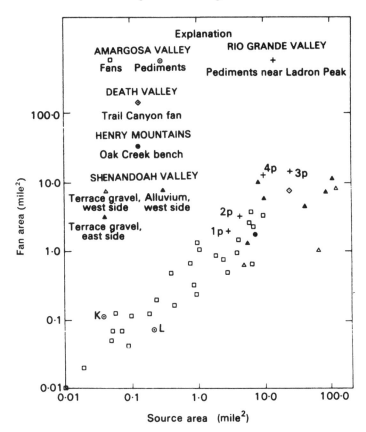

Fig. 13.6. The relationship between area of a fan or pediment and its source area in adjacent mountains (from Denny, 1967).

plain surfaces cut into rock (pediments) are therefore characteristic of areas with a high ratio of lowland to highland; and extensive alluvial fans are characteristic of areas with a low ratio of lowland to highland. These conclusions are in agreement with Lustig's (1969) analysis of the Basin and Range Province of U.S.A. Denny thus shows that, given a dynamic equilibrium, alternations of cut and fill must occur at each place, even in the absence of environmental fluctuations. The mechanism by which this chiefly occurs is by stream piracy, because downcutting

351

streams are able to capture aggrading streams in the same or neighbouring catchments, as has been shown for the Henry Mountains by Hunt, Averitt and Miller (1953).

In comparing this dynamic equilibrium model with a model in which alternations of cut and fill are produced by small changes in the environment, the most important discriminant is whether changes from cut to fill or vice versa are more or less simultaneous within an area. If they are simultaneous, then environmental changes would appear to be critical; if they are not simultaneous, then it is likely that dis-equilibrium is more important. In practice the two models are very closely linked, because an environmental change will be able to trigger a change between cut and fill in an unusually large number of streams.

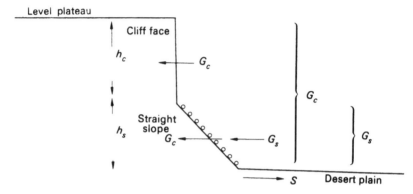

Fig. 13.7. Idealized model showing relationships between cliff and debris slope heights in equilibrium.
G_c = rate of breakdown of rock to joint block debris.
G_s = rate of breakdown of joint blocks to sand.
S = rate of transport of sand from base of slope.

An equilibrium model is also appropriate for considering the relative importance of the two steep slope components in desert landscapes. In a simple model derived from the 'talus weathering ratio' of Schumm and Chorley (1966), we can imagine a vertical cliff of height h_c, and a straight slope of height h_s, with a sand plain below (fig. 13.7). Rock breakdown is assumed to take place in two phases; (1) sound rock to joint blocks at a rate of (horizontal) retreat G_c. This process is considered to take place on both the free face and on the straight slope. (2) Joint blocks to sand at a rate of (horizontal) retreat G_s. Debris is removed at the base of the steep slopes, in the form of sand, at a total transport rate S.
In equilibrium, therefore:
Total rate of production of joint blocks = Total rate of breakdown of joint blocks = Total rate of transport of sand from base of straight slope.

In symbols:

$$G_c(h_c + h_s) = G_s \cdot h_s = S,$$

(13.1)

or

$$\frac{h_s}{h_c} = \frac{G_c}{G_s - G_c}.$$

(13.2)

It is immediately apparent that an equilibrium form can only exist if $G_s > G_c$. Where this is not the case ($G_s < G_c$) and the cliff breaks down readily, it will tend to be eliminated from the landscape, and the debris slope will accumulate a somewhat thicker layer of material to reduce the rate of bedrock break-down to an equilibrium value. Slope profiles in which cliff faces are absent are quite common in semi-arid areas, especially for igneous rocks and for clays and shales. In the latter cases the breakdown products consist of desiccation blocks rather than joint blocks (Schumm, 1964).

At the other extreme, where $G_s \gg G_c$, are those rocks which either shatter to sand on impact after slab failure or rockfalls, or are subject to grain release which effectively by-passes the joint block stage. In this case the straight slope becomes insignificant, as can be seen from the expressions above. Cliff-dominated steep slopes of this sort are very characteristic of sandstones with soluble cements in the south-western U.S.A. Between these two extremes are rocks for which the two breakdown rates are suitably matched to produce steep slopes with cliff and debris slope components, but the presence of both in a profile is most common where resistant rocks overlie less resistant rocks, especially shales (Schumm and Chorley, 1966).

In this widespread case, of a massive jointed rock overlying a less resistant rock which breaks into small fragments, the rates G_s, G_c are determined by the differing properties of the two rocks. Joint blocks are only effectively produced over the depth of the upper rock layer, which we will denote by h_1; but they are allowed to disintegrate to sand sizes over the full length of the length of the debris slope. Thus equation 13.1 must be replaced by:

$$G_c \cdot h_1 = G_s \cdot h_s$$

(13.3)

Two cases may be distinguished. The contact between the two rock types may be on (1) the free face or (2) the debris slope. The rate of joint block removal, G_s, refers in both cases to material from the resistant upper layer; but the rate of cliff retreat, G_c, is determined by the underlying weak rock in case (1) and by the resistant rock in case (2). Thus, in the first case:

$$h_s = h_1 \cdot \frac{G_c}{G_s} \gg h_1;$$

while in the second case:

$$h_s = h_1 \cdot \frac{G_c}{G_s} \sim h_1.$$

It is apparent that many, if not most actual cases will lie between these two extremes, with the result that the junction between free face and straight slope is exactly at the geological contact, and equation 13.3 is satisfied by a thickening of the debris mantle on the straight slope and a consequent reduction in the rate of cliff undercutting, G_c.

The persistence of steep slope elements in the landscape depends strongly on the relative rates of *weathering-limited* rock breakdown and *transport-limited* removal on the plain. Where the rates of cliff retreat and block breakdown are high, then the equilibrium values of h_c, h_s are

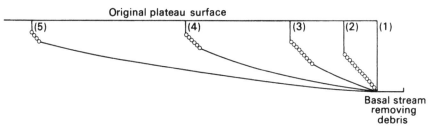

Fig. 13.8. Development of desert slopes from a plateau, under conditions of basal removal from a fixed point.

relatively small, so that the steep slopes will tend to be eliminated more quickly in favour of lower gradient *transport-limited* slopes. In comparing periglacial and desert slopes, the much greater rate of cliff retreat through frost action is the critical difference, leading to rounded forms with moderate relief under periglacial conditions and to plains around steep residuals in desert conditions. Desert areas are, in other words, characterized by long-continuing parallel retreat of the steep slope elements.

These models allow a generalized slope development to be formulated, which is controlled by the following factors: (1) The ratio of cliff height to debris slope height remains constant until both are eliminated (but see also fig. 16.2). In fig. 13.8 the ratio has been taken as 1·0. (2) The rate of sand removal at the foot of the steep slope is proportional to the height of that slope. As the steep slope is reduced in height, the gradient of the basal concavity normally becomes lower to compensate for the reduction in debris carried. Successive profiles are therefore produced by the erosional regrading of the previous desert plain surface, and not merely by its passive extension. (3) The profiles shown in fig. 13.8 refer to slope conditions between major streams. Cut and fill associated with the streams are therefore not included, although it has been suggested above that the action of streams may contribute to the overall form of the break in slope. (4) Fig. 13.8 has been sketched for the conditions of a broad plateau and a fixed stream which is exporting the debris reaching it from the lower concavity. If debris accumulates without stream removal, then the lower part of the concavity will be depositional, although the upper part will continue to be regraded as shown in the figure.

FULLY ARID LANDFORMS

This analysis of desert landforms has concentrated mainly on semi-arid rather than fully arid forms, although much of it applies to both. However, three important differences are found in really arid areas: first, wind removal is relatively important; second, climatic change is less effective; and third, dune systems become an important landform.

At very low rainfalls, the total rate of fluvial sediment production is less than in semi-arid lands (Langbein and Schumm, 1958) and ultimately falls to zero in completely dry areas. Wind transport is, however, increased in absolute terms by the removal of vegetation which both slows the wind speed and binds the soil. Relative to water transport, wind transport increases even more in very dry areas, and becomes the dominant transporting agent. Material in the silt size-range can be carried very long distances once in suspension in the air, even out of the desert area, so that wind not only redistributes material, but can be the major agent in removing debris altogether.

At annual rainfalls of 100 mm or less there is so little vegetation that minor environmental fluctuations can scarcely alter its geomorphic effect, with the result that cycles of cut and fill are not apparent except for major climatic changes, perhaps corresponding in magnitude to main glaciations. Where, as is usual under very arid conditions, through-flowing drainage is absent, the desert plains are predominantly areas of accumulation and they remain so more or less indefinitely.

The slow rates of chemical weathering characteristic of arid regions together with the export of fine material by wind, produces a dominance of sand and coarser material in deserts. Arid areas are therefore commonly surfaced with coarse debris, which becomes areally sorted by the wind. In some areas fine sand accumulations produce dune fields, and large areas of the Sahara especially, show evidence of present or fossil dune patterns, which are one of the most striking macro-features of the area.

CHAPTER 14

TROPICAL LANDFORMS

High temperatures are the main diagnostic property of tropical areas. Mean annual temperatures commonly lie between 20 and 30 °C (70–90 °F), with little annual range; and frosts are absent. This definition of 'tropical' has greater geomorphic significance than one based on geographic area, that is the zone between the tropics, and in any case describes properties which are usual at low elevations in this intertropical area.

Rainfall amounts and distributions vary greatly in the tropics, from extremely arid deserts to rain-forest areas with over 3,000 mm per year. Landforms show a correspondingly wide variation and in limiting cases tend towards forms which have been discussed in previous chapters. Under arid conditions, high temperatures have the least influence on landforms, and the forms are similar to those discussed in chapter 13; whereas under cool moist tropical conditions, the forms tend toward humid temperate forms (chapter 11). There remain two modal types of tropical landscape which are discussed in this chapter.

(1) Under conditions of moderate rainfall (500–1,000 mm), annual rainfall is less than potential evapo-transpiration for several months of the year, and the vegetation cover is somewhat open with perennial species showing some adaptations to a dry season. This is the savanna zone which covers the greater part of the tropics.

(2) Where rainfall is higher (> 1,500 mm) and less seasonal, there are no months in which plant growth is restricted and a close cover of evergreen rain-forest is normal.

THE BALANCE BETWEEN PROCESSES

Under rain-forest conditions, removal of material in solution is probably the major process of landscape lowering. Solute concentrations in streams are usually somewhat lower than in temperate streams (Strakhov, 1967; Douglas, 1967) for reasons which have been discussed in chapter 9; but the greater amounts of runoff more than compensate for the lower concentration, and total rates of chemical removal are higher than in any other climatic environment: they are perhaps two to five

356

times greater than in humid temperate areas. Under savanna conditions, mean rates of chemical weathering are lower, ranging from the very low rates of chemical removal typical of arid areas to the rather high rates associated with tropical rain-forest. However, near the savanna–rain-forest boundary, rates of chemical removal are typically a little higher than in temperate areas (figs. 9.10 and 9.11).

Soil wash rates vary widely under tropical conditions. In dry savanna areas, wash is very effective in landscape lowering; but in areas of higher rainfall the increased vegetation cover reduces wash rates to a minimum which occurs near the savanna–rain-forest boundary. At still higher annual rainfalls the trend is reversed and removal by wash processes becomes increasingly effective because rainfall intensity is increasing whereas the controlling influence of vegetation cover is already at a maximum (fig. 8.18). On moderate gradients, therefore, wash lowers the surface more rapidly than chemical solution at both low (<1,000 mm) and high (>2,500 mm) rainfalls, whereas at moderate tropical rainfalls (1,000–2,500 mm) chemical removal is the more effective.

Mass movements are everywhere the most important group of processes on steep slopes, and under given conditions of gradient and regolith the more intense rainfalls of the tropics will lead to higher rates of degradation by mass movements. Slopes in dynamic equilibrium are, however, likely to react to higher rainfalls by tending to absorb the effect of the change, according to Le Chatelier's principle. This means that not only will the rate of degradation be somewhat greater, but also that the threshold rainfall required to produce widespread sliding is higher, in accordance with the observations of Ruxton and Berry (1957). These two requirements appear at first to be in conflict, but because higher rainfalls are also associated with higher rates of chemical weathering, a slightly lower slope angle (than in a comparable situation at a lower rainfall) will satisfy both requirements: that is, the lower angle increases the threshold for movement, and the higher rate of chemical removal requires a higher rate of mass movement degradation to maintain the dynamic equilibrium (see chapter 9, p. 268). However, this influence will be relatively slight compared to differences in stable angles due to lithology and slope base conditions.

Soil creep processes in wetter areas and rainsplash in drier areas will each tend to be the dominant process near hilltops and both will tend to make them convex. Rates of temperate soil creep and semi-arid rainsplash have been shown to be comparable (chapters 8 and 10) and ineffective, except near the divide, compared to wash and solution processes. Soil creep rates have not yet been measured in deep tropical soils although Kojan's (1967) measurements in California suggest that continuous creep rates in deep soils might be very much higher than in the thinner soils of temperate areas. The relative magnitudes of creep or splash rates near the divides, and of the wash, solution or mass movement processes lower down the hillslope will determine the relative size of the convexity in tropical slope profiles.

The relative rates of wash, solution and mass movements depend not only on the climate but also on the local gradient. At steep gradients mass movements are always the most important processes, although the critical angle at which they become important varies with bedrock and regolith conditions. On very gentle slopes chemical removal is the most effective process, because its rate depends only slightly on gradient. Only at slopes so low that water movement is impeded does the rate of chemical removal decline. At intermediate gradients (5–25°) wash is the most important process under some climates; in savanna areas wash is clearly dominant on all but the lowest gradients whilst in rain forest areas of moderate rainfall (1,500–2,500 mm) wash will be more effective than solution only on rather steep slopes, if at all. In very wet tropical areas (> 3,000 mm) wash is dominant over a wider range of slopes once more.

DEEP WEATHERING

Evidence of the amount of climatic change for the tropics is at least as uncertain as for temperate areas, but generally suggests that the Pleistocene included periods which were cooler and wetter than today. Estimates show a probable temperature lowering of 5–6 °C which is a somewhat smaller lowering than for temperate areas and suggest that rainfalls increased somewhat at the same time (Grove, 1968). There were probably some periods which were drier than today, though not necessarily warmer. Areas which have always remained arid, for example in the central Sahara, show many wind-produced forms and minimal water action: chemical weathering has always been exceedingly slow. Fringe desert areas, for example the Lake Chad area, show evidence of fossil watercourses and lakes, but still little weathering. Central semi-arid areas show the alternations which characteristically produce pediment-like landforms (chapter 13), but in tropical areas there is an important difference which in part distinguishes savanna landscapes from the temperate semi-arid landscapes described in chapter 13. During wetter or colder periods, tropical semi-arid lands show much higher rates of weathering than under temperate conditions (for example in Arizona). The extensive plains which occur on any large land mass are therefore weathered to much greater depths than under temperate semi-arid conditions. In areas which are today near the savanna–forest boundary, weathering has continued at high rates with few interruptions, so that the weathered layer is commonly many metres or tens of metres deep. In wetter (forest) areas conditions for rapid weathering have persisted for a very long time, but plains may not be weathered to such great depths if drainage through the regolith is impeded by a permanently high water table, so that part of the flow must go over the surface and not reach chemical equilibrium.

Deeply weathered plains are a widely-described feature of tropical areas, but they do not seem to be related solely to the higher rates of weathering which are common in the tropics. At least as important is the time which has been available for weathering. Plains are a feature of any continental interior which has been reasonably stable tectonically during the Tertiary, simply because average gradients must be very low. In savanna and other semi-arid areas the contrast between mountain and plain is very marked and this serves to accentuate the plains, which are present in all climates. More important than this topographic stability is the relative climatic stability of the tropics, which has allowed weathering at relatively high rates to continue throughout almost the entire Tertiary era. In this the tropics are unique amongst the geomorphic zones, because cool temperate areas have been disturbed by glaciation and arid areas have exceedingly low rates of weathering.

One of the results of the deep weathering process which profoundly affects the landscapes of dissected tropical plains is the formation of duricrusts. Aluminium, iron and silicon oxides are concentrated by weathering as residuals, and also move as colloids or in solution (Ruxton, 1958) to increase the concentrations below the surface of desert plains. These concentrations become extremely tough when they are exposed by later dissection, and are then able to act as resistant caprocks. Duricrusts of this sort are widespread in tropical Africa and Australia, mainly in savanna areas, where seasonal water movement is thought to play an important part in their formation.

Ferruginous lateritic (ferricrete) and siliceous (silcrete) duricrusts are the most widespread forms of duricrust, and their presence or absence has been attributed to the influence of climate (Stephens, 1961) and to parent material (Woolnough, 1927). On a world scale, silcretes are mainly confined to areas which today have rainfalls of 100 to 250 mm; and laterites to areas with 250 to at least 1,500 mm. Each type of crust shows a tendency to become thicker but less compact and concentrated as the rainfall increases. In transitional zones the two types of concentration may both be present (Wopfner and Twidale, 1967; Mabbutt, 1967) and the dependence on lithology is most marked. Similar relationships apply to the distribution of calcareous accumulations in the United States (Leopold, Wolman and Miller, 1964, p. 122) and in semi-arid areas of Africa. The process which these observations seem to outline is as follows. In regions which are not subject to continuous leaching and have seasonal or annual water deficit, there is some re-deposition of dissolved or colloidal material in the weathered layer (chapter 9) when soil water becomes super-saturated through evaporation. There is nevertheless a steady loss of material from the superficial zone of water circulation where weathering is concentrated. As time passes the more soluble substances are gradually removed: first the calcium salts, then the silica and finally even the iron and aluminium compounds. At different stages each in turn will dominate the material re-deposited in the soil, while the sesquioxides mainly accumulate as a residual in the source area, at a p_s of 0·3–0·6 (pp. 261–6).

At a given site, therefore, in zones with an annual water deficit, there will tend to be a progression in time from a predominance of lime accumulation to silica accumulation to iron or aluminium accumulation. Similarly, at a given time, drier areas have slower rates of net removal so that there will be a spatial transition from lime accumulation in the driest areas to iron accumulation in the wettest areas. In transitional areas the effect of lithology will be most marked, with, for example, silcretes on siliceous rocks and ferricretes on basic rocks. It appears that effective duricrust formation requires periods of the order of magnitude of the entire Tertiary, so that the younger and more temperate North American deserts show mainly calcareous crusts, whereas the older deserts of Australia have silcretes in areas which are today at least as dry.

EQUILIBRIUM SLOPES

Tropical slope forms depend very much on the history of weathering in an area. Where relief has been developed by incision into an earlier plain which had previously been weathered to great depths, then the properties of the weathered material are the most important factors in subsequent slope development. Such slopes are considered in the next section, under the heading of 'multicyclic slopes'. In other areas the relief is too young to have permitted deep weathering, or else downcutting has been so great that former deep-weathered material has been completely eroded. In such cases, present depths of weathering are in equilibrium with current processes of mechanical removal. These slopes are considered in this section, for the examples of rain-forest at about 2,000 mm annual rainfall, and savanna at about 800 mm.

In areas of very rapid downcutting, valleys are typically V-shaped, with straight slope sides and in extreme cases a cliff face. These forms are found under all conditions of climate and lithology, and the stable slope angles are determined by stability under conditions of *weathering-limited* mass movements. Wentworth (1943) describes a process of alternate weathering and sliding of steep slopes in Hawaii to maintain average angles of 45–48°, composed of upper bedrock slopes of about 60° and lower taluvial slopes at 35°, on basalts. Simonett (1967) found straight slope angles of 36° maintained on granites in New Guinea in an area of simple V-shaped valleys. These values from areas of high rainfall do not differ appreciably from stable angles in semi-arid areas of 33–36° (fig. 13.5).

Where downcutting is less intense and slopes have become stable to mass movement processes, the whole slope will be evenly lowered by solution, summits will become rounded through the action of creep and rainsplash, and the lower parts of the slope will be moulded by wash processes. In the humid tropics (2,000 mm) solution is the more important process, so that the resulting profile remains almost straight in its lower parts, with only a slight basal concavity. This is because solution produces more or less parallel retreat, and wash, which would tend to produce a concavity, is of less importance (fig. 14.1a). In the drier savanna areas (800 mm) the relative importance of the two processes is reversed and a marked wash concavity develops. The lower rate of chemical solution produces thinner soils, and the overall rate of removal of debris is higher than in more humid areas. For both of these reasons the relative size of the summit convexity is less in savanna areas than in the humid tropics (fig. 14.2a). At this stage of development soils are generally red, stony, and thin.

As downcutting slows down still more, and overall relief and average gradients are reduced, the rain forest and savanna forms look more divergent (Young, 1968). Under rain-forest, parallel slope retreat resulting from chemical removal is now more rapid than valley lowering so that a flat valley floor begins to form (Chamberlain and Chamberlain, 1910), in which water movement is more restricted, and although weathering continues, chemical removal is much reduced. The development of a flat valley floor also allows the valley river to impinge on the slope base and so remove any vestigial concavity due to wash. Creep is the only process which is able to produce curvature in the upper parts of the slope, which is therefore completely convex (fig. 14.1b). Soils change abruptly from freely draining types on the convex slopes to swampy valley bottom soils with a very high water table, but of similar textures (Young, 1968, p. 11). Ground water laterites may form in these valley bottoms.

Under savanna conditions, the dominant wash concavity persists as relief is lowered. Steep slopes may persist near the top of slopes, in the same way as in more arid regions. Surface grain size differences may accentuate the concavity, as they do in arid areas, but the greater rates of weathering generally lessen this effect in savanna areas. Summit convexities are less narrow than in arid areas because of the greater soil thicknesses (of savanna) and the less rapid removal by wash (fig. 14.2*b*). Soils are much better developed than in arid areas, and it is from savanna areas that tropical catenas were described (Milne, 1947), and in which they are best developed. Soils (though not the total depth to unaltered bedrock) thicken downslope (Ruxton and Berry, 1957), especially on the lower concavity, and in this part of the slope profile they are also progressively finer downslope (Young, 1968; Bunting, 1965, chapter 17), as may be expected in association with a slope formed by wash processes. The degree of weathering also tends to be greater

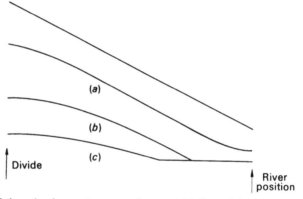

Fig. 14.1. Ideal slope development sequence from an initially straight slope under rain-forest conditions (2,000 mm at 25 °C).

downslope, partly because the soil is in equilibrium with lower rates of surface lowering downslope (chapter 9) and partly because of lateral transfer and re-deposition downslope. Conditions on the gentler slopes and clay bottomlands appear to be very suitable for the gradual concentration of sesquioxides which may form duricrusts if subsequently dissected.

The sequences of slope development described above can be compared in a three-component diagram (fig. 14.3*a*), which shows the relative rates of lowering by the three major processes operating on slopes which are stable to mass movements; creep or rainsplash, wash and solution. The curve shows the way in which the proportions vary with rainfall under typical tropical conditions at about 25 °C. As rainfall increases, it shows the transition from wash dominated slopes at <500 mm annual rainfall; to wash plus creep slopes of savanna areas at 750–1,000 mm; to creep plus solution slopes of rain forest areas at 1,500–2,500 mm; and finally to wash dominated slopes again in very humid areas at 5,000 mm. In fig. 14.3*b* the actual rates of removal are shown (from the data of part two), and it is apparent that areas near the savanna–rain forest boundary (1,000–1,500 mm) are evolving at the slowest rates.

MULTICYCLIC SLOPES

Vast plains in Australia and Africa were subjected to weathering during long periods of the Tertiary, producing such deep weathering that subsequent incision is often entirely within the weathered zone, or at least profoundly influenced by it. Depths of weathering are commonly 10–50 m and may exceed 200 m locally, and surfaces with these depths of

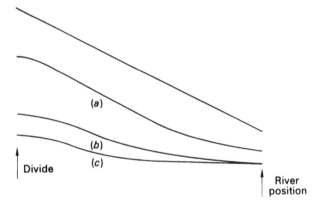

Fig. 14.2. Ideal slope development sequence from an initially straight slope under savanna conditions (800 mm at 25 °C).

weathering on them are typically assigned to the late Cretaceous or early Tertiary (Jennings and Mabbutt, 1967; King, 1962). From these generalized observations we can calculate average rates of removal in solution, assuming that the weathering has been continuous, and has reduced the parent material by 50 per cent of its original substance.

Over a period of 10^8 years, the average rate of rock removal (somewhat greater than actual surface lowering) in solution, assuming no mechanical removal, has been about 100 mm per million years. On the same assumptions, the rate of accumulation of a residual layer, 20 per cent of which is iron oxides, can be calculated as:

Maximum thickness = 5 . (depth of rock weathered) . (proportion of Fe oxides in parent material).   (14.1)

This relationship assumes that negligible iron oxides have been removed

363

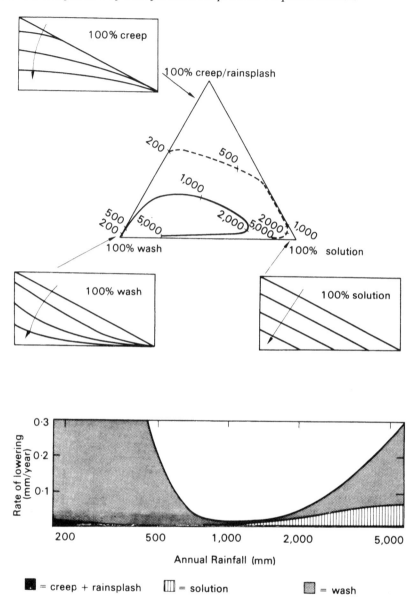

Fig. 14.3. (*a*) Triangular diagram showing relative rates of removal by creep, wash and solution on moderate (10°) slopes. Solid line shows values for the tropics (25 °C), and broken line for temperate areas (10 °C): values on each line show annual rainfalls in millimetres. Diagrams at each corner show idealized slope development sequence for each 'pure' process. (*b*) Estimated total rates of removal by each process in the tropics (25 °C) at different annual rainfalls, for moderate (10°) slopes.

during weathering, to produce a maximum possible thickness of iron enrichment in the residual soil. The most variable term in the above equation is the proportion of iron oxides in the parent material, which may vary from 2–15 per cent from acid to basic rocks. Corresponding to these extreme percentages, 30 m of weathered rock, during the course of the Tertiary, will yield 3–24 metres of iron-enriched soil according to the parent material which therefore exerts a strong influence, not on the type of enrichment and so potentially the type of duricrust developed, but on its thickness. Very high rainfall areas may, however, show an absence of duricrusts, despite deep weathering, as has been suggested above (p. 359). Alternatively, an Fe-rich duricrust to a small extent, and Si- and Ca-rich crusts commonly, are produced by re-deposition, preferentially in basins. Thick (50 m) Secondary crusts also appear to form, particularly in $CaCO_3$, as a result of deflation of dust from areas of primary deposits.

Younger plain surfaces have commonly been assigned to late Tertiary, Miocene or Pliocene ages, so that the times available for weathering have been about one-fifth of that for the older plains. On the simplest assumption, that weathering proceeds at a uniform rate, depths of weathering will also be reduced to one-fifth. Sesquioxide accumulation horizons will be affected by both the less complete weathering in the shorter time period, which will tend to reduce its thickness; and by the presence of a surrounding Fe-rich plateau, which may supply material for re-deposition and so tend to increase its thickness. The older duricrusts appear, however, to be highly resistant to subsequent removal, and usually remain *in situ* until they break up during incision and are incorporated as detritus at the lower plain level. Sesquioxide accumulations and duricrusts on late Tertiary surfaces therefore appears to be very thin, except perhaps in areas of iron-rich parent rocks.

Multicyclic landscapes in which deep weathering plays an important role are therefore mainly confined to areas in which early Tertiary (or older) plains are being eroded. The dissection of younger plain surfaces conforms much more to the equilibrium slope model above because of the shallower weathering, unless the surface of the younger plain is still within the older weathered material. Multicyclic landscapes tend to develop in two ways, according to the extent to which duricrusts are developed. (1) In areas of thick and tough duricrust, the indurated layer acts as a cap-rock which preserves the former surface, and allows effective lowering only along the fringes of the plateau, as the duricrust is undermined. A stepped landscape tends to develop, dominated by duricrusted remnants. (2) In areas where the duricrust is poorly developed, and readily eroded, incision tends to produce planation down to a new plain level, or to the unweathered bedrock, whichever is the higher. This landscape is a plain dominated by isolated rock knobs, which were more resistant to sub-surface weathering, and form inselbergs after exhumation by incision.

Duricrust landscapes and inselberg landscapes may exist side by side: for example inselbergs may be exhumed around the fringes of a duricrust plateau. There are, however, large areas which are dominated by one or other type of landscape, and the determining factor has commonly been identified as lithology (e.g. Thomas, 1968). Areas of basic rock which is not apparently resistant itself are able to produce thick duricrust layers during deep weathering, which protect the rock from subsequent erosion; whereas areas of acid rocks, especially massive granites and gneisses, have low iron contents, and consequently poorly developed duricrusts which erode to form inselberg landscapes.

Stepped duricrust landscapes usually occur in savanna areas with somewhat sparse vegetation and in which wash is an important process on moderate slope gradients. Dissection of a plateau in these areas tends to produce a stepped landscape, with a sharp break of slope at the base of steep slopes, and the duricrust merely emphasizes this form and provides coarse boulders to veneer the debris slope which, in an area of deep weathering would otherwise be less well defined. Successive stages of incision produce a series of steps with the lower, less duricrusted levels having less steep slopes between them.

Inselberg landscapes can arise in two distinct ways; and the residual hill forms tend to be somewhat different in the two cases. (1) they may arise through exhumation of a variably weathered soil–bedrock boundary; and (2) they may be produced by semi-arid slope retreat in sound rock until only isolated residuals are left. The second type of residual, which is called a 'butte' here, is very characteristic of the American south-west, where deep weathering is rare. Buttes form in a variety of lithologies, including granites, basalts, and sedimentaries. In shales they are usually associated with a cap-rock layer of limestone or sandstone, but in massive sandstones they commonly occur within a single lithologic unit. Buttes generally have a flattish top and a cliff face and/or a debris-covered straight slope (chapter 13). Where the debris slope is present, the topographic form of the butte seems to be distinctive, but where it is absent, as in some sandstones, the cliff face may be somewhat rounded thus producing a form similar to that of exhumed inselbergs.

Tropical inselbergs have been widely attributed to exhumation (Linton, 1955; Ollier, 1960; Thomas, 1965) and they are thought to form where the rock is more resistant to weathering, principally by having wider spaced joints. Small differences in primary porosity have been seen (chapter 9), to have considerable effects on the rate at which weathering begins and sub-surface bedrock may have a relief almost as great as the depth of weathering. Weathering in any case produces rounded cores, even on a large scale; and this is usually emphasized in tropical inselbergs by dilation jointing which tends to produce convex spalls, so that inselbergs are typically dome shaped. The common association of inselbergs with massive acid rocks may be due to the poor development of duricrusts during deep weathering; on the other hand the development of inselbergs on plutonic rocks is favoured by their massive jointing and lateral variability in joint spacing.

The planation process by which deep-weathered material is removed from residual hills is not very different from that occurring in duricrusted areas, but the rates of planation are higher and the step form is less pronounced in the absence of a plentiful supply of coarse debris. The two ways in which residual hills may be formed are mutually exclusive at any one location, but there is little doubt that each is effective in some areas.

SUMMARY

Few process studies have been carried out in tropical areas, so that part of our knowledge is based on extrapolation from temperate and semi-arid areas. It is important to keep in mind that much of the literature on the tropics has confused the effects of differences in process which occur in the tropics with the properties of any large land mass; that is, to possess large plain areas which are subjected to denudation and weathering for very long times. Tropical processes differ mainly in the larger importance of chemical solution, but it is only two to five times greater than in temperate areas. This difference modifies slope forms, especially in producing flat floored valleys in the humid tropics, but the differences may be considered slight. It is the combination of higher weathering rates with long times which produces the most distinctive tropical landscapes, namely, duricrusted plains and inselberg plains.

PART FOUR: SYNTHESIS

CHAPTER 15

SLOPE PROFILES

In this chapter, the most important ideas contained in parts two and three will be brought together in a general discussion of slope profile change through time. In order that these more recent developments may be placed in a historical perspective, the chapter begins with a summary of the classical models of W. M. Davis and Walther Penck. This is followed by a synthesis of modern attitudes to the problem, devoted particularly to an appraisal of the influence of climate and rock type on the pattern of slope profile development. In the last section we briefly explore the topic of valley asymmetry in an attempt to examine the role of (micro) climate at a radically different scale.

CLASSICAL MODELS

The diagram of fig. 15.1, supposedly contrasting the development of slope profiles through time according to Penck and Davis, is perhaps the most well-known figure in slope geomorphology. It is taken from Davis' (1932) *Piedmont Benchlands and Primärrumpfe* at a time when Davis' ideas appear to have become very confused. The left side of the illustration, purporting to demonstrate Penck's ideas, is in fact a great misrepresentation of the German geomorphologist's beliefs; and the right side of the diagram, whilst little different from Davis' initial model (Davis, 1899), appears to be inconsistent with some of his statements only two years before (Davis, 1930). In view of the superficial treatment often given to summaries of the Penckian and Davisian models in the past, a fairly lengthy treatment is devoted to them here.

The ideas of W. M. Davis

The earliest lengthy account of slope profile development by Davis is found in his classic *Geographical Cycle*. It is characterized not only by an impressive literary style, but, unfortunately, also by a marked vagueness. Nowhere is there any explicit reference to the effects of a particular process on the geometry of slope profiles, and the various denudational processes operating on hillslopes are merely lumped together as *the agencies of removal.*

The transportation of the weathered material from its source to the stream in the valley bottom is the work of various slow acting processes, such as the surface wash of rain, the action of ground water, changes of temperature, freezing and thawing, chemical disintegration and hydration, the growth of plant roots, the activities of burrowing animals. All these cause the weathered rock waste to wash and creep slowly downhill, and in the motion thus ensuing, there is much that is analogous to the flow of a river.   (Davis, 1954, p. 266)*

When the graded slopes are first developed they are steep, and the waste that covers them is coarse and of moderate thickness; here the strong agencies of removal have all they can do to dispose of the plentiful supply of coarse waste from the strong ledges above. In a more advanced stage of the cycle, the graded slopes are moderate, and the waste that covers them is of finer texture and greater depth than before; here the weakened agencies of removal are favoured by the slower weathering of the rocks beneath the thickened waste cover, and by the greater refinement (reduction to finer texture) of the loose waste during its slow journey. In old age, when all the slopes are very gentle, the agencies of waste removal must everywhere be weak, and their equality with the process of waste supply can be maintained only by the reduction of the latter to very low values.   (Davis, 1954, pp. 268–9)*

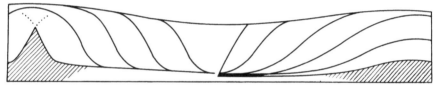

Fig. 15.1. Widening of valleys and change in slope profile according to W. M. Davis (right) and Davis' interpretation of W. Penck (left) (from Davis, 1932).

The essential features of the change in slope profiles through time are depicted in the right-hand side of fig. 15.1. The most noticeable major change is the early replacement of an initially straight-sided slope by a convexo-concave sigmoid profile. (The same point is emphasized by Bryan, 1940, p. 258 in his summary of Davis' scheme.) After this there is a progressive decrease in the average steepness of this profile during the course of the cycle, without any major change in its geometry.

   Subsequently Davis recognized that, although this scheme may represent the pattern of slope development in the 'normal' cycle, that is, in areas of humid temperate climate, it was inapplicable to more arid areas where slope retreat occurs. By 1930, Davis appears to have become very confused, and clearly incorporates elements of slope retreat in the sequence of slope development in humid areas also. The sequence of illustrations in fig. 15.2 demonstrates his attitude to the evolution of humid slopes at that time. It presents a rather different picture to that

* From Geographical Essays by W. M. Davis; Dover Publications, Inc., New York, 1954.
  Reprinted through permission of the publishers.

depicted by the right-hand side of fig. 15.1. It is, of course, possible that this stems simply from a perspective, rather than a profile, representation, but Davis' writings, as well as his diagrams, now explicitly refer to slope retreat.

The continued removal of the finer soil from the valley-side slopes causes them to recede from the banks of the graded stream to which they previously descended; narrow strips of valley floor will be developed at their base, back of whatever flood-plain is simultaneously formed by the stream. As these strips widen, the concavity of the profile across the valley bottom is given broader expression.

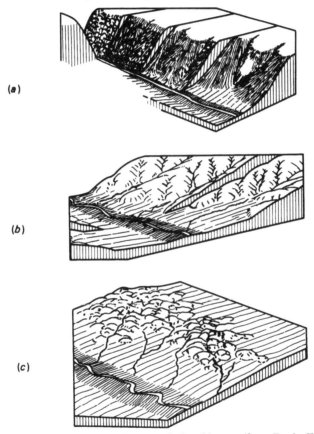

(*a*)

(*b*)

(*c*)

Fig. 15.2. Three stages in the cycle of erosion in humid areas (from Davis, W. M., 1930, *Journal of Geology*, **38**, University of Chicago Press).

The lateral valley-floor strips are, like the flood plain, underlain with degraded rock which may be called the valley-floor basement. It has a somewhat ragged and vague surface, except that where cut by lateral shifts of the stream it may be more even and better defined. The lateral strips will be everywhere covered by detritus, partly derived from local subsoil weathering, partly washed down from the valley sides; and the detrital cover will, in time, constitute a large part of the valley floor.

The lateral swinging of the graded stream broadens the floodplain and its rock basement, and thus the valley-floor strips may be encroached upon; but in spite of this, they gain width by the retreat of the valley-side slopes.

(Davis, 1930, p. 136)

In the words of King (1953, p. 738),

In his 1930 analysis Davis has gone a long way from his simple belief in 'graded' hillslopes of the Geographical Essays. His exposition ends indeed with a diagram [fig. 15.2c] of the ultimate landscape form under the erosion cycle which is far removed from his earlier concept of the classical peneplain. It is none other than the pediplain, typical senile landscape familiar in semi-arid environments, utterly unlike the peneplain of broad convexities in that it is composed of numerous broad, concave pediment or pediment-like surfaces from which rise steep-sided residuals, not gentle monadnocks.

In the same paper Davis continues:

As the various processes of arid and of humid erosion are thus seen to differ in degree and manner of development rather than in nature, and as their differences in degree and manner are wholly due to differences in their climates, so the forms produced by the two erosions may be shown to differ in the degree to which certain elements of form are developed rather than in the essential nature of the elements themselves.

(Davis, 1930, p. 147)

As King so eloquently puts it:

Davis here stood upon the threshold, with the unification of epigene landscapes before him. But the shackles of his earlier thought, nearly 50 years upon him, were not to be broken so easily, and he never entirely accepted real identity of landscape evolution under humid and arid influences.

(King, 1953, p. 724)

Two years later, Davis (1932) reverts completely to his original ideas on slope flattening and the illustration of fig. 4 of his article *Piedmont benchlands and Primärrumpfe* (our fig. 15.1), seemingly inconsistent with his sketches of the 1930 paper (our fig. 15.2), appear emphasizing this.

In view of these apparent inconsistencies it is not surprising that many geomorphologists find Davis almost as confusing as Penck. Today his 1930 ideas are often overlooked and, notwithstanding King's comments, Davis' name is still associated with the concept of slope flattening and the development of a peneplain.

The ideas of Walther Penck

The development of slope profiles through time, under conditions where streams are no longer cutting down into a landmass, was discussed at length by Penck (1953) on pages 134–44 of his *Morphological Analysis*, in a section entitled, significantly enough, the *flattening* of slopes.

Penck begins with a straight, steep rock face (*steilwand* or *felswand*) standing above a stream at its base. Why Penck assumed a uniform slope, as a starting point, after all the attention he paid to waxing and waning development is not clear. Perhaps this was purely for reasons of expediency. The fact that he did assume an initially straight rock face is fortunate since it makes it possible to compare his ideas with those of others. In the light of this, Davis' representation (fig. 15.1) of the Penckian scheme as beginning with a convexo-concave profile is even more perplexing. Accepting this starting point, Penck then argued as follows.

Weathering converts the surface of the rock face to loose rubble which, assuming the initial slope to be steeper than the angle of stability of loose rock debris, falls to the slope base and is removed by the

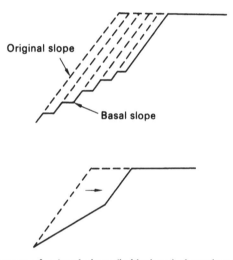

Fig. 15.3. The emergence of a basal slope (haldenhang) through retreat of a cliff face (steilwand) according to W. Penck (based on Penck, 1953).

stream. Since 'the whole surface of the cliff has the same exposure and succumbs equally in every part to the process of reduction' (p. 134), continual weathering and rockfall result in the retreat of the cliff face at an unchanging angle as depicted in fig. 15.3. Penck demonstrated, step by step, that the effect of the retreat of the cliff is to produce a new straight slope, the *haldenhang* or basal slope, presumably, although not explicitly stated, at the angle of repose of the loose debris. Although Penck's reasoning is not mathematical, his derivation of the *haldenhang* appears to be identical to the later ideas of Bakker and Le Heux (1952) (specifically the case where $c = 0$) treated in chapter 6. Penck, however, went further than the Dutch authors for the *haldenhang* itself was believed to develop in a similar way.

From its first appearance, the basal slope develops independently, since the rock on it, too, is being reduced. Here, however, the mere loosening of pieces of rock from

373

the general fabric is not enough to produce denudation, as it is on the steep cliff face. A far greater degree of reduction, i.e. far greater mobility, is required for rock derivatives to migrate on the very much smaller gradient, and for this, much longer periods of time are needed. The development of the basal slope is therefore very much slower; but it proceeds in the same direction as that of the cliff face above.

It is still being assumed that the rock is all of the same composition and also has the same exposure. In unit time a layer of definite thickness, everywhere the same, is loosened from the basal slope. But it is only when a multiple of that unit time has elapsed that the loosened material is sufficiently mobile to migrate spontaneously.

(Penck, 1953, p. 136)

The retreat of the *haldenhang* creates a new slope of even gentler gradient, the *abflachungshang*. It should be emphasized that Penck does not discuss the factors controlling the angle of the *abflachungshang*, although implicitly he links it to the stability of the reduced rock debris. Nevertheless, perhaps we can remark that the sequence so far described closely resembles the model suggested by Carson and Petley (1970) and discussed in chapter 7. But Penck's development does not end here. The process is continually repeated, and the retreat of the *abflachungshang* leaves a slope of even gentler gradient which in turn retreats itself.

The denudation on all inclined flats (*geneigten flächen*) proceeds now in such a way that the flats retreat parallel to themselves at a constant gradient, and at their foot a flat of smaller gradient grows upwards increasingly, so long as no river erodes deeper there, hindering the formation of the gentler foot-flats.

(Penck, 1925, p. 89)

The net result of this continual slope replacement is shown in fig. 15.4. According to Penck, 'if left undisturbed, a slope of any gradient whatsoever, provided it is uniform, becomes a slope system concave in profile' (Penck, 1953, p. 140). In this statement he is clearly, although perhaps not deliberately, repeating similar ideas proposed by Davis.

When the migration of divides ceases in late maturity, and the valley floors of the adjusted streams are well-graded, there is still to be completed another and perhaps more remarkable sequence of systematic changes than any yet described: this is the development of graded waste slopes on the valley sides.

(Davis, 1954, p. 266)*

A graded waste sheet may be defined in the very terms applicable to a graded water stream; it is one in which the ability of the transporting forces to do work is equal to the work that they have to do. This is the condition that obtains on those evenly-slanting, waste-covered mountain sides that engineers call the 'angle of repose' but that should be called, from the physiographic standpoint, the angle of first-developed grade.

... rivers normally grade their valleys retrogressively from the mouth headwards. So with waste sheets; they normally begin to establish a graded condition at their base and then extend it up the slope of the valley side whose waste they 'drain'.

(Davis, 1954, p. 267)*

* From *Geographical Essays* by W. M. Davis; Dover Publications, Inc., New York, 1954. Reprinted through permission of the publishers.

Davis went on to argue that successive angles of grade, developed later than the angle of repose, similarly extend upslope and eventually produced a concave lower slope profile. Equating the *haldenhang* with Davis' angle of first-developed grade, there is clearly a great deal of similarity between the two ideas.

So far there has been no mention of the upslope convexity, but on

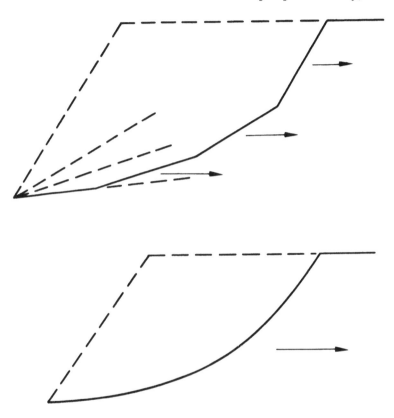

Fig. 15.4. The successive retreat of inclined valley flats (geneigten flächen) and the development of a concave profile according to W. Penck (based on Pcnck, 1953).

pages 141–4 of *Morphological Analysis*, Penck proposes a mechanism by which the upper parts of slope profiles gradually become rounded. His arguments are very similar to those offered in chapter 11, and will not be discussed here. The net result of the rounding of the summits and the development of the *geneigten flächen* is clearly a convexo-concave profile, a profile produced not by successive waxing and waning development, but during a period of constant base level.

The attempt by Davis (1932) to depict graphically Penck's views on slope development in the left-hand side of fig. 15.1 is clearly very mis-leading. This was stressed independently by Tuan (1958) in a rather obscure article, and more recently by Simons (1962) in an article pre-viously mentioned. Unfortunately the misinterpretation of Penck's ideas by Davis has been repeated on many occasions and fig. 15.1 must rank as one of the most misleading diagrams in the history of geo-morphology. Subsequent to the original article by Davis, the diagram was reproduced in Bryan's (1940*b*) well-known paper on *The Retreat of Slopes*, Cotton's (1948) *Landscape*, King's (1953) famous *Canons of Landscape Evolution*, and, in a slightly different form, by Dury (1959) in *The Face of the Earth*. The article by King (1953) was particularly mis-leading since he attempted to equate Penck's ideas with his own on pediplanation, and identify the pediplain with the end-product of the Penckian model, *Endrumpfe*. Actually the two ideas are very different. The pediplain, as defined by King, is a surface produced by the con-tinual retreat of two elements, a scarp and a talus slope. Now although

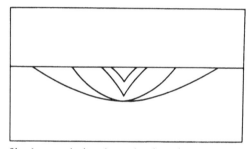

Fig. 15.5. Slope profile changes during the cycle of erosion according to W. Penck (from Penck, 1953).

these two elements may coincide with the *steilwand* and *haldenhang* of the Penckian system, the emergence of *Endrumpfe* is due to the retreat of a succession of *geneigten flächen*, not merely two of them.

Admittedly the picture of slope development on the left hand side of fig. 15.1 resembles very closely King's ideas on scarp retreat and the emergence of a pediplain. This sequence is, however, a complete mis-representation of Penck's ideas as we have already emphasized. Penck's own illustration of the pattern of slope development (Penck, 1925), depicted in fig. 15.5, is radically different to Davis' attempted portrayal of it. The difference between these two figures (fig. 15.1 and fig. 15.5) is very important. As Simons (1962) notes:

His [Penck's] entire argument rests on the premise that the steep parts of a slope must be denuded more rapidly than the gently inclined parts. For a whole valley-side to retreat in the manner suggested by Davis' diagram, all parts of the slope, steep and gentle, would have to be denuded at the same rate, so that each small part or unit of the slope would retreat at the same speed. Penck, *in direct contradiction of the parallel retreat hypothesis*, spoke of 'the known law that intensity of denudation, which is equal to the rate of development of the slope units, increases with their steepness'. (Simons, 1962, p. 3)

Enough has been said to demonstrate that the classic antithesis created by Davis, between the Penckian system and his own, has been very misleading. Penck, like Davis, believed in the flattening of slopes through time. The real father of the school of parallel retreat is probably Bryan (1940); its greatest proponent, Lester King. Indeed, accepting some of Davis' statements in his 1930 paper on *Rock Floors in Arid and in Humid Climates*, it would be possible to invert the entire classical Davis/Penck dichotomy, but perhaps enough damage has already been done.

A MODERN APPROACH

The models of slope profile development conceived by Davis and Penck have been presented primarily to add historical perspective to the more modern ideas of this text. The casual treatment of denudational processes in both models renders them of little use to the modern geomorphologist. We shall have little cause to refer to them again.

In parts two and three, a large amount of evidence was compiled to show that most, even if not all, slopes possess rounded convex summits, and are separated from stream channels at their bases by shallow concave elements. The upslope convexity and the concave base are both products of processes that are essentially transport-limited. The *upslope convexity* stems from a combination of weathering, creep and rainsplash. The *concave base* appears to result from the retreat (not necessarily at an unchanged angle) of the valley-side slope through surface wash and solution; it may also be partly, or even wholly, a depositional feature. The remnant part of the slope profile between the upslope convexity and the concave base is, on the other hand, primarily a reflection of weathering-limited processes, particularly instability movements, but also, in semi-arid areas, surface wash in a weathering-limited form. We shall call this part of the profile, between the upslope convexity and the basal concavity, the *main slope*. In the deductive treatment which follows, we shall argue that slope development is perhaps most profitably approached by separate analysis of (a) the form, and (b) the behaviour, of the main slope.

In the simplest case the main slope is a single slope unit. It may be a vertical sandstone cliff, a straight, debris-mantled slope, or a smoothly concave, rill-dissected scarp. Our ideas on main slope behaviour are most conveniently illustrated by these simple cases. There are three

377

possible systematic ways in which the main slope may behave under denudation: retreat, decline and shortening. These are shown in fig. 15.6. The process of retreat is the mechanism by which a basal concavity is produced; the shortening of the main slope is due to the development and down-slope extension of the hilltop convexity. Neither a concave base nor an upper convex slope is produced by decline alone.

On many slope profiles, the main slope is not a simple, single-unit slope, but consists of a number of distinct units. Sometimes this is the result of lithologic variations in the rock mass; often it is due to an instability sequence (chapter 7) in which one slope unit is retreating

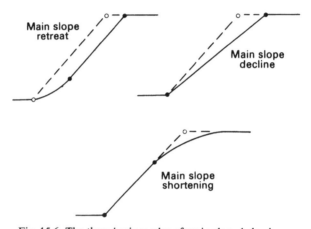

Fig. 15.6. The three basic modes of main slope behaviour.

under landslides, and is being replaced by another, more stable, gentler slope. The most common example of an instability sequence is the rockwall–talus slope profile. This occurs in all climates wherever slopes are cut in strong, closely-jointed rock. It is especially common in semi-arid areas. In cold regions, a three-unit (rockwall–talus–taluvium) main slope profile is probably more common than a two-unit suite. This is also true for humid temperate areas. Such multi-unit main slopes are not restricted to closely-jointed rocks, e.g. double-segment profiles in the London Clay (Hutchinson, 1967) noted in chapter 7, but they are most common in such rocks.

In order to appreciate the development of a complex (multi-unit) profile, we need to know more than simply whether the main slope, *as a whole*, retreats, declines or shortens. We need to assess the extent of *internal adjustment* in the geometry of the main slope itself. Since a complex main slope is essentially a succession of retreating slope units, the important issue is whether or not these units retreat at the same rate.

In the situation in which the individual units do retreat at the same rate, there is clearly no internal change in the geometry of the main slope, and a multi-unit main slope may be treated as a simple slope. Internal alteration will take place only if units retreat at different rates, that is, in effect, if the upper, steeper units retreat more rapidly than the gentler units below. The result is that the upper units gradually disappear from the main slope profile and are replaced by the lower units. This internal change in the geometry of the main slope is termed *slope replacement*, following Savigear (1960). A simple example of slope replacement is the retreat of a rockwall and the accumulation at its base, *without any subsequent removal*, of talus derived from rockfall; eventually the talus slope, extending upwards, completely obliterates the original rockwall. This is radically different to the retreat of a rockwall-talus slope as envisaged by Lester King (1953); according to King, the talus slope and the rockwall retreat together at the same rate, so that they may be regarded as a simple main slope.

From the discussion above, it may be appreciated that the evolution of *any* slope profile undergoing denudation may be approached through a study of three distinct points:

(1) the *form* of the main slope, that is, whether the main slope is simple (single-unit) or complex with more than one unit;

(2) the *internal alteration in geometry* of the main slope over time, that is, whether or not slope replacement occurs;

(3) the *overall behaviour* of the main slope, that is, whether it retreats, declines, shortens, or combines more than one of these changes.

This is illustrated in figs. 15.7–15.9. The differences between the figures are meant to demonstrate the effect of lithology, particularly the *disintegration sequence* of the rock mass, on slope development. The contrasts between (*a*) and (*b*) elements of each figure are meant to indicate the differences between semi-arid and humid temperate climatic conditions respectively. These figures are simply a graphic summary of the main ideas and observations discussed in parts two and three of this book. The sequences are therefore not described in detail here, although a brief résumé of these figures follows.

1. *Valley-side slope development during initial phase of active stream down-cutting*

i. Valley-wall is shown vertical in fig. 15.7*a*; this is based on the assumption that, in this type of rock in a semi-arid environment, the weathering-limited processes act very slowly *relative* to stream down-cutting.

ii. Valley-walls in fig. 15.8 are steep (45–75°) but not vertical. This is based on the assumption that weakening of the jointed rock mass accompanies stream incision and, in effect, the rock mass acts as a densely-packed cohesionless system; as a result, rock avalanches (chapter 6, p. 123) move debris off valley-walls, lowering

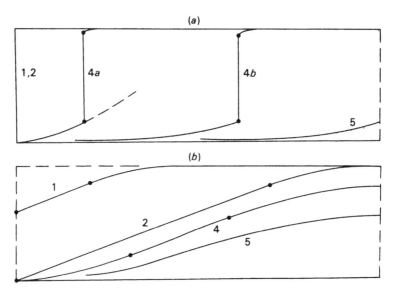

Fig. 15.7. Slope profile development in massive sandstone rock in (a) semi-arid and (b) humid temperate environments.

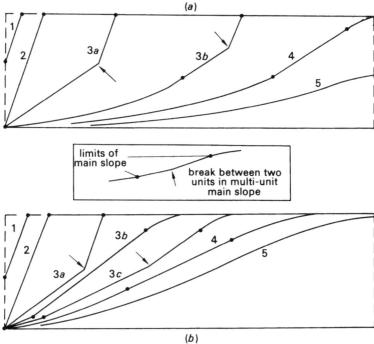

Fig. 15.8. Slope profile development in strong, closely-jointed rock in (a) semi-arid and (b) humid temperate environments.

them at a constant angle which may vary from 70–75° (chapter 6, p. 124) to 45° (chapter 11, p. 318)

iii. Valley-wall in fig. 15.9*a* is steep due to genuine cohesion and capillary cohesion (chapter 4, p. 76) of the clay mass in semi-arid areas; slightly concave profile reflects rill- and gully-erosion which accompanies sliding on walls during stream down-cutting. This is the *badland scarp* referred to by Schumm and Chorley (1966, p.18).

iv. Valley-walls in fig. 15.7*b* and 15.9*b* are gentler (20–30°) and also show greater development of the summit convexity. Gentler main slope assumes rapid weathering of these rocks in humid environment and loss of cohesion in surface soil mantle; angle of maximum stability during stream down-cutting phase is, therefore, low (chapter 7, p. 179). Development of convexity during stream down-cutting is due to rainsplash and creep, and has been outlined previously (fig. 10.12).

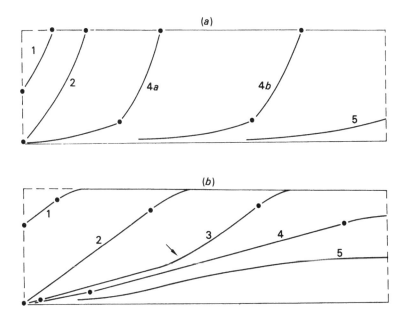

Fig. 15.9. Slope profile development in a clay mass in (*a*) semi-arid and (*b*) humid temperate environments.

2. *Valley wall shown at instant when active stream down-cutting has halted and, henceforth, slope processes dominate.*
3. *Slope replacement of initial valley-wall.* Whether or not slope replacement occurs, and the complexity of the process if it does so, depends on the instability sequence (chapter 7, p. 183) of the slope. This in turn reflects the disintegration sequence of the rock mass and the climatic setting.

i. In sandstone (fig. 15.7), it is assumed that there is no replacement of the initial slope by another unit. In (*a*), loose sand grains which fall to the cliff base (granular disintegration or impact-disintegration after slab-failure) are easily washed away by surface water. In (*b*), it is assumed that the sandy soil-mantled slope is already at the angle of ultimate stability; this assumption is made because once the rock has weathered to a sandy mantle, little further weathering (and change in shear strength) is likely.

ii. Similar statements might be made about a clay mass in which the disintegration sequence is also rather simple. Note, however, that one phase of slope replacement is shown in (*b*). This is the change from a temporary angle of stability to an ultimate angle of stability for the clay mass, described in chapter 7, p. 180. Note also that in some cases this change may occur through slope decline rather than slope replacement.

iii. The complexity of the disintegration sequence in jointed rocks is reflected in the development of the slope profiles. In (*a*), a one-phase instability-sequence (rockwall replaced by talus-mantled slope) is shown; it is assumed, following King (1953), that washing-away of fine debris off the talus slope prevents replacement from occurring. As a result, the rockwall-talus slope retreats as one slope, leaving behind a pediment. In (*b*), three phases of instability are shown, following Carson and Petley (1970), although there may be more. The rockwall is replaced by the talus-mantled slope, and this, in turn, is replaced by the taluvial slope. The taluvial slope is unlikely to stand at the ultimate angle of stability of the mantle, and there may be further replacement or some decline (from 3c to 4) in steepness.

4. *Modification of the stable main slope.* Once stable, the main slope is slowly obliterated from the slope profile through retreat (solution and soil wash) and shortening (creep and rainsplash) as discussed in chapters 8–10.

i. In semi-arid conditions (*a*), retreat is assumed to be dominant. It is due to strong surface wash of loose material that is only poorly, if at all, bound together by vegetation. Solution is probably less important, although (figs. 9.17–9.18) re-deposition of solutes may accentuate the pediment concavity at the slope base. Only a small amount of shortening of the main slope is shown, and this is due more to rainsplash than creep. The amount of summit convexity is, however, very variable, and is particularly dependent on the soil type as shown by Schumm (1956*b*) in the South Dakota Badlands.

ii. In humid temperate areas (*b*), shortening is assumed to be the most important geometric change; it is due to both creep and rainsplash. Surface wash is much less effective on well-vegetated slopes, and, therefore, retreat is much smaller than in semi-arid areas. The basal concavity in humid temperate areas thus remains an embryonic pediment of the most rudimentary type. Solution, however, is probably very strong, and may be an important mechanism of retreat in these areas.

iii. In cold areas and the humid tropics, available evidence suggests that slope development is intermediate between extremes (*a*) and (*b*) in figs. 15.7–15.9. Cold areas seem to resemble semi-arid areas in *basic* slope form; and, if one accepts Jahn's (1960) ideas, this may be attributed to the relatively important role of surface wash processes. Slope profiles in the humid tropics appear to be more closely aligned to those of humid temperate areas. On many profiles, however, the basal concavity is more strongly-developed and abuts more sharply against the main slope than in humid temperate areas; if one accepts figs. 9.14–9.15, this may be a reflection of the stronger role of solution in the tropical environment.

5. *Obliteration of main slope and emergence of convexo-concave profile.* Extension of the ideas in 4 above indicates that the eventual profile should be dominantly concave in semi-arid areas and dominantly convex in humid temperate areas. Cold areas and humid tropical areas should show intermediate forms.

The graphic summary above has been essentially qualitative. The lengthy treatments in chapters 6–10 have shown, however, that slope development may be approached quantitatively. In the early stages of development, the important features of slope profiles are the *form* and *internal adjustment in geometry* of the main slope; this has been shown to be a function of instability sequences. The number of instability sequences, the steepness of the threshold slopes and the mode of change from one threshold slope to another, all depend (chapters 6 and 7) on a complex interaction of weathering and soil mechanics which can only be successfully approached through a quantitative examination. Similarly the *overall behaviour* of the main slope (retreat, decline, and shortening) is also best examined quantitatively. In chapters 8–10, it was shown that the characteristic form (Kirkby, 1971) to which slopes, once stable, tend to attain, is very dependent on the type of process operating on them. Earlier, in chapter 5, it was argued that many slope processes are most profitably approached in terms of the exponents in equation 5.4, with $f(x) \propto x^m$:

$$C \propto x^m \cdot \left(-\frac{\partial y}{\partial x} \right)^n$$

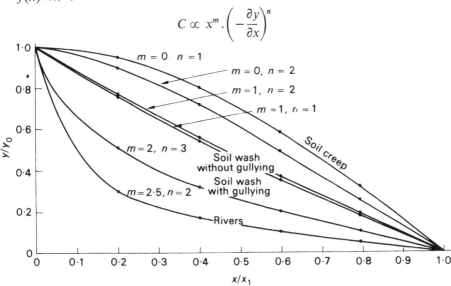

Fig. 15.10. The characteristic form of profiles associated with different denudational processes (from Kirkby, 1971).

where $n \geqslant 0$. The full value of this approach can now be seen. Using this equation, it was shown (chapter 5) that the characteristic form of the slope profile approximates to (equation 5.5):

$$-\frac{dy}{dx} \propto x^{(1-m)/n}$$

assuming that contour curvature (discussed in chapter 16) is negligible. Processes characterized by values of $m < 1$ result in shortening and pro-

383

duce upslope convexities; processes characterized by $m > 0$ result in retreat and produce basal concavities. This is shown in fig. 15.10 which is, essentially, a more quantitative expression of stages 4–5 in figs. 15.7–15.9. The important point is that, *in terms of slope profile development*, the most meaningful single aspect of a process is its classification, in terms of the exponents m and n, in equation 5.6. In this respect, rainsplash should be grouped with creep; solution should be grouped with surface wash. Unfortunately this is not the type of grouping that has been attempted by geomorphologists in the past.

Our ideas on the significance of climate clearly conflict with those of Lester King (1957, 1962) who argues that 'the four-element hillslope (crest, scarp, debris slope, pediment) is the basic landform that develops in all regions of sufficient relief and under all climates wherein water flow is a prominent agent of denudation' (p. 139). There are two main areas of conflict between this statement and our views. In the first place, King maintains that all main slopes, comprising a scarp and a debris slope, are complex. In contrast, we have emphasized that the character of the main slope depends very much on the weathering characteristics of the rock mass; some scarps may be simple, others may contain more than two component units. Admittedly King qualifies his claim, to some extent, pointing out that 'a weak bedrock, readily breaking down under weathering, tends to eliminate the scarp face' (p. 139), but even this statement hardly acknowledges the general importance of the character of the rock mass as outlined in previous chapters. Notwithstanding this, it is probably true that the scarp–debris slope sequence is the most common of all types of main slope profile. Secondly, where King continually emphasizes the retreat of the scarp–debris slope sequence, we have suggested that the development of a complex main slope may proceed in a variety of ways, involving decline and shortening, as well as retreat, and, in addition, an internal change involving slope replacement. The relative importance of retreat and shortening, as relates to the main slope profile, and the relative importance of retreat and replacement, as far as internal changes in the main slope are concerned, depends very much on the climatic setting. King clearly recognized this, judging by some of his comments, but in his attempt to stress the universality of pediplanation, he biased his general statements towards the retreat of slopes far more than observations outside semiarid areas permit. Notwithstanding these differences, we nevertheless lean far more towards King's attempt to formulate a general framework for the study of slope development than the idiographic approach of Büdel (1948), Peltier (1950) and others in their construction of numerous 'morphogenetic' regions.

THE ROLE OF CLIMATE AS INDICATED BY VALLEY ASYMMETRY

It has long been recognized that valley cross-sections in some areas are markedly asymmetric in that the form and average steepness of slope on one side of the valley differs radically from that on the other side. Moreover, since this asymmetry appears to be especially common in west–east aligned valleys, where differential insolation between opposite slopes is at a maximum, this has led many geomorphologists to believe that the influence of climate is not confined to latitudinal differences, but also exists within small areas. At this scale, it should be remembered, the micro-climate of adjacent or opposing slopes will be influenced not only by the direction in which they face, but also by considerations of local topography, most noticeably valley width and depth. This obviously complicates attempts to ascribe asymmetry purely to the rather consistent variations in direct-beam solar radiation which are associated with slopes of different aspect. Since the variance in other controls, such as lithology, is likely to be much smaller at this scale, it provides an excellent opportunity for a direct study of the influence of climate on slope profile development. Some work on valley asymmetry in cold areas has already been discussed in chapter 12; in this chapter we shall confine our attention entirely to information shed by asymmetry studies on the contrast between humid and dry conditions.

We should emphasize here that it is probably only under rather rare circumstances that valley asymmetry is a product solely of micro-climatic contrasts between opposite slopes. Even at this scale, other controls still exist to complicate the picture. In the first place, valley asymmetry may reflect asymmetric undercutting by streams and, in fact, relate little to processes operating on the slopes themselves; and, secondly, even when asymmetry is produced by differential denudation on opposite slopes, these differences may result from factors other than micro-climate; although these will not usually remain consistent over a large area.

Two mechanisms of asymmetric lateral corrasion by streams have long been recognized, although little empirical work has actually been undertaken to test the validity of the ideas. The first is the supposed influence of the Coriolis force on the movement of streams. Over a century ago, Bauer (1860) argued that, as a result of the Earth's rotation, streams in the Northern Hemisphere would be continuously deflected to the right (and to the left in the Southern Hemisphere) and right-hand valley walls would be subject to much greater undercutting than left-hand walls. Since then numerous workers have echoed this opinion. Summaries of early literature may be found in the articles by Gilbert (1884), Penck (1894) and Fabre (1903), and more recent articles are listed by Emery (1947). The role of the Coriolis force is still controversial (Currey, 1964; Dinga, 1969), but it seems very unlikely that it could be strong enough to produce noticeable asymmetry. So far, no evidence has been produced which really substantiates the concept. The second mechanism, the preferred undercutting of one valley side, rather than another, by the uniclinal or homoclinal shifting of a stream axis down the dip of an inclined bedding plane, is a much more important process. It was emphasized long ago by Powell (1874), working in the Uinta Mountains of Utah, and has been acknowledged in other areas by many workers since then. In areas of alternating weak and strong rocks, such as shales and sandstones, the effect of stream shifting is especially pronounced, but even in areas with little contrast in the sedimentary strata, it may still produce a distinct, though less obvious, asymmetric effect, although this is not substantiated by Kennedy (1969).

Structural conditions may also produce valley asymmetry through contributing to differential rates of weathering and denudation on opposite valley walls. An example of this is cited by Hack and Goodlett (1960) working in the Appalachians of western Virginia. The predominant sandstones and inter-bedded shales in this area dip steeply to the south-east. Considerable asymmetry exists as a result, with a consistent tendency for south-east-facing slopes to be steeper than north-west-facing counterparts. The soil mantle on the down-dip sides of the mountains is much moister than on the north-west-facing side, and this is attributed to percolation inside the mountain masses along bedding planes. The contrasting moisture conditions are reflected in the vegetation pattern with yellow pine forest on north-west-facing slopes and oak on south-east-facing slopes. Hack and Goodlett argue that the contrast in moisture conditions is reflected also in differences in the type of debris transport process taking place on the slopes. The moist south-east-facing slopes suffer little from surface wash and, although mantles are subject to soil creep, little denudation has taken place. Surface wash is much more pronounced on the drier, more easily-eroded north-west-facing slopes, and, as a result, these slopes have been flattened much more than those opposite. The complete sequence of reasoning by these workers contains a number of hidden assumptions, and the evidence they present is certainly not unequivocal, but undoubtedly structural conditions are very important. In addition to its effect on the relative importance of creep and soil wash, it might also be expected that contrasting moisture conditions on north-west and south-east slopes would produce different responses to more rapid forms of instability. At times of prolonged storms, movement of water into the mantle on the down-dip side of hills might be expected to give excess pore pressures, leading to more instability there and, ultimately, to a lower angle of stability.

Indeed asymmetric instability of valley slopes, arising from structurally-controlled contrasts in moisture regimes, is probably much more important generally than is indicated by reports of workers on valley asymmetry.

Despite the importance of asymmetric lateral stream movement and structural influences on the intensity of denudation processes, many examples of valley asymmetry clearly relate to contrasts in micro-climatic conditions due to differences in aspect. In the study by Hack and Goodlett, for instance, the contrasting steepness of north-east and south-west slopes, presumably independent of structure, appears to be readily explicable in terms of differences in exposure. North-east-facing slopes are sheltered more than south-west-facing slopes from drying by the wind and the sun's rays, and, as a result, retain more moisture. Again the more moist slope is consistently steeper than the drier slope opposite, and, as noted previously, Hack and Goodlett ascribe this to the greater intensity of surface wash processes on the drier slope.

Similar conclusions were reached by Emery (1947) in a study of valley asymmetry in west-draining valleys cut in poorly consolidated sedimentary rocks on a coastal terrace in southern California. North-facing valley walls were consistently steeper than south-facing slopes, and, again, the difference appeared to be linked to differences in soil moisture due to contrasting exposure. Vegetation density is much less on the drier south-facing slopes and, presumably, denudation through surface wash processes is much greater. Emery argued, as did Hack and Goodlett, that the effect of more intense erosion on the south-facing slopes is a decrease in steepness there. In this particular case it is unfortunately difficult to separate micro-climate effects from those of structure, since the strata dip fairly consistently, although at a shallow angle, to the south and south-east, and a strong possibility of uniclinal shifting, steepening north-facing slopes, exists.

Unfortunately the conclusions of both of these studies, while probably valid in general terms, are of rather limited value in understanding the contrasting development of slope *profiles* under moist and dry conditions. Both studies examine only differences between opposite slopes in *steepness* (maximum or average) and pay little attention to differences in the actual *geometry* of the slope profile. More recent studies have partly taken this factor into account (Kennedy, 1969) but there does not appear to be any consistent or pronounced variation in the length of the basal concavity in all areas in which asymmetry is present.

In previous chapters it has been argued that surface wash processes do not produce flattening of straight slopes; the effect is retreat of the main slope and extension of the basal concavity. Admittedly this may produce a decrease in average steepness of valley-side slope, but, in the context of slope profile development, it would be misleading to label this as slope decline. Studies of hillslope asymmetry must clearly pay more attention to slope *profiles* if they are to be used in an attempt to understand slope development, and, so far, few studies have done so.

An exception to this is the rather briefly discussed work by Walker (1948) on the slopes of valleys cut in the Hogback and Gros Ventre ranges in western Wyoming. North-facing slopes here are relatively straight in profile, and forest-covered; stream channels closely hug the base of the slope. South-facing slopes, in contrast, are covered with loose bunch grass, a response to the warmer soil, and are heavily gullied. Slope profiles on south-facing slopes reflect this gullying. The steepest

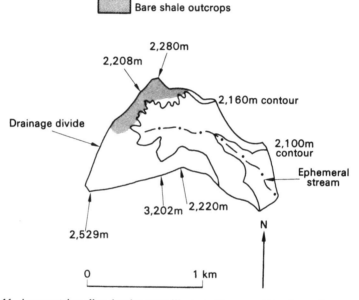

Fig. 15.11. Asymmetric valley development illustrated by a small basin on the eastern rim of the Mesa Verde National Park, Colorado (based on U.S. Geological Survey, 1967).

part of the slope, at about 30°, is as steep as the main slope on the north-facing side, but it occupies a much smaller part of the profile; beneath it occurs a long sweeping concave basal slope. This is attributed to the much greater surface wash on the less-vegetated south-facing slopes. The effect of this has been twofold. The south-facing slope has retreated much farther from the stream, pushing the divide north, and developing the long basal slope beneath it. Associated with the gullying, alluvial fans have developed and the stream has been pushed against the north-facing valley-side. Soil creep is the major process operating on the north-facing slopes, and little, if any, retreat of this slope has taken place. In some ways these observations are akin to those of Emery and of Hack and Goodlett noted earlier; the difference is that Walker emphasizes that the longer south-facing slope stems primarily from retreat under soil wash rather than flattening of the main slope.

Similar asymmetry is developed in the Mancos Shales rim around the Mesa Verde of south-western Colorado. South- and west-facing slopes are commonly bare, heavily-dissected by rills and gullies (fig. 15.11) and seem to have retreated more than opposite slopes. Typical profiles in these bare areas are concave, determined by the gully network, except for the uppermost headslopes leading into the gullies; these are straight slopes cut in the shale at angles of 38–42°. North- and east-facing slopes, while comparable in *average* steepness, lack the sharp concavity associated with gullying on their opposite counterparts.

Potentially, studies of valley asymmetry could prove to be extremely useful in understanding general problems of slope development. This will only be realized, however, if more attention is paid to asymmetry in profile, rather than mere contrasts in average steepness. On a number of occasions previously, we have argued that dry soil conditions favour soil wash, main slope retreat and the development of extended concave basal slopes, whereas moist soil conditions favour mass-movement, shortening of the main slope and convex divides. Walker's work at the micro-climatic scale, just as Schumm's (chapter 8), provides further support for this. It will be interesting to see whether or not future micro-scale studies add to this.

CHAPTER 16

SLOPES IN DRAINAGE BASINS

This book has concentrated on the slope profile as the basic unit for studying landforms, both for its conceptual simplicity and because the influence of contour curvature in modifying profile form has appeared to be of secondary importance. In this chapter, profiles are considered in a more general way for two reasons: first, to assess the importance of contour curvature in modifying slope profiles and in shifting their positions laterally – an assessment which will go some way to justifying the fact that these influences have been ignored so far; and second, to consider the set of profiles in a drainage basin as a spatially distributed set which together determine the topography of the basin, and in particular its drainage texture. The first of these approaches is analysed mainly in terms of the continuity equation, which has already been introduced in chapter 5 and developed for simple profiles in chapter 15. The second approach is analysed more in terms of basin morphometric parameters and slope profile measurements.

THE EFFECT OF CONTOUR CURVATURE ON PROCESSES AND PROFILES

Where contours are curved, the flow lines of water and sediment, which are at right angles to the contours, are not parallel but converge or diverge (fig. 16.1). For the case where the contours are concave outwards, called a hollow by Hack and Goodlett (1960), flow lines converge (fig. 16.1b) so that there are, other things being equal, greater flows of water and sediment passing successive points downslope than in the straight contour situation. Conversely, on a spur or nose, where contours are convex outwards (fig. 16.1c) the flow lines diverge and flows of water and sediment tend to be less than in the corresponding slope position where the contours are straight.

Water flow is greatly concentrated in hollows, and soil moisture is also increased (assuming that the soil has uniform thickness and properties). Where soil properties vary water flows through the soil are similar, but moisture conditions in the soil vary appreciably with the soil properties. For the case of uniform soil properties the water discharge

390

per unit width, q, is approximately related to soil water content by an empirical power law (Kirkby and Chorley, 1967):

$$q \propto (w - w_0)^n, \tag{16.1}$$

where w = water content in mm, w_0 = constant minimum value, and n is a constant exponent (about 4). Under average conditions, the mean flow is equal to the product of the rainfall excess and to the area drained per unit contour length, a (fig. 16.1):

$$\bar{q} = \bar{i} \cdot a, \tag{16.2}$$

where $\bar{i}$ is the mean rainfall excess. Equation 16.1 therefore shows that the moisture content under average conditions is:

$$w - w_0 \propto (\bar{i} \cdot a)^{1/n}. \tag{16.3}$$

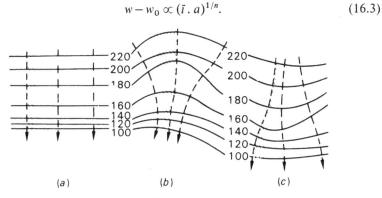

(a) (b) (c)

Fig. 16.1. Contours (solid lines) and flow lines (broken lines) at right angles to the contours.
(a) Straight contours: $a = x$:
(b) Contours concave outwards (hollow): $a > x$:
(c) Contours convex outwards (nose): $a < x$:
where x = distance from the divide along flow line,
a = area drained per unit contour length.
Shaded areas show contribution to a in each case to unit contour length on the 100 contour.

There is thus a general tendency for soil moisture content to increase downslope, but the tendency is much greater in hollows, where a is increasing faster than x.

During high intensity rainfall, flows and moisture contents increase above their minimum values, by amounts which are approximately expressed as:

$$\left.\begin{array}{l} q \propto (i \cdot t)^n \cdot (1 + x/\rho) \\ w \propto i \cdot t(1 + x/n.\rho), \end{array}\right\} \tag{16.4}$$

where i is storm rainfall excess intensity, ρ is the radius of curvature of the contours (positive in hollows), and t is time elapsed (Kirkby and

391

Chorley, 1967, p. 16). These expressions show that hollows not only have higher basal moisture contents, but also increase their moisture contents during storms to a greater extent than areas with straight contours.

The greater moisture contents of hollows have three important effects. First, moisture contents are higher so that hollows are commonly areas of swamps and seeps in humid areas, and of denser vegetation in somewhat arid areas. Second, the higher moisture conditions lead to higher actual evapo-transpiration rates than at similar distances from the divide outside hollows, because the water deficit is lower in the hollows. Third, the higher basal moisture contents and faster increase in moisture content during storms produces saturation of the soil more frequently in hollows than elsewhere, so that overland flow is also more frequent. In addition more frequent saturation in the hollows means that they are more likely to contribute to peak stream flows (even without overland flow). The opposite effects may be observed on spurs, but those are less important.

Sediment transport is influenced by contour curvature in two main ways; (1) by the convergence or divergence of the sediment flow lines which tend to produce more sediment in hollows and less on spurs, as with water flow; and (2) by the way in which the contour shapes influence the processes acting, mainly on account of the increased water flow in hollows. Where the increased water flow is able to transport more than the additional sediment supplied, then the hollow will become a locus of erosion, so that incipient hollows tend to grow larger and become a stable feature of the landscape. Conversely, where water flows are not large enough to carry the additional sediment, incipient hollows tend to fill in and the landscape to remain very regular in outline.

Some of these effects can be seen most clearly by examining the continuity equation for debris movement, for the case where contours are curved. In the notation of chapter 5 (p. 107),

$$\frac{\partial S}{\partial x} - \frac{S}{\rho} + \mu \cdot \frac{\partial y}{\partial t} = 0. \tag{16.5}$$

where S is the rate of debris transport per unit contour length, ρ is the contour radius of curvature (positive in hollows), y is the elevation of the surface at distance x from the divide, t is time elapsed, and μ is the volume of soil produced from each unit volume of rock. Both of the first two terms are influenced by the contour shape; the first through the modification of the debris transport rate in areas of curved contours, and the second term, additionally, through the convergence or divergence of the sediment flow lines.

Equation 16.5 shows clearly the effect of altering the radius of curvature without changing the rate of sediment transport. In hollows (radius of curvature positive) the rate of lowering will be reduced; and on spurs the rate of lowering will be increased. Such effects are, however, rarely to be seen except under rather simple conditions of accumulation; as with scree slopes, where the amount of debris received on the slope is determined by the cliff above. Where the scree slope has convex contours, the rate of thickening is less than where the contours are straight. In an equilibrium situation, areas of the cliff with higher rates

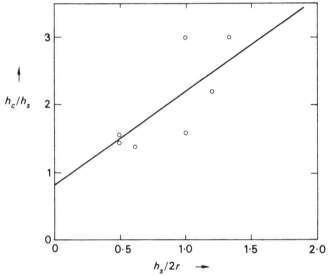

Fig. 16.2. The relationship between butte radius, r, and the proportions of cliff height, h_c, to debris slope height, h_s. Best-fit line indicates a ratio $h_c/h_s = 0.85$ for a straight cliff ($1/r = 0$), and a debris-slope angle of $32°$, compared to a measured angle of $31° = β$.

$$\frac{h_c}{h_s} \propto \left(1 + \frac{h_s}{2r} \cdot \cot β\right).$$ Data for Monument Valley.

of retreat are associated with convex fans of scree; convexities and concavities in the cliff outline are associated with smaller and larger thicknesses of scree material respectively.

A similar simple effect of contour curvature may be seen for the two-stage model of arid slope retreat described in chapter 13 (p. 353). If a sequence of small buttes in similar lithology are compared, the smaller buttes have relatively smaller debris slopes. In the notation of chapter 13 and fig. 13.7:

$$\frac{h_c}{h_s} = \frac{G_s - G_c}{G_c} \cdot \left(1 + \frac{h_s}{2r} \cot β\right), \tag{16.6}$$

where h_c, h_s are cliff and scree heights respectively, G_c, G_s are rates of retreat of bedrock (in cliff and scree) and scree debris respectively, r is the radius of the (vertical) butte cliff, and β is the angle of the debris slope.

Fig. 16.2 shows the best-fit relationship between h_c and h_s for a series of buttes in Monument Valley, Arizona, which is consistent with $G_c = 0.54 \cdot G_s$ and with $\beta = 32°$, compared to the field value of $31°$. For these rather small buttes, the effect of diminishing size is to double the ratio of cliff to debris slope, enhancing the delicate appearance of the narrow buttes.

The influence of contour curvature on the rate and type of process acting is more variable than the simple effect of convergent or divergent debris flow-lines described above. Where removal is weathering-limited (chapter 6), the rate of lowering is directly determined by the rate of weathering; which is, to a first approximation, independent of contour shape. The same remark applies equally to the rate of chemical removal, but it may be expected that the higher water content in hollows will, by increasing the rate of evapo-transpiration, cause increased re-deposition of colloids etc. in temperate areas (chapter 9) and hence lower net rates of removal in hollows. Hollows may therefore tend to become less hollow. On spurs differential rates of chemical solution will generally show a reverse effect, though to a smaller extent.

Slope failures are affected by contour curvature in two opposing ways. In hollows the greater moisture contents and more frequently saturated conditions favour failure, but the shape of the hollow gives some basal support to potential slide masses, particularly of the deep-seated variety (chapter 7). On spurs the opposite occurs. The influence of greater moisture content appears to greatly outweigh the increased basal support in hollows, which commonly appear to be preferred locations for slides to occur.

Slow soil movements of the wash and creep types can be analysed in terms of the continuity equation (16.5) for the transport-limited case. Debris transport rates can be approximated by expressions of the form:

$$S = f(a) \cdot \left(-\frac{\partial y}{\partial x} \right)^n, \tag{16.7}$$

where f is some function of a, the area drained per unit contour length and n is a constant exponent of the slope, $-\partial y/\partial x$ (Kirkby, 1971). Of critical interest is whether hollows and spurs tend to grow into stable features if an initial slight irregularity is present; or whether, instead, the slope processes tend to smooth out initial slight hollows so that they do not develop into stable features. Analysis of equations 16.5 and 16.7 shows that initial hollows will begin to grow if, and only if,

$$\frac{df(a)}{da} > \frac{f(a)}{a}, \tag{16.8}$$

which is also a *sufficient* condition for the profile to become concave (appendix C and Kirkby, 1971). This means that hollows can only *begin* to form in areas where the slope tends to become concave in profile. This analytical solution appears, however, to be an over-simplification because the initial hollow enlargement consists of the formation of rills which are themselves regarded as essentially unstable (chapter 8) because subject to seasonal infilling. In practice therefore, hollows are only able to *start* to develop along permanent channels. Once begun, however, hollows can and do *develop* upslope into areas of convex slope profile, even though they could not be *initiated* there.

If an expression of the type shown in equation 16.7 can be used to describe debris transport on a stable slope, and the area drained per unit contour length, a, does not vary through time, then an approximate form for the characteristic slope profile, after all trace of the initial form has been obliterated, is:

$$-\frac{dy}{dx} \propto \left[\frac{a}{f(a)} \right]^{1/n}, \qquad (16.9)$$

subject to the boundary conditions $y = y_0$ at $x = 0$; $y = 0$ at $x = x_1$. Equation 16.9 may be compared with equation 5.5 for the case $a = x$. This form shows that the equilibrium profile towards which a hollow develops is at a lower elevation than a neighbouring straight-contour slope all the way up to the divide: the effect of the hollow is to increase the values of a more rapidly downslope. Closer analysis shows that if $f(a) \propto a^m$, then the slope profile will tend to become convex throughout if $m \leqslant 0$; concave throughout if $m \geqslant 1$, and convexo-concave if $0 < m < 1$ (appendix C). In the last case the transition from convex to concave will occur nearer to the divide in the hollows than on straight-contour slopes.

For wash and creep processes therefore, hollows can only begin in areas of concave profile, but they ultimately extend headwards towards the divides. The profiles of the hollows differ in scale rather than in type from the neighbouring profiles where contours are straight: the straight-contour profile is compressed laterally so that the point of inflection of the profile is nearer to the divide in hollows, and farther downslope on spurs, than on straight-contour slopes.

River profiles may also be considered as a form of slope profile in which the contours are extremely hollow, concentrating the flow with maximum effect. Since fluvial sediment transport can be approximately represented by equation 16.7, in the form

$$S \propto a^{2 \, to \, 3} \cdot \left(-\frac{\partial y}{\partial x} \right)^2$$

then the long profile tends to a form which can be computed from the continuity equation (16.5). Depending on the exponents in the expression above, the characteristic form for the profile may be approximately expressed as:

$$y = y_0 - B \cdot x^{\frac{1}{2}} \quad \text{or} \quad y = A - B \cdot \log x, \qquad (16.10)$$

where A, B are suitable constants. Both of these forms have been used as empirical fits to river profiles (Tylor, 1875; Jones, 1924).

In making simple analyses of the ways in which contour curvature is able to influence process and slope form, it has been seen that slight curvatures have only rather slight influences, and that only major curvatures are able to produce appreciable effects on the profile, as for example in the case of river profiles – and even this will be questioned below. To a first approximation, the effect of contour curvatures is to produce mainly a compression or extension of the horizontal scale, so that in hollows the point of inflexion at which the profile changes from convex to concave, occurs somewhat closer to the divide; and conversely, on spurs the point of inflexion is moved downslope.

It has been assumed so far that contour curvatures are fixed as hill-slopes develop, but this is certainly not always true. Neighbouring profiles may wear down at different rates so that the contour lines change their forms, altering not only their radii of curvature but also the positions of lines of greatest slope. This means that it may not be feasible to talk about the development of a *single* profile at all, except in rather symmetrical situations like a spur axis or a valley axis; and even in these situations the radii of curvature usually change through time. In some situations profile positions, and hence divides, can move indefinitely: for example, in the case of a ridge developing by soil creep between two streams at fixed, but unequal elevations (fig. 16.3). But this is not the typical situation because the higher stream would normally be steeper and hence cut down at a higher rate than the lower stream. More usually all points in a basin are eroding with reference to a *common* boundary condition provided by the stream level at their common outflow point. Each slope profile which can be defined through time can be shown to tend towards a characteristic form in which the rate of lowering is proportional to elevation above the outflow point (if this is fixed in elevation) (Kirkby, 1971). Because all profiles extend from a common summit point (or a small number of summit points) to a common out-flow point, the constant of proportionality must be the same for all profiles. In other words, the basin tends towards a condition in which contours change their elevation values but not their forms, so that lines of greatest slope become fixed in map position, and contour curvatures no longer vary through time. This theoretical proposition, which has been proposed in a slightly different context by Sprunt (1970, oral communication), has far-reaching implications for basin development which will be examined below. What is important at this point is to note that there are theoretical reasons to suppose that divides and lines of greatest slope will not migrate indefinitely, but will settle into equilibrium positions within a basin which reflect the processes acting just as fully as do the slope profiles.

MEASURING SLOPES IN BASINS

Study of the set of slopes which form a drainage basin is usually severely limited by the large volume of indigestible data which even a small basin

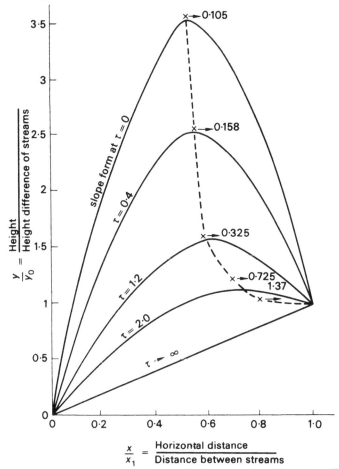

Fig. 16.3. Dimensionless slope profile showing the development of a hillside by soil creep when the base-levels are at different elevations on the two sides of a divide. Broken line shows migration of divide at marked relative rates. τ is a measure of time elapsed.

contains. These data must be compressed into a set of representative profiles, or some sort of generalized profile, or into a series of morphometric parameters which together describe the basin without losing too many of its inter-relationships.

Profile measurements are obviously necessary, and it is clear that contour curvature is sufficiently important for it to be measured down the profile together with slope gradients. Such a profile is, however, somewhat awkward to present, and at least three possibilities are open: (1) to present the contour radius of curvature as an independent piece of information, so that the profile consists of a curve for elevation (or gradient) and a curve for contour radius of curvature (or its reciprocal). This is complete, but difficult to compare with other profiles. (2) to draw the profile in terms of the area drained per unit contour length, *a*, instead of in terms of horizontal distance direct. This approach partly integrates the contour information, but lacks the visual effect of a simple profile. There is also a question whether the value of *a* measured from surface contours is effective in shaping the profile, or whether a 'contributing area' should be used instead – this is a particular problem with hollows and river profiles, as may be seen below. (3) to draw the profile directly in terms of distance from the divide and to express the relationship between distance and measured area in terms of a single constant which is the mean ratio of 'measured area drained per unit contour length' to 'distance from the divide'. This ratio is a true constant equal to 1·0 for straight contours; takes values less than one for spurs; and greater than one for hollows and stream courses. For spurs the ratio is commonly more or less constant, as is shown for a number of examples in fig. 16.4, in which the radius of curvature is plotted against distance from the divide. A constant ratio of ρ/x in this figure is associated with a constant ratio of:

$$\lambda = \frac{a}{x} = \frac{\rho/x}{1 - \rho/x}. \qquad (16.11)$$

For hollows the relationship of *a* to *x* is less simple: in small hollows, the radius of curvature initially decreases as the hollow deepens, but this tendency is reversed once a channel is formed. In a channel, the area drained per unit contour length may be considered equal to the drainage area divided by the channel (or bottomland) width. Since basin length $\propto$ (Area)$^{0.6}$ (Hack, 1957) and channel width $\propto$ (Discharge)$^{0.4 \text{ to } 0.5}$ $\propto$ (Area)$^{0.4 \text{ to } 0.5}$ (Leopold and Maddock, 1953); the area drained per unit width,

$$a \propto (\text{Area})^{0.5 \text{ to } 0.6} \propto x^{0.8 \text{ to } 1.0}$$

Hence, for a moderate range of drainage areas, it is reasonable to

398

assume a *linear* increase of 'area drained per unit contour length' with distance downstream. In the hollow or river case the constant of proportionality, λ, is clearly greater than 1·0. Where a simple presentation is desired, therefore, the third possibility for describing a slope profile, in which the normal profile is associated with single ratio, λ, may

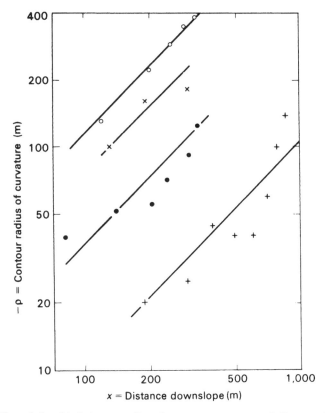

Fig. 16.4. The relationship between radius of contour curvature and distance downslope from a divide, on a series of spurs (noses). $\lambda = \dfrac{a}{x} = \dfrac{\rho/x}{1-\rho/x}$

○ Dartmoor, south-west England $\lambda = 0\cdot54$.
× Dartmoor, south-west England $\lambda = 0\cdot44$.
● Chalk Downs, Dorset, England $\lambda = 0\cdot28$.
+ Cotswolds, Oxon, England $\lambda = 0\cdot10$.

be preferred to the more complex but more complete possibilities (1) and (2).

Slope profiles are normally plotted on arithmetic scales, with or without vertical exaggeration, and this method has obvious visual advantages; but other forms of plotting may be preferred in special circumstances. Because slope is perhaps more important than elevation, it is sometimes plotted direct (usually as the sine or tangent of the angle) against distance (usually horizontal, but may be distance measured *along* the slope): this method emphasizes differences in gradient at the expense of smoothness, and distinguishes clearly between zones of convex, straight and concave slope. An alternative method consists of

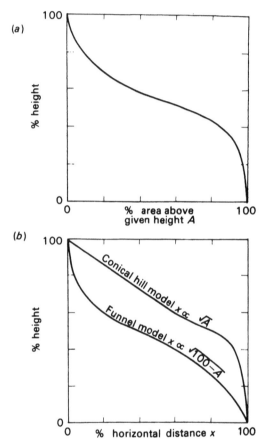

Fig. 16.5. (*a*) Hypsometric curve for an area. Hypsometric integral = 56 per cent. (*b*) Clinographic curves for the same area. Upper curve is based on conical hill model: lower curve on funnel model.

a log–log plot of 'elevation below the divide' against 'distance', a method which has been found to give straight lines for many sections of slope profiles (Hack and Goodlett, 1960), especially near the divides. Figs. 16.9–16.11 show examples of such plots.

Generalized profiles may be obtained from an amalgam of individual slope profiles, but attempts have also been made to use the areal data within a basin to summarize the slopes, perhaps into some sort of synthetic profile. The hypsometric curve is the most frequently-used device, consisting of a cumulative frequency curve showing the total area above each given altitude. Strahler (1952*a*) concentrated on the drainage basin as the descriptive unit, and expressed area and height in dimensionless terms, so that area and height are expressed as percentages of total basin area and total basin relief respectively (fig. 16.5*a*). A critical parameter of the hypsometric curve is the area beneath it, the 'hypsometric integral', which is equal to the volume of material in the basin above its outflow point, as a proportion of the volume of material in a basin-shaped prism extending over the full relief range of the basin. There are two main difficulties with the hypsometric curve as a generalized slope profile: (1) The hypsometric integral settles down to a value of about 50 per cent as a basin attains 'maturity' and thereafter changes little. This may show a property of basins, or may show that the method is insensitive. It clearly shows that the highest points in a basin are reduced along with the rest of the basin. (2) If the hypsometric curve is considered as a frequency distribution, it is scarcely remarkable that the form of the curve is very similar to that for a normal distribution, even allowing for the high degree of auto-correlation amongst the elevations of points in a basin. From this viewpoint, the curve cannot be expected to look at all like a single slope profile, since the individual basin profiles are all over different elevation ranges. It is also doubtful whether the distribution should be referred to its *range* (i.e. the total basin relief), as this is an unreliable statistic. It appears preferable, except in the rare cases (e.g. Schumm, 1956*a*) where the *original* basin relief is known, to define the shape of the curve by its skewness; high hypsometric integrals being associated with negative skewness and low integrals with positive skewness (Evans, 1970).

An additional problem in using the hypsometric curve as a generalized profile is that its dimensions are wrong, since it is a plot of elevation against area [length^2]. This may be remedied by using the square roots of the areas, starting either from the highest or lowest point. In the first case, the basin is imagined as a cone with a single summit point, and the area within each contour is considered as the area of a circle, so that the horizontal distance for that elevation is the radius of the circle, proportional to the square root of the area. This is the clinographic curve (Hanson-Lowe, 1935), but the second case, starting at the lowest point, is equally valid, in considering the basin as a circular funnel with a single outlet point. The two methods give distinctly different results (fig. 16.5*b*), and both suffer from the problem described above, that they are probability distributions rather than profiles.

Morphometric parameters may refer to slope gradients, slope lengths and drainage channel texture, or shape and directional properties. Because the slopes and channels form dual systems, it is often possible to use slope and network parameters more or less interchangeably. Simple measures of gradient are the relief ratio (basin relief/basin length) and the ruggedness (basin relief × drainage density), but both are somewhat indirect measures. An effective method of sampling a map is to count the number of contour crossings along a random traverse (often a grid line). Correcting for random contour orientations, the mean tangent slope

$$= \frac{2}{\pi} \cdot \frac{\text{contour vertical interval}}{\text{mean contour separation along traverse}}.$$

All of these methods yield mean slopes, but it has been argued (Melton, 1957; Carson and Petley, 1970) that maximum valley side slope (excluding cliffs) is the most significant parameter, especially where there are appreciable straight-slope segments. Such straight-slope angles can only be obtained with any accuracy by field measurement; and a simple mean of the maximum slope angles is likely to be misleading, as the distributions commonly have several modes (chapter 15).

Channel network properties have been studied intensively since Horton's (1945) classic paper on (amongst other things) stream ordering, but the conclusion of much recent work is that constant or near constant ratios between the numbers, lengths and areas of streams of consecutive orders are properties of nearly all networks (Shreve, 1967; Leopold and Langbein, 1962), and so have little geomorphic significance. Shreve (1966, 1969) has suggested that the *link*, or section of stream between successive nodes (junctions and/or ends) should be taken as the basic unit of study, and stream order given less prominence. On such a scheme, the measure of drainage texture need no longer be drainage density (total channel length/total area) with the dimension $[L^{-1}]$, but may instead be based on *link* length, which has the advantage of sampling from a large population (of links) with a known distribution (Gamma according to Shreve, 1969). The best measure may be mean *link length*, with dimension $[L]$, but a more convenient measure is *link frequency* (total number of links/total area), with dimension $[L^{-2}]$, which still correlates closely with drainage density, as is shown in fig. 16.6. As may be expected, the relative spread increases at lower densities, suggesting that inverse sampling should be adopted, the sample area being enlarged until it contains a given number of links (say 50–100).

Drainage texture fixes the whole scale of a drainage basin, so that link length, or the reciprocal of drainage density is directly proportional to some average slope length. Basin variables which seem to be independent of scale are those describing orientation and shape. Measures of basin shape have always been notoriously variable within small areas, and do not differ appreciably between areas, but some measure of contour shape appears to be of value, because of the importance of contour curvature, particularly in hollows. A series of measurements of contour radius of curvature may therefore provide a distribution which characterizes an area. Preliminary studies suggest that a still simpler measure, the proportion of an area in which the *contours* are concave (hollow) shows reasonable consistency within an area, and

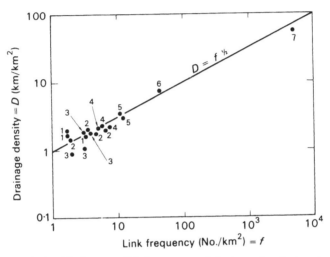

Fig. 16.6. The relationship between *link* frequency and drainage density based on square sample units (not drainage basins). A *link* is the section of stream between adjacent nodes.

Area	Sample area	Source
1 Chilterns, southern England	4 km²	
2 Cotswolds, southern England	4 km²	
3 Dartmoor, south-west England	4 km²	1:25,000
4 Exmoor, south-west England	4 km²	O.S. maps
5 New Forest, southern England	4 km²	
6 Spanish Fork Quad, Utah	1·5 km²	Melton, 1957
7 Tucson Quad, Arizona	0·02 km²	

ranges from 20–40 per cent in southern England. A similar measure, of the proportion of an area in which *profiles* are concave, may also be a useful measure of shape: these two parameters are discussed further below.

Stream orientation may not be significant in an area of homogeneous bedrock, but such areas are few; and orientation is widely considered to depend on geological structure and denudational history (Davis, 1909; Sparks, 1960, chapter 6). It is therefore surprising how few quantitative studies have been made (Lubowe, 1964; Schick, 1965), but workers have been held back by the difficulty of handling orientation data. Work has commonly been limited to stream directions, which have been analysed on the basis of rose diagrams (fig. 16.7), sometimes with the overall flow direction removed so that the total resultant flow vector is zero. A critical, but little-discussed element has been the choice of a unit stream length, the direction of which forms an element in the rose diagram, and decides the scale on which structural control is to be tested.

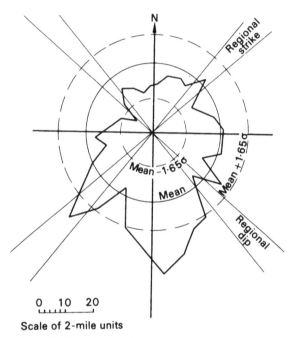

Fig. 16.7. Streamflow directions for south-west Scotland, based on two-mile stretches of stream classed in 10° intervals (from Kirkby, 1963, fig. 2).

We have seen that, although a spatial set of slope profiles provides one of the most meaningful generalizations of hillslopes in an area or a drainage basin, and is the one most directly linked to the approach of this book, it is at certain times either too complete or not complete enough. When it is desired to summarize the information in the profiles, the most significant parameters are mean slope length (drainage density or link length), the distribution of straight slope angles (or its modal value or values), and the mean slope or total relief of the area. Additional information to that contained in the profiles consists of data on areal distributions, summarized in hypsometric curves and in proportions of contour and profile concavity; and in orientation data.

BASIN SLOPE MODELS

One of the simplest models of a basin assumes that the set of profiles in a basin may be arranged as a time sequence, so that the spatial set of forms can be considered as showing the development of hillslopes in the basin; steeper profiles developing into less steep. This idea is implicit in the work of Young (1963b, pp. 12/13), and fig. 16.8 shows such a sequence

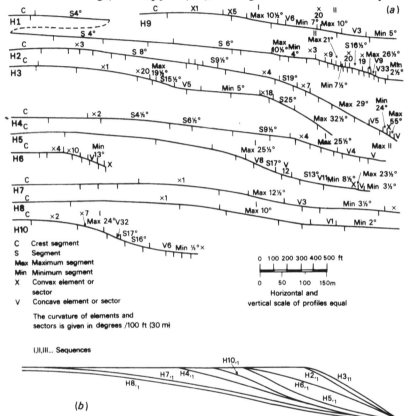

Fig. 16.8. (a) Ten surveyed slope and soil profiles from the Heddon Basin, Exmoor, England. (b) Eight of the sequences reduced to the same height to suggest a possible sequence of slope development through time (from Young, 1963b, figs. 4 and 5).

constructed from slopes in the Heddon Basin, Exmoor, north Devon. Substitution of space for time in this way is intuitively justifiable in certain cases, for example in the case of a cliff which is progressively protected from marine erosion through the growth of a spit in front of it (Savigear, 1952) so that successive profiles have been protected for longer and longer periods; but for the set of profiles in a basin no such progression exists. Carried to its extreme, basin slopes might be thought

405

to develop towards the gentlest profile of all, the valley axis profile, with its relatively long concavity, but such a conclusion appears improbable. We prefer to view the set of profiles as a *space* set, acted on by a range of different processes dictated by local conditions of geology and contour curvature, so that each profile becomes characteristic of the particular set of processes operating on it. In this view the set of *profiles* reflects the set of *processes* operating and there need be no tendency for profiles to develop one into another.

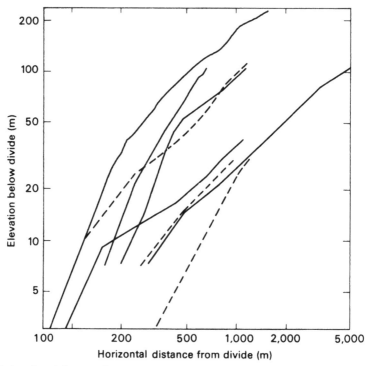

Fig. 16.9. A series of slope profiles and river profiles, taken from the Dartmoor granite area of the 1:25,000 map SX 58.

A second model of slope profiles considers them as a set with variable horizontal and vertical scales. Stretching of the scales in this way can be accomplished graphically if vertical fall from the divide is plotted against horizontal distance from the divide on log–log paper. Changes of scale are now represented by translational movement of the profiles in the x- or y-directions, *without* rotation. Fig. 16.9 shows a series of profiles plotted in this way, with data taken from the 1:25,000 series map (sheet SX 58) for a part of Dartmoor. It is found that by superimposing the profiles, a single composite profile may be constructed from all the plotted points of the individual profiles, as is shown in fig. 16.10. The inset shows the profile corresponding to this log–log plot.

Differences in vertical scale correspond to local differences in relief, but differences in horizontal scale appear to correspond to differences in process. Profiles for which the horizontal distances are less, generally correspond to hollows and stream profiles; whereas profiles for which

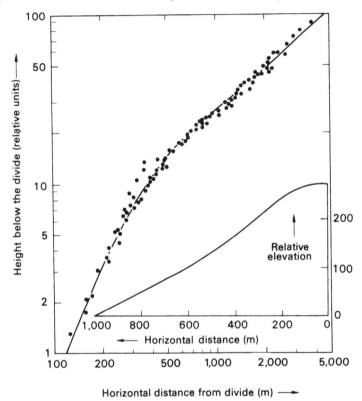

Fig. 16.10. Composite slope profile from data in fig. 16.9 for Dartmoor data. Plotted points correspond to all contour points on profiles shown. Inset shows composite profile on arithmetic scales. Horizontal scales have been chosen as suitable for slopes with straight contours.

the horizontal distances are more, usually correspond to straight slopes and even more to noses or spurs. These tendencies are in agreement with the ideas expressed above (p. 395) about the general effect of contour curvature on profiles, but it is a great empirical convenience if the transformations involved are even approximately linear, as seems to be the case from the evidence typified by figs. 16.9 and 16.10. Examples of other profiles generalized in this way are shown in fig. 16.11.

Some theoretical justification for what is essentially an empirical model for generalizing slope profiles can be obtained by analysing the forms characteristic of a set of processes under a range of conditions of contour curvature. For a range of slope processes which can be de-

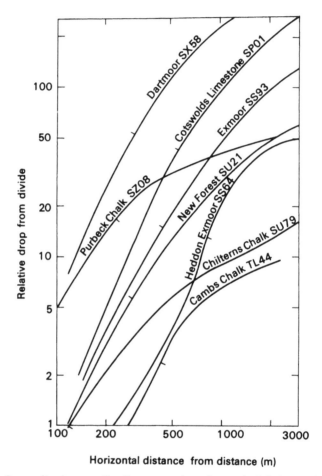

Fig. 16.11. Composite slope profiles from a number of areas in Britain. Vertical scale is relative only, and does not indicate relief differences between areas. Numbers refer to 1:25,000 map sheets. Ticks show points of inflexion of profiles.

scribed as a function of area drained per unit contour length and slope gradient (equation 16.7 above), the slope profile tends towards a characteristic form which may be approximated by equation 16.9 above. If it is further assumed that area drained per unit contour length (and hence also radius of contour curvature) is directly proportional to distance from the divide, which is also approximately true (p. 398, and equation 16.11 above), then equation 16.9 can be solved in the form:

$$y_0 - y \propto \int \left\{ \frac{a}{f(a)} \right\}^{1/n} . da. \qquad (16.12)$$

Now the right-hand side of this equation is a function of a alone, and a is proportional to horizontal distance, x; so that changes in the scale of $(y_0 - y)$, the fall from the divide, and in x should accommodate all variations of relief and contour curvature for a given overall process, specified by $f(a)$ and the exponent n.

Slope development theory therefore not only provides an approximate justification for the empirical method of generalizing for stable slopes in an area, but also shows how some deductions about process may be made from the generalized profile. With both the function, $f(a)$ and the exponent n unknown, it is not possible to deduce the process, even assuming that the process obeys a law which can be expressed by equation 16.7 and suitable boundary conditions; but it provides useful comparisons between slopes if an analysis is made on the assumption that the slope exponent, $n = 1$, and that $a \propto x$. In this case, equation 16.9 gives:

$$f(a) \propto \frac{x}{\text{slope at } x}, \qquad (16.13)$$

where the slope referred to is the ground slope, and not the slope of the curve. Values of $f(a)$ calculated in this way vary somewhat for the composite profiles of fig. 16.11, but all can be approximated to the form:

$$f(a) \propto u^m + x^m, \qquad (16.14)$$

for constant u and m, with the exponent m varying from 2 to 3.5. The first, constant term in this expression may be interpreted as a creep or solifluction component, and the distance term as a wash or fluvial sediment transport component. In the case where the distance exponent, $m = 2$, the expression for $f(a)$ corresponds to the slope profile:

$$y = A - \tfrac{1}{2} B. \log (u^2 + x^2), \qquad (16.15)$$

which has a convexity near the divide, the magnitude of which is dependent on the value of the constant u, and becomes concave downslope, tending towards the form $y = A - B . \log x$ for large distances, which has already been seen to be an empirical fit to many stream long profiles (equation 16.10). In these ways the empirical slope and river profile generalizations may be combined and related to a theoretical background which allows some interpretation of the profiles obtained in terms of the processes acting.

Synthesis

In matching profiles by changing the horizontal scales, it has been assumed that area drained per unit contour length is directly proportional to distance from the divide, but no mention has been made of the constants of proportionality involved with the real profiles. In fact, other factors besides contour curvature influence the ratio, especially the rate of downcutting of the stream at the slope base which, as it increases, reduces $\lambda = a/x$; but some consistent trends may be seen. Spurs and straight slopes, as a group, have lower values of λ than do river profiles, but only by a factor of about $\times 2$, instead of the tens or

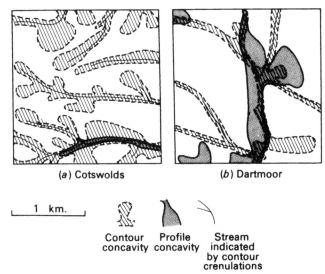

(*a*) Cotswolds   (*b*) Dartmoor

1 km.

Contour   Profile   Stream
concavity   concavity   indicated
by contour
crenulations

Fig. 16.12. Examples of areas with different proportions of contour concavity (hollows) and profile concavity.

	Contour concavity (% of square area)	Profile concavity (% of square area)
(*a*) Cotswolds	39	3
(*b*) Dartmoor	22	18

hundreds of times which is suggested by the surface topography. If the evidence of the profiles is to be reconciled with the evidence of the topography, it is necessary to introduce the idea of a *contributing area* which is effective in producing stream flows, and constitutes only a small percentage of the total surface drainage area. In a hydrologic context, Betson (1964) has calculated contributing areas as peak stream runoff divided by peak rainfall intensity, and has obtained values of less than 10 per cent for basins with a good soil and vegetation cover. Similar contributing proportions give reasonable lateral shifts for slope and river profiles in the areas shown in fig. 16.11.

To learn more about the slopes of a basin, we must look at the areal

410

distribution of the profiles to see how long actual slopes are, and how the contour concavities and convexities affect the slope lengths, and so interact with the form of the profiles found. It might be supposed that areas with mainly convex contours would have low values of area drained per unit contour length, even close to stream courses, so that the slope profiles would also be predominantly convex. That is to say the amount of contour concavity might be positively correlated with the amount

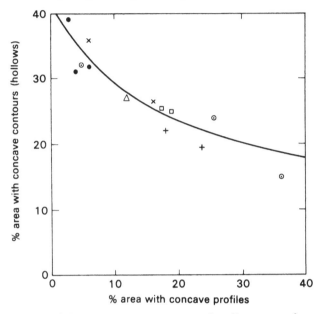

Fig. 16.13. The relationship between contour curvature and profile curvature for unglaciated temperate valleys, mainly in S. England. Data from 1:25,000 maps and from Hack and Goodlett (1960).
Symbols:
● Jurassic Limestone (Cotswolds).
○ Chalk (Chilterns).
△ Sandstone (New Forest).
□ Sandstone (Exmoor).
+ Granite (Dartmoor).
× Appalachians.

of profile concavity as these parameters vary within an area and between areas. In fact, however, the reverse appears to be true: areas with mainly convex contours have a *lower* proportion of convex slope profiles.

Fig. 16.12 shows two examples taken from the 1:25,000 maps of southern England, which show typical distributions of contour and profile concavities. In the Cotswold area slope profiles are predominantly convex, even in the upper reaches of small valleys, but the

411

TABLE 16.1. *Linear slope dimensions*

Area	1:25,000 map sheet	D.D. (km^{-1})	1/2 D (m)	Mean straight slope length (m) A	Dist. to point of inflection (m) B	Ratio A/B	% profile concavity in area
Exmoor	SS 93	2·2	230	650	440	1·5	18
New Forest	SU 21	3·3	150	620	285	2·2	12
Dartmoor	SX 58	1·6	310	820	300	2·6	21
Cotswolds	SP 01	1·6	300	575	550	1·0	4
Chilterns[1]	SU 79	1·6	310	455	115	3·9	30

[1] Chilterns data refer to areas with broad valley bottoms in this table, and in fig. 16.14.

hillsides are covered with a network of spoon-shaped hollows. On the other hand, slopes in Dartmoor commonly have concave bases and hollows are much rarer, with hilltops having rounded convex contours being very characteristic of the area. Fig. 16.13 shows that this inverse relationship is common, and the best-fit curve has been drawn to pass through the extreme case of a perfect pediment, on which contours are all convex and profiles all concave.

It is not clear why this inverse relationship exists, but it appears from

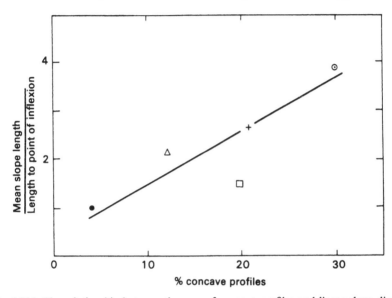

Fig. 16.14. The relationship between the area of concave profiles and linear slope dimensions. Symbols as in fig. 16.13.

maps similar to those shown in fig. 16.12 that hollows are not usual on concave slope profiles, except in actual stream channels. Since concave wash slopes are associated with overland flow, it might be inferred that hollows are associated with sub-surface throughflow and conditions of local moisture seepage, but this is no more than a suggestion. The areal proportion of concave profiles may be related to the linear dimensions of the slope profiles, as may be seen from the data in table 16.1 and fig. 16.14. Where the average length of slope profiles is much longer than the average distance to points of inflexion, then a high proportion of the average profile, and hence of the area, is concave in profile.

Table 16.1 also shows drainage densities, measured from contour crenulations in the various areas, and it may be seen that $1/2D$, which is nominally the mean slope length, has no relationship to the actual mean slope lengths. This difference may be partly explained by opera-

413

tional difficulties of defining an appropriate network for drainage density, but probably also has some geomorphic significance. If slope profiles are traced on a map (down lines of greatest slope), it is found that most points lie on profiles which terminate in major valleys, and only a few in small lateral channels which have very restricted valley development. It may therefore be useful to distinguish between a 'valley density', defined as the reciprocal of twice the mean slope length; and the drainage density defined in the usual way.

If the model for profile development of stable slopes based on the continuity equation and empirical slope process equations (referred to on pp. 394–5 and 408–9) is applied to *all* profiles in the landscape, including stream profiles, it leads to conclusions about basin erosion rates independent of the particular slope profile forms. For each profile, the characteristic form towards which it tends can be written in the form (see appendix B):

$$y = Y(x) \cdot T(t), \tag{16.16}$$

where Y, T are functions of distance, x, and time, t, respectively. The rate of lowering can then be expressed as proportional to elevation multiplied by a function of time. For processes in which the rate of transport (at capacity) is a function of distance multiplied by the nth power of the slope, the time function is either a constant ($n = 1$) or else is inversely proportional to time elapsed ($n > 1$). It has been shown above that slope forms can be closely approximated by transport laws in which the exponent n is equal to 1, and so this model is explored below, in the form that the rate of lowering is a constant proportion of the elevation. It has been suggested above (p. 396) that the constant of proportionality will be the same all over a drainage basin; and by extension, this argument may be applied on a continental scale. Of course, qualifications must be made about the attainment of characteristic profile forms all over a continent; and about constancy of process over a range of climates and lithologies – and of course these conditions are never met – but the model does provide a yardstick with which to compare rates of downcutting, as has been done by Ahnert (1970).

If this model is applied to drainage basin data, the rate of downcutting should be related to the mean basin elevation, and fig. 16.15 shows the range of data for a set of large river basins in a wide variety of lithological and climatic environments. If rivers from glaciated areas are excluded, the median best-fit line conforms to a linear proportionality, as supposed in the model, and the mean rate of lowering corresponds to a rate of reduction of 10 per cent per million years. In looking at the extreme values, the excluded glaciated areas are areas

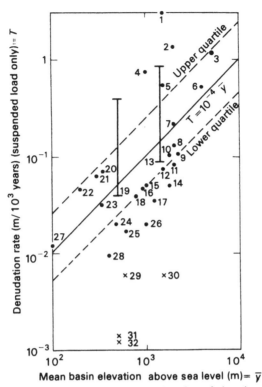

Fig. 16.15. Suspended solids denudation rate as a function of elevation, mainly for large rivers in America and Asia. Median best-fit line refers only to basins not glaciated during the Pleistocene. Data from Dole and Stabler, 1909; Schumm and Hadley, 1961; Maner, 1958; Holeman, 1968.

● Non-glaciated basins.
× Pleistocene-glaciated basins.
Key to rivers plotted:

1. Lo (China).
2. Yellow or Hwang (China).
3. Kosi (India and Nepal).
4. Ganges (India).
5. Red or Yuan (North Vietnam and China).
6. Brahmaputra (East Pakistan and Tibet).
7. Indus (West Pakistan).
8. Yangtse (China).
9. Colorado (U.S.A.).
10. Mekong (South-East Asia and China).
11. Rio Grande (U.S.A. and Mexico).
12. Missouri (U.S.A.).
13. Small basins in east Wyoming (U.S.A.).
14. Salt (U.S.A.).
15. Red (U.S.A.).
16. Mississippi (U.S.A.).
17. Cheyenne (U.S.A.).

18. Tennessee (U.S.A.).
19. Small basins in Oklahoma and Texas (U.S.A.).
20. Roanoke, Virginia (U.S.A.).
21. Santee, North Carolina (U.S.A.).
22. Savannah, Georgia (U.S.A.).
23. Amazon (South America).
24. Nile (north-east Africa).
25. Susquehanna, Pennsylvania (U.S.A.).
26. Brazos, Texas (U.S.A.).
27. Pearl, Missipi (U.S.A.).
28. Potomac, Virginia and Maryland (U.S.A.).
29. Hudson, New York (U.S.A.).
30. S. Joaquin, California (U.S.A.).
31. Kennebec, Maine (U.S.A.).
32. Mississippi at Minneapolis, Minnesota (U.S.A.).

which are patently out of equilibrium with current processes, and contain large proportions of fresh rock outcrops, which are subject to low rates of mechanical removal. At the upper extreme, the rivers of South-East Asia all plot somewhat high, but the highest of all are the Yellow and Lo rivers, which flow through large areas of highly erodible loess. There thus seems to be some connection between extreme positions with respect to the median line and the erodibility of the material in the

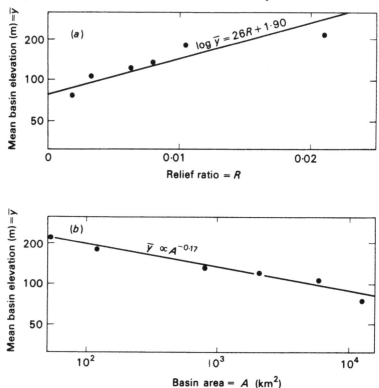

Fig. 16.16. The relationships between mean basin elevations and (*a*) relief ratio, (*b*) basin area for the River Thames basin, England. The best-fit slopes correspond closely to values found for variation of denudation rates:

In (*a*) 26 corresponds to 27·35 (Schumm and Hadley, 1961).
In (*b*) −0·17 corresponds to −0·15 (Brune, 1948).

basin. Climatic differences, in general agreement with the results of Langbein and Schumm (1958), can also be detected but are much less in magnitude.

Other models for basin erosion rates have been proposed, notably in terms of relief ratio (Schumm and Hadley, 1961) and in terms of basin area (Brune, 1948), and all have a predictive value because all are interdependent. Fig. 16.16 illustrates the extent and nature of this interdependence for the River Thames basin, England. It may be seen that the best-fit coefficients relating mean basin elevation to relief ratio and to basin area correspond closely to those found in the studies referred to. What this means is that the three measures, mean basin relief, relief ratio and basin area are all equivalent to one another for basins which have reached their characteristic forms; so that any one correlation will give an estimate of denudation rate.

Denudation is accomplished through the slope gradients, so that, for basins which have not reached characteristic form, a measure of slope appears most likely to predict denudation rate. Relief ratio is such a measure and appears to take into account differences in lithology and erodibility better than a measure based on elevation alone; but appears to require a scale factor since large basins tend to erode *faster* than small basins with the same relief ratio. Correlations like:

$$T = 1 \cdot 5 \times 10^4 . R^3 . A^{\frac{1}{2}}, \tag{16.17}$$

seem to be appropriate, where T is the denudation rate in m per 1,000 years, R is the relief ratio and A is the basin area in km^2.

The elevation model is a convenient one for comparing rates on the world scale, where little is known about detailed topography. The rates shown in fig. 16.15 seem to be in accord with world rates of erosion. A total rate of 14×10^9 tons per year (Stoddart, 1969) is spread over an effective area (excluding internal drainages and ice-caps) of $1 \cdot 0 \times 10^8$ km^2, at a mean elevation of 800 m, giving a mean rate of world denudation of 9 per cent per million years, in close agreement with the median line of fig. 16.15. The implications for peneplanation times are also reasonable: a massif initially at 1,000 m elevation will be reduced to 900 m after one million years; 370 m after ten million years; 50 m after thirty million years; and 7m after fifty million years. The span of twenty to thirty million years suggested by this evolution for the production of a surface of low relief corresponds to geological estimates of the times required for peneplanation (Schumm, 1963).

The model discussed above has assumed that river profiles are a special case of slope profiles, obeying the same process laws, and has treated the whole basin as a single slope unit. This approach breaks down where the focus of interest is the interaction of slopes and rivers; and in this case a two-phase model is required; slopes and rivers being considered separately. Perhaps the most important instance where a two-phase model is necessary is in the estimation of drainage density, where the point of change from slope to river is critical in determining the position of stream heads and so the density of the whole network. Other cases where the interaction of slope and river is important are (1) in considering the formation of flood plains and the lateral migration

417

of rivers as they alternately undercut opposing slopes and modify their forms; and (2) in considering the formation of asymmetric valleys where the differing sediment supplies from opposite slopes themselves help to determine the pattern of stream lateral movement as it cuts down. Such models have not so far been satisfactorily worked out, but seem necessary to explain some of the features of a basin.

DRAINAGE TEXTURE

Some measure of the scale of topography appears to be vital to a knowledge of how basins look; and some understanding of the factors affecting the scale seems essential to an understanding of the topography. Choice of a network which is geomorphically relevant has provided many problems, and has limited comparisons between workers, but there do seem to be some consistent differences between areas. If drainage density is to be related to other factors, it is important to consider whether the set of 'explanatory' factors relates to the same set of conditions as the drainage pattern it attempts to explain.

The available evidence suggests that drainage density and the channel pattern adjust through time initially very fast, then more slowly as soil and landforms adjust to changing conditions; and eventually, hardly at all, by the time the landscape has approached a dynamic equilibrium or characteristic form, even though erosion is still continuing. Evidence for the first phase is shown most clearly in the dissection of raised shorelines after earthquakes (Morisawa, 1964; Kirkby and Kirkby, 1969) which show that drainage nets are mainly present within a year, and show relatively small increases over the next few years (fig. 16.17a). Over a period of tens of thousands of years, Ruhe's (1952) data, analysed by Leopold, Wolman and Miller (1964), showed a gradual approach to an apparent steady state value by comparing the drainage densities of successively older till sheets in Iowa (fig. 16.17b). There is a rather rapid increase of drainage density for about the first 20,000 years, and then a levelling off at a density of about 0.34 km^{-1} (0.85 miles^{-1}). However, this interpretation depends critically on the dating and lithological similarity between the till sheets, especially of the Tazewell till dated by Leopold *et al.* (1964, p. 424) at 17,000 years B.P. Without this point, the other tills fit just as well a model of linear increase of drainage density through time, as is shown in fig. 16.17c; and in this model there is no levelling off of the rate of increase. There is therefore an increase of drainage density for at least 20,000 years, and possibly for a much longer time period as the drainage density changes along with the soils and landforms.

If drainage patterns are thought to change and increase in density

Slopes in drainage basins

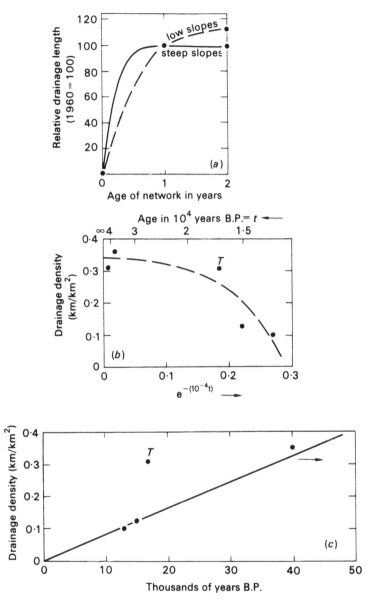

Fig. 16.17. Changes in drainage density over time. (*a*) Rapid adjustment after uplift of a lake floor. Data from Morisawa (1964); (*b*) and (*c*): slow adjustment to changing conditions on till sheets of different ages in Iowa. Data from Ruhe (1952), analysed by Leopold, Wolman and Miller (1964); (*b*) is an exponential plot implying an approach to an equilibrium drainage density; (*c*) is an arithmetic plot implying a continuing linear increase in drainage density.

419

until maturity, then the time scale of continuing change must be thought of as millions rather than tens of thousands of years. Once maturity is reached, evidence of change is scarce but Schumm's (1956) evidence from rapidly eroding badlands suggests that channel networks and drainage density are changing to a relatively slight degree while measurable lowering goes on.

Accepting these ideas of rates of change of drainage density, one must accept present drainage densities (defined in terms of presently wetted channels) as being in equilibrium with hydrologic conditions over periods of a few years only, and it is desirable to adopt the concept of a dynamic drainage density changing in response to individual flows. Over this time period the topography is essentially fixed and the observed flows are produced by the input rainfall as modified by the fixed environment. Over a longer time period, drainage density may be defined in terms of the eroded channels visible in the field. The relevant

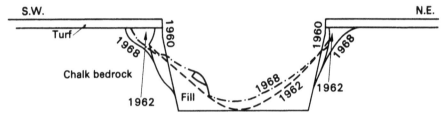

Fig. 16.18. Filling of an artificial ditch at Overton Down, Wiltshire, by bank collapse resulting from frost-heave. Based on Jewell *et al.* (1963) and Crabtree (1971).

time scale appears to be the period for which a channel remains identifiable in the absence of flow. Under cultivation an unused channel can disappear within a year, but natural slope processes will take somewhat longer, especially under a continuous vegetation cover. For example the experimental earthwork at Overton Down, Wiltshire, included a trench which at first filled rapidly with frost shattered material from the banks but still remains a distinct depression, although not at all similar to a freshly eroded channel (fig. 16.18) (Jewell *et al.*, 1963; Crabtree, 1971). A period for initial filling of at most ten years therefore seems appropriate. Over a longer time the edges of the former channel will become progressively rounded until they survive only as a shallow depression.

Repeated extension of a channel followed by infilling will, over periods of hundreds or thousands of years, produce a permanent depression or small valley which should show as a contour crenulation on a map of moderately large scale. Because crenulations show repeated use of a channel they correspond, more nearly than any other observable feature, with a geomorphically meaningful drainage density in a *graded time* interval. This is only true however, if climate and vegetation conditions have remained constant but if these conditions have changed in such a way as to reduce the drainage density a crenulation network may survive as an anachronism for an exceedingly long period, perhaps thousands or tens of thousands of years. In the context of southern England, it has been argued that the crenulation network is much too extensive to be related to present hydrology. This is argued mainly for valleys in limestone (e.g. Sparks, 1960, pp. 161–6), but has also been applied to other lithologies. Gregory (1966) describes valleys in southeast Devon for which the crenulation density is $1 \cdot 39$ km^{-1}, compared to a density of current eroded watercourses of only $0 \cdot 77$ km^{-1}, and considers that this difference may be due to a reduction in discharge.

Although the crenulation network need no longer be related to current runoff conditions, even in exceptional floods, it still matches the hydrology of the relatively long period over which the valleys were formed. Over a *graded* or *cyclic* time span, the hydrological conditions are responsible for forming the valley network, so that this network is of the greatest interest in a study of slope and valley forms. This emphasis on the 'crenulation' network contrasts with the hydrologist's view of eroded channels as formed by the sequence of storms in the immediate past and delivering water in *steady* time spans from a valley topography which is considered as fixed.

At least two types of criterion may be applied to the analysis of stream-head position and so drainage density. (1) Horton (1945) estimated the critical distance from the divide at which channels are just able to erode. If this model is generalized to stream head *areas*, it may be possible to balance the tractive force of the flow against strength of the soil and vegetation surface. (2) A channel may also be considered to be maintained as a balance between infilling from the side slopes and stream erosion along the valley axis. Above a stream head, a channel is excavated only in very rare storms, and the valley bottom is filled from the side slopes, producing a depositional concavity in side slope profile. At successive points down-valley, stream erosion becomes more frequent, and the valley axis tends to be cut down, producing a more V-shaped valley; and this form continues until it gives way to a flood plain in the valley bottom.

Empirical explanations of differences between areas have been most satisfactorily made for semi-arid areas (Melton, 1957), where increases in vegetation cover from place to place produce consistent increases in permeability and therefore decreases in contributing area. Fig. 16.19 shows the relationship between drainage density and percentage bare (unvegetated) area. Chorley and Morgan (1962) have shown that differences in rainfall intensity may make a significant difference to drainage density, but the importance of other factors has not been demonstrated.

Although the flows determine the drainage density in the long term, the drainage density determines the flows in the short term. Within a single basin Gregory and Walling (1968) have related the totol length of wetted channels to total basin discharge for the same flood, and Weyman (personal communication) has found comparable relationships for a very small catchment (fig. 16.20*a*). Comparing different basins, Carlston (1966) has found that mean annual flood per unit area is proportional to the square of drainage density (fig. 16.20*b*) and

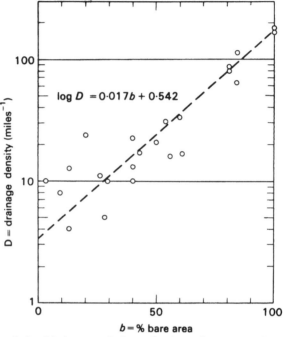

Fig. 16.19. The relationship between drainage density and percentage bare (unvegetated) area; mainly for semi-arid areas (from Melton, 1957, fig. 11).

Rodda (1969) has related mean annual floods in U.K. to drainage density by the relationship:

$$\log Q = 1{\cdot}08 + 0{\cdot}77 \log A + 2{\cdot}92 \log r + 0{\cdot}81 \log D, \qquad (16.18)$$

where r is the mean annual daily maximum rainfall. These time and space relationships both depend partly on the response of the network; given channel inputs are routed through it; but they differ in the relationship between drainage density and inputs. In comparing storms for one basin, higher rainfall intensities saturate a longer distance of valley axis, which can then contribute rapidly to the strʌam flow. In comparing areas, high drainage density areas are generally areas of greater overland flow, so that similar rainfalls will lead to higher peak channel

422

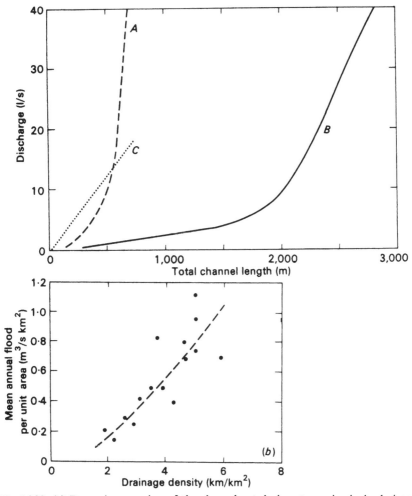

Fig. 16.20. (*a*) Dynamic expansion of the channel net during storms in single drainage basins.

A: south Devon: 3·3 km² (after Gregory and Walling, 1968).
B: south-east Devon: 13·6 km² (after Gregory and Walling, 1968).
C: Somerset: 0·11 km² (Weyman, 1970, personal communication).
(*b*) Spatial variation of drainage density with peak discharge for a series of basins, each represented by one point. Basins are from the Appalachians, with drainage areas of 10–150 km². Data points from Carlston (1963).

inputs in areas of high drainage density. In both cases channel input tends to increase with drainage density.

The problems of drainage scale and the way in which it is controlled have only been touched on here, and could be the subject of a book on their own. The texture of the landscape also has a subtler significance than that of a simple scale factor. At very high drainage densities the landscape becomes like a badlands in which stream channels dominate the appearance of the surface. Elevations, however, are not reduced in proportion with the slope length, and badlands are commonly in areas of moderate to high relief. As a result the stream networks of Mesa Verde or the Book Cliffs, for example, combine to form a skyline which appears to the observer as a slope profile, but is in fact made up of an intricate visual superimposition of short sections of divides.

In summary, slope profiles are valuable units of study, even when groups of them are considered together in the context of a drainage basin. Slopes within a basin, which no longer show marked straight segments, appear to have features in common which allow the whole basin to be characterized by a generalized profile which can, at least to a first approximation, take into account differences in contour curvature and rate of stream downcutting by simple changes of horizontal and vertical scale. The areal pattern of slope profiles can be characterized by hypsometric or similar curves, but it is difficult to relate them to slope properties. An inverse relationship has been demonstrated between areas of contour curvature and areas of profile curvature and it may be that this is an unexplained part of the total equilibrium form of the basin, which is thought to approach a state in which points initially on a contour remain level throughout a basin as it develops, while their elevation values are reduced at a rate proportional to their height above sea level.

Perhaps the most informative single parameter of a basin is one which describes its texture, for example drainage density or link frequency. Although stream profiles may to some extent be considered as a special case of slope profiles, profile development theory at present can only partly explain the development of a drainage network and so the characteristic texture of an area. The channel network and its texture, however, provide the crucial link connecting hillslope and river studies. Not only is it a spatial link, the channel head defining the point at which slope processes give way to river processes and the channel bank defining the lateral extent of the slopes; but it is also a temporal link. Hillslopes should be studied essentially as a sediment transport system which is considered in *graded* or *cyclic* time spans, and in which the rivers are only a means of removing debris; whereas rivers should be studied as a water transport system which is considered in *steady* or *graded* time spans, and in which the slopes are only a topography which modifies the water flowing into the channels. Over the *graded* time spans in which both systems are meaningful, the dynamic equilibrium which maintains the density of the channel network provides the essential link between these two systems and these two viewpoints.

DERIVATION OF SLOPE STABILITY FORMULAE

---

1. *The critical height of a vertical bank* (*after Lohnes and Handy, 1968*)
To prove:

$$H'_c = \frac{4c}{\gamma}\left\{\frac{1}{\cos\phi - 2\cos^2 a.\tan\phi}\right\} - z \qquad (6.7)$$

where
$$a = 45 + \phi/2.$$

At the time of limiting equilibrium, the shear force along the incipient failure plane in fig. 6.4 is just balanced by the shear strength along it. We have, from the diagram, therefore:

$$(T=)\,W.\sin a = \frac{c.x}{\cos a} + W.\cos a.\tan\phi\,(=S).$$

Now, from the diagram, $W = \tfrac{1}{2}.x.(H'_c + z).\gamma$, and, substituting in the equation above, we have:

$$\tfrac{1}{2}.x.(H'_c + z).\gamma.\sin a = \frac{c.x}{\cos a} + \tfrac{1}{2}.x.(H'_c + z).\gamma.\cos a.\tan\phi$$

which simplifies to:

$$(H'_c + z) = \frac{2c}{\gamma}\left\{\frac{1}{\cos a\,[\sin a - \cos a.\tan\phi]}\right\}.$$

Now,
$$\cos a\,[\sin a - \cos a.\tan\phi]$$
$$= \cos a.\sin a - \cos^2 a.\tan\phi$$
$$= \tfrac{1}{2}.\sin 2a - \cos^2 a.\tan\phi$$
$$= \tfrac{1}{2}[\cos\phi - 2\cos^2 a.\tan\phi].$$

Therefore,

$$H'_c + z = \frac{2c}{\gamma}\left\{\frac{1}{\tfrac{1}{2}[\cos\phi - 2\cos^2 a.\tan\phi]}\right\}$$

and
$$H'_c = \frac{4c}{\gamma}\left\{\frac{1}{\cos\phi - 2\cos^2 a.\tan\phi}\right\} - z \qquad (6.7)$$

425

Appendix A

2. *The critical height of a vertical bank (after Terzaghi, 1943, p. 153)*
To prove: the equation above (6.7) is identical to

$$H'_c = \frac{4c}{\gamma} \tan a - z. \tag{6.8}$$

We need to show that:

$$\tan a = \frac{1}{\cos \phi - 2 \cos^2 a \,.\, \tan \phi}$$

Now,

$$\cos \phi = \sin(90 + \phi) = \sin 2a, \quad \text{and}$$

$$\tan \phi = -\cot(90 + \phi) = -\cot 2a;$$

therefore, with respect to the equation above, we have:

$$\text{RHS} = \frac{1}{\sin 2a + 2 \cos^2 a \,.\, \cot 2a}$$

$$= \frac{1}{2(\sin a \,.\, \cos a + \cos^2 a \,.\, \cot 2a)}$$

$$= \frac{\sin 2a}{2(\sin a \,.\, \cos a \,.\, \sin 2a + \cos^2 a \,.\, \cos 2a)}$$

$$= \frac{2 \sin a \,.\, \cos a}{2[2 \sin^2 a \,.\, \cos^2 a + \cos^2 a (\cos^2 a - \sin^2 a)]}$$

$$= \frac{\sin a \,.\, \cos a}{(2 \sin^2 a \,.\, \cos^2 a + \cos^4 a - \sin^2 a \,.\, \cos^2 a)}$$

$$= \frac{\sin a \,.\, \cos a}{\sin^2 a \,.\, \cos^2 a + \cos^4 a}$$

$$= \frac{\sin a \,.\, \cos a}{\cos^2 a (\sin^2 a + \cos^2 a)}$$

$$= \tan a = \text{LHS}$$

3. *The general relationship between critical height and slope angle: the case of a plane failure surface passing through the toe of the slope (after Culmann, 1866)*

A. The general equation for the situation depicted in fig. A1 is:

$$H_c = \frac{2c}{\gamma} \frac{\sin i}{\sin(i - a)[\sin a - \cos a \,.\, \tan \phi]} \tag{A 3.1}$$

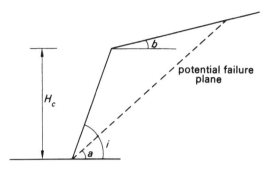

Fig. A1. The Culmann method of stability analysis.

In the critical condition, we know (the full derivation of equations A3.1 to A3.3 is provided by Carson (1971)) that:

$$a = \tfrac{1}{2}(i+\phi) \qquad\qquad (A\ 3.2)$$

and, in this case, the previous equation simplifies to:

$$H_c = \frac{4c}{\gamma} \cdot \frac{\sin i . \cos \phi}{[1-\cos(i-\phi)]} \qquad\qquad (A\ 3.3)$$

When $i = 90°$, this reduces further to:

$$H_c = \frac{4c}{\gamma} \cdot \frac{\cos \phi}{[1-\sin \phi]} \qquad\qquad (A\ 3.4)$$

It is more usual to see equation A3.4 written in the form:

$$H_c = \frac{4c}{\gamma} . \tan(45+\phi/2). \qquad\qquad (A\ 3.5)$$

The two terms $\cos \phi/(1-\sin \phi)$ and $\tan(45+\phi/2)$ are, of course, identical, both being equal to $\sqrt{[(1+\sin \phi)/(1-\sin \phi).]}$

Equation A3.5 is, it will be recalled, equation 6.8 for the situation in which no tension cracks occur ($z = 0$). Note the following points in relation to equation A3.3:

    (*a*) if $i \leqslant \phi$, then H_c is infinite;

    (*b*) if $\phi = 0$, then $H_c = \dfrac{4c}{\gamma} \dfrac{\sin i}{[1-\cos i]}$

    (*c*) if $\phi = 0$ and $i = 90°$, then $H_c = 4c/\gamma$.

    B. The case in which tension cracks have developed in the bank *prior to failure* has been discussed by Terzaghi (1943, p. 153–4). The situation is shown in fig. A2. In this analysis, let us define W as the

427

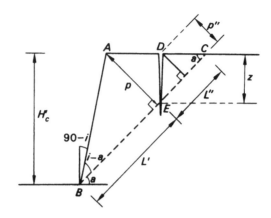

Fig. A2. Modified Culmann method: slopes with tension cracks.

$$P = AB \cdot \sin(i-a)$$

$$\frac{H'_c}{AB} = \cos(90-i) = \sin i$$

$$\therefore P = \frac{H'_c \cdot \sin(i-a)}{\sin i}$$

weight of the wedge ABC, W'' as the weight of the wedge CDE, and $W' = W-W''$. Similarly, let the lengths BC, EC and BE be denoted by L, L'' and L', where $L' = L-L''$. The height of the tension crack is denoted by z.

From the diagram,

$$W = \tfrac{1}{2} \cdot L \cdot p \cdot \gamma$$

$$= \tfrac{1}{2} \cdot L \cdot H'_c \cdot \frac{\sin(i-a)}{\sin i} \cdot \gamma,$$

$$L'' = z/\sin a$$

and

$$W'' = \tfrac{1}{2} \cdot L'' \cdot p'' \cdot \gamma,$$

where

$$p'' = z \cdot \cos a,$$

thus

$$W'' = \tfrac{1}{2} \cdot z^2 \cdot \cot a \cdot \gamma.$$

At the time of limiting equilibrium, the shear force along BE is just balanced by the shear strength, and we have:

$$W' \cdot \sin a = cL' + W' \cdot \cos a \cdot \tan \phi.$$

Consider, for the sake of simplicity, the case in which $i = 90°$.

$$W = \tfrac{1}{2} \cdot L \cdot H'_c \cdot \cos a \cdot \gamma,$$

$$L = H'_c/\sin a,$$

so that
$$W = \tfrac{1}{2}.(H'_c)^2.\cot a.\gamma.$$

Substituting $W-W''$ for W', and denoting H'_c by H, for convenience, in the equation above, we obtain:

$$\tfrac{1}{2}.\cot a.(H^2-z^2).\gamma.\sin a = cL'+\tfrac{1}{2}(H^2-z^2).\gamma.\cot a.\cos a.\tan \phi$$

and, putting $L' = L-L'' = H/\sin a-z/\sin a = (H-z)/\sin a$, we have:

$$\tfrac{1}{2}.\gamma.\cot a.(H^2-z^2).\sin a =\frac{c(H-z)}{\sin a} +\tfrac{1}{2}(H^2-z^2).\gamma.\cot a.\cos a.\tan \phi$$

On rearranging, this simplifies to:

$$\frac{c(H-z)}{\sin a} = \tfrac{1}{2}.\gamma.\cot a.(H^2-z^2).(\sin a-\cos a.\tan \phi)$$

$$= \tfrac{1}{2}.\gamma.\cot a(H-z).(H+z).(\sin a-\cos a.\tan \phi)$$

and we obtain:

$$H+z = \frac{2c}{\gamma}.\frac{\tan a}{(\sin a-\cos a.\tan \phi).\sin a}$$

$$= \frac{2c}{\gamma}.\frac{1}{\cos a(\sin a-\cos a.\tan \phi)}$$

or, reverting to our previous notation,

$$H'_c = \frac{2c}{\gamma}.\frac{1}{\cos a(\sin a-\cos a.\tan \phi)} -z.$$

Now, returning to equation A 3.1, in the case where $i = 90°$,

$$H_c = \frac{2c}{\gamma}\frac{1}{\cos a(\sin a-\cos a.\tan \phi)}$$

and we obtain:

$$H'_c = H_c-z$$

in which the presence or absence of the superscript $'$ indicates, respectively, the critical height with or without tension cracks.

4. *Circular arc analysis for clay slopes: conventional method of slices (after May and Brahtz, 1936)*

To show:
$$F_s=\frac{\sum\limits_{B}^{A}[c'.l+(W.\cos a-u.l).\tan \phi]}{\sum\limits_{A}^{B} W.\sin a} \qquad (7.14)$$

Appendix A

In relation to fig. 7.11, the Factor of Safety with respect to shearing strength, for the slope as a whole, is given by:

$$F_s = \frac{\sum\limits_B^A S}{\sum\limits_B^A T}$$

Now, $\qquad\qquad S = c'.l + (N - u.l).\tan\phi$

and, resolving normal to the base of the slice,

$$N = W.\cos a,$$

where N is the *total* normal force at the base; substituting above, we have:

$$S = c'.l + (W.\cos a - u.l).\tan\phi.$$

The shear force (T) is found by resolving forces parallel to the base of the slice:

$$T = W.\sin a.$$

We therefore have:

$$F_s = \frac{\sum\limits_B^A [c'.l + (W.\cos a - u.l).\tan\phi]}{\sum\limits_B^A W.\sin a} \qquad (7.14)$$

5. *Circular arc analysis for clay slopes: modified method of slices (after Bishop, 1955)*

To show:

$$F_s = \frac{\sum\limits_B^A \left\{ \left[c'.l + \left(\dfrac{W}{\cos a} - u.l \right).\tan\phi \right] \Big/ \left[1 + \dfrac{\tan a.\tan\phi}{F_s} \right] \right\}}{\sum\limits_A^B W.\sin a}. \qquad (7.15)$$

We again use the formula

$$F_s = \frac{\sum\limits_B^A S}{\sum\limits_B^A T} = \frac{\sum\limits_B^A [c'.l + (N - u.l)\tan\phi]}{\sum\limits_B^A W.\sin a}$$

but this time, N is determined differently, and inter-slice pressures are included. If we resolve forces vertically in fig. 7.12, we obtain:

$$N.\cos a = W^* - \frac{S}{F_s}.\sin a$$

where $W^* = W + X_n - X_{n+1}$; the insertion of F_s emphasizes that not all the available shear strength is necessarily utilized in resisting the weight of the slice. Substituting for N in the expression for S, we obtain:

$$S = c'.l + \left(\frac{W^*}{\cos a} - \frac{S.\tan a}{F_s} - u.l\right).\tan \phi$$

or

$$S = \frac{c'.l + \left(\dfrac{W^*}{\cos a} - u.l\right).\tan \phi}{1 + \dfrac{\tan a.\tan \phi}{F_s}}$$

and the factor of safety is thus equal to

$$F_s = \frac{\sum\limits_B^A\left\{\left[c'.l + \left(\dfrac{W^*}{\cos a} - u.l\right).\tan \phi\right]\middle/\left[1 + \dfrac{\tan a.\tan \phi}{F_s}\right]\right\}}{\sum\limits_B^A W.\sin a}$$

Bishop (1955) has shown that, with an estimated loss of accuracy of less than one per cent, it can be assumed that $X_n = X_{n+1}$, or $W^* = W$, and this expression is then equal to equation 7.15. The essential difference between the modified and the conventional method of slices is not, therefore, in the treatment of the inter-slice pressures, but, rather, in the derivation of the normal force acting on the base of the slice.

6. Stability analysis for planar slides

In the situation given by fig. 7.4 (planar slides on an infinite slope), the angle a ($= \theta$ in fig. 7.4) is constant for all slices. Therefore we may consider one slice alone and ignore the summation signs in equation 7.15. In this case, equation 7.15 becomes:

$$F_s = \frac{c'.l + (W.\cos a - u.l).\tan \phi}{W.\sin a}$$

which is identical to the solution by the conventional method of slices (May & Brahtz, 1936) given by equation 7.14. Note that in the situation given by fig. 7.4, and retaining the notation of fig. 7.11,

$$W = \gamma.z.l.\cos a,$$

and the formula for the Factor of Safety reduces to:

$$F_s = \frac{c' + (\gamma . z . \cos^2 a - u) . \tan \phi}{\gamma . z . \sin a . \cos a}$$

which is identical to equation 7.4 for the case of limiting equilibrium ($F_s = 1$). The full derivation is provided by Carson (1971).

DERIVATION OF CHARACTERISTIC FORM EQUATIONS

These equations for characteristic slope form are derived from the continuity equation in the form of equation 5.3, with the weathering factor, μ, equal to $1\cdot0$; for *transport-limited* removal in which the capacity rate of transport takes the form of equation 5.4 in which there is no significant threshold slope angle; and for a fixed divide at $x = 0$ and fixed basal removal point at $x = x_1$, $y = 0$. These results, and some others for characteristic forms, are quoted, though not as fully derived, in Kirkby (1971), for a wider range of conditions including those of threshold slope angles, *weathering-limited* removal and curved contours.

For the *transport-limited* case with $\mu = 1\cdot0$, the continuity equation (5.3) becomes:

$$\frac{\partial C}{\partial x} = -\frac{\partial y}{\partial t},$$

where C, the capacity rate of transport takes the form assumed in equation 5.4:

$$C = f(x).\left(-\frac{\partial y}{\partial x}\right)^n.$$

Let us seek a solution to these equations of the form:

$$y = Y(x).T(t), \tag{B1}$$

where Y is a function of x alone, and T is a function of t alone.

Then:

$$\{f'(x).(-Y')^n - n.f(x).Y''.(-Y')^{n-1}\}.T^n = -.YT'$$

or

$$\frac{f'(x).(-Y')^n - n.f(x).Y''.(-Y')^{n-1}}{Y} = -\frac{T'}{T^n} \tag{B2}$$

where dashes denote differentiation. The left-hand side is now a function of x alone; and the right-hand side of t alone: each must therefore be equal to a constant, which we will denote by λ. Since slopes decline through time, λ will be positive. Different values of λ produce a series

433

of solutions, and for the case $n = 1$, the general solution is a linear combination of these solutions. After long times elapsed the solution with the smallest value of λ (satisfying the boundary conditions) progressively dominates the solution, and this appears to be the *characteristic form* towards which the slopes tend. It can be shown rigorously to provide the characteristic form for $n = 1$, and is assumed to behave similarly for other values of n. Thus, for the characteristic form:

$$f'(x).(-Y')^n - n.f(x).Y''.(-Y')^{n-1} = \lambda Y. \tag{B 3}$$

Integrating with respect to x:

$$f(x).(-Y')^n = \lambda \int_0^x Y.dx,$$

or

$$-Y' = \left[\frac{\lambda \int_0^x Y.dx}{f(x)} \right]^{1/n}. \tag{B 4}$$

To obtain an approximate solution,

let

$$\int_0^x Y.dx = \theta.Y_0.x \quad \text{for some} \quad 0 < \theta < 1,$$

where Y_0 is the value of $Y(x)$ at $x = 0$, the divide.

Then:

$$-Y' = \left\{ \frac{\lambda.\theta.Y_0.x}{f(x)} \right\}^{1/n}, \tag{B 5}$$

so that, approximately,

$$Y(x) \approx Y_0 - (\lambda.\theta.Y_0)^{1/n}.I(x),$$

where

$$I(x) = \int_0^x \left\{ \frac{x}{f(x)} \right\}^{1/n}.dx. \tag{B 6}$$

The value of $\lambda.\theta$ is chosen with reference to the basal removal condition, $Y(x_1) = 0$, so that:

$$Y(x) \approx Y_0 \left(1 - \frac{I(x)}{I(x_1)} \right), \tag{B 7}$$

which is equivalent to equation 5.5.

If $f(x)$ is chosen in its simplest form, as proportional to x^m where m is a constant exponent, then equation (B 7) can be solved as:

$$Y(x) \approx Y_0 \left\{ 1 - \frac{\int_0^x x^{(1-m)/n}.dx}{\int_0^{x_1} x^{(1-m)/n}.dx} \right\}$$

$$\approx Y_0 \left\{ 1 - \left(\frac{x}{x_1} \right)^{([1-m]/n+1)} \right\}, \tag{B 8}$$

which is equivalent to equation 5.6.

If the exponent m is put equal to 1 in equation (B 8), the approximate solution for the characteristic form is a straight slope, whereas the exact solution is a slightly concave slope. For this case especially, it is worth obtaining a more accurate, though still approximate solution to equation (B 4) by applying to it the inequality, valid over $0 \leqslant x \leqslant x_2 \leqslant x_1$:

$$Y_0 \gtrless Y(x) \gtrless Y(x_2). \tag{B 9}$$

In this way we obtain upper and lower bounds for $Y(x)$ which show how far our approximations may be in error, and allow a more accurate approximate solution to be obtained by taking the arithmetic mean of the upper and lower bounds. Thus:

$$\left\{ Y_0^{(n-1)/n} - \frac{n-1}{n} . \lambda^{1/n} . I(x) \right\}^{n/(n-1)} \geqslant Y(x) \geqslant Y_0 - (\lambda Y_0)^{1/n} . I(x). \tag{B 10}$$

For the case where $n = 1$, the left-hand side of this inequality is replaced by its exponential limit, namely:

$$Y_0 . e^{-\lambda I(x)}.$$

Equation B 10 allows equation B 7 to be corrected somewhat, although the correction varies with the exponent, n. The closer approximations

are: for $n = 1$:

$$Y(x) \approx \tfrac{1}{2} Y_0 \left\{ 1 - 1.28 \frac{I(x)}{I(x_1)} + e^{-1.28 I(x)/I(x_1)} \right\} \tag{B 11}$$

and for $n = 2$:

$$Y(x) \approx Y_0 \left\{ 1 - 1.17 \frac{I(x)}{I(x_1)} + 0.17 \left(\frac{I(x)}{I(x_1)} \right)^2 \right\}. \tag{B 12}$$

From these corrected forms follow other corrections to equation B 8; but these have not been incorporated in constructing fig. 5.7, in which the curve for $m = 1$ should more accurately be slightly concave throughout its length. It is correctly drawn in fig. 8.19 for $m = 1$; $n = 2$.

The remaining curves in fig. 8.19 have been derived from equations B 7 or B 8; except for (c) which has been derived with greater accuracy from equation B 12.

For (a): $m = 1.5$; $n = 1.25$. Therefore $1 - \dfrac{Y}{Y_0} = \left(\dfrac{x}{x_1} \right)^{0.6}$.

For (b): $m = 2.0$; $n = 2.0$. Therefore $1 - \dfrac{Y}{Y_0} = \left(\dfrac{x}{x_1} \right)^{0.5}$.

For (c): $m = 1.0$; $n = 2.0$. Therefore $1 - \dfrac{Y}{Y_0} = 1.17 \left(\dfrac{x}{x_1} \right) - 0.17 \left(\dfrac{x}{x_1} \right)^2$.

For (d): $f(x) = 0.04 + x^2$; $n = 2.0$.

Therefore:

$$I(x) = \int_0^x \left(\frac{x}{0\cdot04 + x^2}\right)^{\frac{1}{4}} . \, dx;$$

and the curve shown in fig. 8.19 is obtained after integrating $I(x)$ numerically in the range, $x = 0$ to 1.

CONDITIONS FOR PROFILE CONVEXITY OR CONCAVITY

Slope profile continuity equations allow some *general* deductions to be made about profile convexity or concavity, especially near the divide or near the basal removal point. We can also comment on the tendency of the profile as a whole, by examining the characteristic forms towards which the profiles tend as the initial form of the profile is gradually eliminated. We can thus say whether the profile will *tend to become* convex or concave throughout its length.

For the *transport-limited* case, with or without chemical removal, the continuity equation 5.3 can be written as:

$$\frac{\partial C}{\partial x} = T,$$

where T is the rate of *mechanical* surface lowering, and C, the capacity rate of transport, takes the form assumed in equation 5.4:

$$C = f(x) . \left(-\frac{\partial y}{\partial x} \right)^n.$$

Combining these two equations:

$$C = f(x) . \left(-\frac{\partial y}{\partial x} \right)^n = \int_0^x T . \, dx$$

or

$$\frac{\partial y}{\partial x} = -\left\{ \frac{\int_0^x T . \, dx}{f(x)} \right\}^{1/n}. \tag{C 1}$$

Differentiating with respect to x:

$$\frac{\partial^2 y}{\partial x^2} = -\left\{ \frac{\int_0^x T . \, dx}{f(x)} \right\}^{1/n - 1} . \left\{ \frac{T . f(x) - \int_0^x T . \, dx . f'(x)}{\{f(x)\}^2} \right\}.$$

Since $\int_0^x T . \, dx / f(x)$ is positive; the sign of $\partial^2 y / \partial x^2$ is the same as the sign of the expression:

$$\frac{f'(x)}{f(x)} - \frac{T}{\displaystyle\int_0^x T . \, dx} \tag{C 2}$$

Appendix C

Case 1: Near the divide, if there is a non-zero rate of surface lowering ($T_0 \neq 0$), then for small values of x:

$$\int_0^x T \, dx \simeq T_0 . x,$$

and expression (C 2) is:

$$\frac{f'(x)}{f(x)} - \frac{1}{x}. \tag{C 3}$$

If the distance function, $f(x)$ behaves as x^m near the divide, then the divide will be convex ($\partial^2 y / \partial x^2 < 0$) if $m < 1$;
and concave if $m > 1$.

Thus soil creep, solifluction, rainsplash etc, for which $m = 0$ are not the only processes which may be able to produce a divide convexity; and Gilbert's (1909) argument is unnecessarily restrictive.

Case 2: Near the stream or basal removal point, $T = T_1$, the rate of lowering of the basal point, and $\int_0^x T . \, dx$ is equal to the slope length, x_1, multiplied by the average rate of slope lowering, T; so that if $f(x)$ behaves as x^m near the basal point, the expression (C 2) becomes:

$$\frac{1}{x_1}\left(m - \frac{T_1}{T}\right). \tag{C 4}$$

The base of the slope will therefore be concave if the basal point is fixed or is rising in elevation ($T_1 \leqslant 0$), for all $m > 0$. If the basal point is being lowered in elevation ($T_1 > 0$), then we may distinguish the cases of:
(a) Normal slope lowering with some reduction in slope angles ($T_1 < T$).
   In this case the base of the slope will certainly be concave if $m \geqslant 1$.
(b) River incision with $T_1 > T$, in which case a basal convexity is formed

if $m < T_1/T$.

Case 3: If the profile is approaching a characteristic form (in which case T_1 is being assumed as zero), then if chemical removal may be neglected, the rate of surface lowering,

$$T = -\frac{\partial y}{\partial t} \rightarrow -\frac{\partial}{\partial t}[Y(x).T(t)] \quad \text{in the terminology of equation B1}$$

$$= -\frac{\partial T(t)}{\partial t} . Y(x)$$

$$= \lambda . y \quad \text{for some } \lambda \text{ which varies with time.}$$

438

Expression C 2 then becomes:

$$\frac{f'(x)}{f(x)} - \frac{y}{\int_0^x y \, . \, dx}. \tag{C 5}$$

Now, in a characteristic form, y is a monotonic decreasing function of x (that is to say that hillslope profiles go steadily downslope), so that:

$$y_0 \, . \, x > \int_0^x y \, . \, dx > y \, . \, x, \tag{C 6}$$

which is the integral form of inequality B 9. Applying this inequality to equation C 5, a sufficient, though not necessary condition for a profile to *become* concave under the action of a process is:

$$\frac{f'(x)}{f(x)} \geqslant \frac{1}{x}; \tag{C 7}$$

and to *become* convex ($\partial^2 y/\partial x^2 < 0$):

$$\frac{f'(x)}{f(x)} < \frac{1}{x} \, . \, \frac{y}{y_0}. \tag{C 8}$$

Thus if $f(x)$ behaves as x^m, the slope becomes
  concave throughout if $m \geqslant 1$;
  convex throughout if $m \leqslant 0$.
It can also be shown that the slope becomes convexo-concave if $0 < m < 1$.
More general forms of equations C 7 and C 8 are quoted in Kirkby (1971).

NOTATIONS USED

A	Area; Constant.
A_g	Area of gaps and joints.
a	Angle; Acceleration; Area drained per unit contour length.
B_0	Empirical function of solid-fluid friction angle in sediment transport.
C	Capacity rate of debris transport.
$C(z)$	Rate of soil movement at depth z below soil surface.
c	Particle concentration; Pore space concentration; Coefficient; Cohesion.
$\bar{c}$	Cohesion per unit normal stress
c'	Cohesion (effective stress value).
c_e	True cohesion (after Hvorslev, 1937).
c_i	Effective cohesion (after Terzaghi, 1962).
c_r	Residual cohesion (after Skempton, 1964).
D	Diffusivity; Rate of surface lowering through chemical removal as dissolved load; Deviator stress.
d	Grain diameter.
E	Modulus of elasticity; Total storm energy per unit area.
E_n, E_{n+1}	Resultants of horizontal forces between slices.
e	Base of natural logarithms (2·7183).
e	Instantaneous rate of storm energy production; void ratio (pore space/grain space).
F	Friction force; Total amount of water infiltrated.
F_s	Factor of Safety (with respect to shearing strength).
f	Instantaneous rate of infiltration; Dimensionless friction factor.
$f(\ )$	Function of ( ).
G	Rate of straight slope retreat, measured in a horizontal direction.
g	Gravitational acceleration (9·81 m s^{-2}).
H	Accumulated day-degrees above or below given temperature.
H_c	Critical height (no tension cracks).
H'_c	Critical height (with tension cracks).
h	Height; Height of capillary rise.

I_p	Plasticity index.
i	Angle; Index suffix; Intensity of rainfall or rainfall excess.
J	Cumulative frequency.
j	Index suffix; Frequency density.
K	Capillary conductivity of soil; Coefficient of earth pressure; Constant.
k	Constant; Size of irregularities on soil surface (equivalent grain roughness); Solubility in water.
LL	Liquid Limit.
L	Latent heat of fusion of ice (330 kJkg^{-1}).
l	Length.
$M(z)$	Accumulated moisture change at depth z (dimensionless).
m	Mass; Constant exponent; Volumetric moisture content; Stability coefficient.
N	Normal force.
N_s	Stability number.
n	Number of items; Constant exponent; Stability coefficient.
P	Force; Wetted perimeter; Total proportion of *all* bedrock constituents remaining in undissolved residue ($= \Sigma p$).
PL	Plastic Limit.
p	Net rate of percolation; Probability; Proportion of a bedrock constituent remaining in undissolved residue ($0 < p < 1$) on a 'by substance' basis.
p_s	Proportion undissolved residue at the soil surface.
Q	Total water discharge.
q	Water discharge per unit width.
q_u	Unconfined compressive strength.
q_z	Water discharge per unit cross section within the soil.
R	Relief ratio (basin relief/basin length); Annual precipitation.
r_u	Pore pressure ratio.
r	Radius; Hydraulic radius; Daily rainfall.
r_0	Mean rainfall per rain-day.
S_t	Sensitivity.
S	Sediment transport per unit width; Shear strength (force).
s	Slope; Shear strength (stress).
s_r	Residual shear strength; Sediment transport per unit width in a storm of rainfall r.
T	Shear force; Surface tension; Rate of surface lowering by mechanical erosion.
t	Time elapsed.
u	Pore pressure.
v	Velocity; Volume.
v_d	Volume of talus deposited at cliff base.

v_r	Volume of rock removed from cliff.
W	Weight; Rate of weathering of bedrock to form soil.
w	Water content (% water relative to weight of dry soil).
X_n, X_{n+1}	Resultants of vertical forces between slices.
x	Horizontal distance: Horizontal axis direction.
y	Elevation; Vertical distance; Vertical axis direction.
Z	Constant length.
z	Vertical distance; Depth below soil surface or water table; Vertical axis direction.
z_0	Thickness of tension (earth pressure) zone.
α	Angle; Angle of contact; Angle between failure plane and major principal plane.
β	Angle; Slope angle.
γ	Unit weight of soil.
γ_w	Unit weight of water.
Δ	Dimensionless density ratio.
ε	Strain; Porosity.
$\dot{\varepsilon}$	Strain rate.
θ	Dimensionless shear stress; Temperature; Slope angle.
κ	Thermal conductivity.
μ	Dynamic molecular viscosity; Coefficient of friction; Volume of soil produced from a unit volume of unweathered rock.
ν	Kinematic molecular viscosity.
ρ	Soil or sediment density; Radius of contour curvature, measured positive in hollows and negative on spurs.
ρ_w	Water density.
σ	Total normal stress.
σ'	Effective normal stress.
σ_1, σ_2, σ_3	Major, intermediate and minor principal stresses.
τ	Shear stress.
ϕ	Total gravitational and hydraulic potential; Angle of internal friction.
$\phi(\ )$	Function of ( ).
ϕ'	Angle of internal friction (effective stress value).
ϕ_e	Hvorslev's true angle of internal friction.
Φ	Dimensionless sediment transport per unit width.
ψ	Hydraulic potential.

442

BIBLIOGRAPHY

AHNERT, F. 1970. Functional relationships between denudation, relief and uplift in large, mid-latitude drainage basins. *American Journal of Science*, **268**, 243-63.

AKROYD, T. N. W. 1957. *Laboratory Testing in Soil Engineering.*

ANDERSON, H. W. 1951. Physical characteristics of soils related to erosion. *Journal of Soil and Water Conservation*, **6**, 129-33.

ANDERSON, H. W. 1954. Suspended sediment discharge as related to streamflow, topography, soil and land use. *Transactions of the American Geophysical Union*, **35**, 268-81.

ANDRÉ, J. E. and H. W. ANDERSON, 1961. Variation of soil erodibility with geology, geographic zone, elevation and type of vegetation in the North California wildlands. *Journal of Geophysical Research*, **66**, 3351-8.

ANDREWS, J. T. 1961. Permafrost in southern Labrador-Ungava. *Canadian Geographer*, **5**, 34-5.

ANTEVS, E. 1932. *Alpine Zone of Mount Washington, Maine.* Auburn, Maine.

BAGNOLD, R. A. 1954. Some flume experiments on large grains but little denser than the transporting fluids, and their implications. *Proceedings of the Institution of Civil Engineers*, Paper 6041, 174-205.

BAGNOLD, R. A. 1956. The flow of cohesionless grains in fluid. *Philosophical Transactions of the Royal Society of London, Series A*, **249**, 235-97.

BAGNOLD, R. A. 1960. Some aspects of the shape of river meanders. *United States Geological Survey Professional Paper 282-E.*

BAKKER, J. P. and J. W. N. LE HEUX, 1946. Projective-geometric treatment of O. Lehmann's theory of the transformation of steep mountain slopes. *Koninklijke Nederlandsche Akademie van Wetenschappen, Series B*, **49**, 533-47.

BAKKER, J. P. and J. W. N. LE HEUX, 1947. Theory on central rectilinear recession of slopes. *Koninklijke Nederlandsche Akademie van Wetenschappen, Series B*, **50**, 959-66 and 1154-62.

BAKKER, J. P. and J. W. N. LE HEUX, 1950. Theory on central rectilinear recession of slopes. *Koninklijke Nederlandsche Akademie van Wetenschappen, Series B*, **53**, 1073-84 and 1364-74.

BAKKER, J. P. and J. W. N. LE HEUX, 1952. A remarkable new geomorphological law. *Koninklijke Nederlandsche Akademie van Wetenschappen, Series B*, **55**, 399-410 and 554-71.

BAKKER, J. P. and A. N. STRAHLER, 1956. Report on quantitative treatment of slope recession problems. *International Geographical Union, 1st Report on the Study of Slopes.*

BAUER, K. E. von, 1860. Über ein allgemeines Gesetzin der Gestaltung der Flussbetten. *Bulletin of the St Petersburg Imperial Academy of Science*, **2**, 1-49, 218-50, and 353-82.

BAULIG, H. 1950. *Essais de Géomorphologie.* Paris.

443

Bibliography

BAVER, L. D. 1956. *Soil Physics.* Wiley. New York.

BEATY, C. B. 1959. Slope retreat by gullying. *Bulletin of Geological Society of America,* **70,** 1479–82.

BEER, DE, E. 1965. Influence of the mean normal stress on the shear strength of sand. *Proceedings of 6th International Conference on Soil Mechanics and Foundation Engineering,* **1,** 165–9.

BERRY, L. and B. P. RUXTON, 1959. Notes on weathering zones and soils on granite rocks in two tropical regions. *Journal of Soil Science,* **10,** 54–63.

BESKOW, G. 1935. Soil freezing and frost heaving. *Sveriges geologiska undersokning. Afhandlingar och uppsater, Series C,* **375,** 222–42.

BETSON, R.P. 1964. What is watershed runoff? *Journal of Geophysical Research,* **69,** 1541–52.

BIRD, J. B. 1959. Recent contributions to the physiography of Northern Canada. *Zeitschrift für Geomorphologie, N.F.* **3,** 151–74.

BIRD, J. B. 1967. *The Physiography of Arctic Canada.* Baltimore.

BIROT, P. 1968. *The Cycle of Erosion in Different Climates.* (Translated by Jackson, C. I. and K. M. Clayton.) Berkeley.

BISHOP, A. W. 1954. Correspondence on a paper by A. D. Penman. *Géotechnique,* **4,** 43–5.

BISHOP, A. W. 1955. The use of the slip circle in the stability analysis of slopes. *Géotechnique,* **5,** 7–17.

BISHOP, A. W. 1966. The strength of soils as engineering materials. (The Rankine Lecture.) *Géotechnique,* **16,** 91–128.

BISHOP, A. W. and L. BJERRUM, 1960. The relevance of the triaxial test to the solution of stability problems. *Proceedings of the American Society of Civil Engineers Research Conference on the Shear Strength of Cohesive Soils,* 437.

BISHOP, A. W. and D. J. HENKEL, 1962. *The Measurement of Soil Properties in the Triaxial Test.*

BISHOP, A. W. and N. R. MORGENSTERN, 1960. Stability coefficients for earth slopes. *Géotechnique,* **10,** 29–150.

BJERRUM, L. 1954a. Geotechnical properties of Norwegian marine clays. *Géotechnique,* **4,** 46–69.

BJERRUM, L. 1954b. Stability of natural slopes in quick clay. *Proceedings of the European Conference on the Stability of Earth Slopes, Stockholm,* **3,** 101–19.

BJERRUM, L. 1955. Stability of natural slopes in quick clay. *Géotechnique,* **5,** 101–19.

BJERRUM, L. 1967. Progressive failure in slopes of over-consolidated plastic clay and shales. *Journal of Soil Mechanics and Foundations Division, American Society of Civil Engineers,* **93,** 1–49.

BJERRUM, L. and F. JØRSTAD, 1968. Stability of rock slopes in Norway. *Norwegian Geotechnical Institute Publication* **79,** 1–11.

BJERRUM, L. and B. KJAERNSLI, 1957. Analysis of the stability of some Norwegian clay slopes. *Géotechnique,* **7,** 1–16.

BLACK, C. A. (Ed.), 1965. *Methods of Soil Analysis.* Part II, 771–1572. Chemical and Microbiological Properties. American Society of Agronomy.

BLACKWELDER, 1925. Exfoliation as a phase of rock weathering. *Journal of Geology,* **33,** 793–806.

BRACE, W. F. 1964. Brittle fracture of rocks. In *State of Stress in the Earth's Crust,* (Ed. by W. R. Judd), New York.

BRAWNER, C. O. 1966. Slope stability in open pit mines. *Western Miner,* **39,** 56–72.

BREDTHAUER, R. O. 1957. Strength characteristics of rock samples under hydrostatic pressure. *Transactions of American Society of Mechanical Engineers,* **79,** 695–708.

BRICKER, O. P., GODREY, A. E. and E. T. CLEAVES, 1968. Mineral-water interaction during the chemical weathering of silicates. *Advances in Chemistry Series No 73: Trace Inorganics in Water*, 128–42.

BRITISH STANDARDS INSTITUTION, 1967. *Methods of Testing Water Used in Industry.* British Standard 2690.

BRUNE, G. 1948. Rates of sediment production in midwestern United States. *Soil Conservation Service Technical Publication 65.*

BRYAN, K. 1922. Erosion and sedimentation in the Papago country, Arizona, *United States Geological Survey Bulletin 730.*

BRYAN, K. 1940a. Gully gravure: a method of slope retreat. *Journal of Geomorphology*, **3**, 87–107.

BRYAN, K. 1940b. The retreat of slopes. *Annals of the Association of American Geographers*, **30**, 254–67. (In the Symposium: Walther Penck's contribution to Geomorphology.)

BRYAN, K. 1946. Cryopedology – the study of frozen ground and intensive frost action with suggestions on nomenclature. *American Journal of Science*, **244**, 622–42.

BRYAN, K. 1954. The geology of Chaco Canyon, New Mexico. *Smithsonian Miscellaneous Collection*, **122**.

BRYAN, R. B. 1969. The relative erodibility of soils developed in the Peak District of Derbyshire. *Geografiska Annaler*, **51A**, 145–59.

BÜDEL, J. 1948. Die Klima morphologischen zonen der polarländer, *Erdkunde*, **2**, 25–53.

BUNTING, B. T. 1965. *The Geography of Soil.* Hutchinson, London.

BYERLEE, J. D. 1968. Brittle-ductile transition in rocks. *Journal of Geophysical Research*, **73**, 4741–50.

CADY, J. G. 1965. Petrographic microscope techniques. In *Methods of Soil Analysis*, Part I. (Ed. by Black, C. A. *et al.*) American Society for Agronomy, Inc.

CAILLEUX, A. 1952. Recentes variations du niveau des mers et des terres. *Bulletin de la Société géologique de France*, **2**, 135–44.

CAINE, N. 1969. A model for alpine talus slope development by slush avalanching. *Journal of Geology*, **77**, 92–100.

CAINE, N. and J. N. JENNINGS, 1968. Some blockstreams of the Toolong Range, Koscinsko State Park, New South Wales. *Journal and Proceedings of the Royal Society of New South Wales*, **101**, 93–103.

CAQUOT, A. 1934. *Equilibre Des Massifs à Frottement Interne. Stabilité Des Terres Pulverulentes et Cohèrentes.* Gauthier-Villars, Paris.

CARLSTON, C. W. 1963. Drainage density and streamflow. *United States Geological Survey Professional Paper 422-C.*

CARSLAW, H. S. and J. C. JAEGER, 1959. *Conduction of Heat in Solids.*

CARSON, M. A. 1967a. The magnitude of variability in samples of certain geomorphic characteristics drawn from valley-side slopes. *Journal of Geology*, **75**, 93–100.

CARSON, M. A. 1967b. *The Evolution of Straight Debris-Mantled Hillslopes.* Unpublished Ph.D. Dissertation, Cambridge University.

CARSON, M. A. 1969a. Soil Moisture. In *Water, Earth and Man* (Ed. by R. J. Chorley).

CARSON, M. A. 1969b. Models of hillslope development under mass failure. *Geographical Analysis*, **1**, 76–100.

CARSON, M. A. 1971a. Application of the concept of threshold slopes to the Laramie Mountains, Wyoming. *Transactions of the Institute of British Geographers*, Special Publication No. 3.

CARSON, M. A. 1971b. *The Mechanics of Erosion.* Pion, London.

445

Bibliography

CARSON, M. A. and D. J. PETLEY, 1970. The existence of threshold hillslopes in the denudation of the landscape. *Transactions of the Institute of British Geographers,* **49,** 71–95.

CARTER, C. A. and R. J. CHORLEY, 1961. Early slope development in an expanding stream system. *Geological Magazine,* **98,** 117–30.

CASAGRANDE, A., 1931. Discussion: a new theory of frost heaving. *Proceedings of Highway Research Board,* **11,** 168–72.

CASAGRANDE, A. 1947. Classification and identification of soils. *Proceedings of American Society of Civil Engineers,* 783–810.

CHAMBERLAIN, T. C. and R. T. CHAMBERLAIN, 1910. Certain valley configurations in low latitudes. *Journal of Geology,* **18,** 117–24.

CHANDLER, R. J. 1966. The measurement of residual strength in triaxial compression. *Géotechnique,* **16,** 181–86.

CHORLEY, R. J. 1959. The geomorphic significance of some Oxford soils. *American Journal of Science,* **257,** 503–15.

CHORLEY, R. J. 1962. Geomorphology and general systems theory. *United States Geological Survey Professional Paper,* 500-*B.*

CHORLEY, R. J. 1964. Geomorphological evaluation of factors controlling shearing resistance of surface soils in sandstone. *Journal of Geophysical Research,* **69,** 1507–16.

CHORLEY, R. J. 1969. The role of water in rock disintegration. In *Water, Earth and Man.* (Ed. by R. J. Chorley.)

CHORLEY, R. J. and M. A. MORGAN, 1962. Comparison of the morphometric features, Unaka Mountains Tennessee and N. Carolina, and Dartmoor. *Bulletin of Geological Society of America,* **73,** 17–34.

CHOW, V. T. 1959. *Open-channel Hydraulics.* McGraw-Hill, New York.

CLAYTON, R. W. 1956. Linear depressions in savannah landscapes. *Geographical Studies,* **3,** 102–26.

COATES, D. F. 1967. *Rock Mechanics Principles.* Department of Energy, Mines and Resources (Canada), Mines Branch Monograph 874.

COOK, N. G. W. 1965. The failure of rock. *International Journal of Rock Mechanics and Mining Science,* **2,** 389–403.

COOK, N. G. W., HOEK, E., PRETORIUS, J. P. G., ORTLEPP, W. D. and M. D. G. SALAMON, 1966. Rock mechanics applied to the study of rockbursts. *Journal of South African Institute of Mining and Metallurgy,* **66,** 435–528.

CORBEL, J. 1954. L'érosion terrestre, étude quantitative. *Annales de géographie,* **73,** 385–412.

CORBEL, J. 1959. Vitesse de l'érosion. *Zeitschrift für Geomorphologie,* **3,** 1–28.

CORTE, A. E. 1963*a.* Experiments on sorting processes and the origin of patterned ground. *Proceedings of International Permafrost Conference, Lafayette, Indiana,* 130–5.

CORTE, A. E. 1963*b.* Relationship between four ground patterns, structure of the active layer, and type and distribution of ice in the permafrost. *Biuletyn Peryglacjalny,* **12,** 7–90.

CORTE, A. E. 1966. Particle sorting by repeated freezing and thawing. *Biuletyn Peryglacjalny,* **15,** 175–240.

COTTON, C. A. 1948. *Landscape.* New York.

COTTON, C. A. and M. T. TE PUNGA, 1955*a.* Fossil gullies in the Wellington landscape. *New Zealand Geographer,* **11,** 72–5.

COTTON, C. A. and M. T. TE PUNGA, 1955*b.* Solifluxion and periglacially modified landforms. *Transactions of the Royal Society of New Zealand,* 1001–31.

COTTRELL, A. M. 1963. Fracture. *Proceedings of the Royal Society, Series A*, **276**, 1–18.

COULOMB, C. A. 1776. Essais sur une application des règles des maximis et minimis à quelques problems de statique relatifs à l'architecture. *Academie des Sciences, Paris. Mémoirs présentées par divers Savants.*

CRABTREE, K. 1971. Overton Down Experimental Earthwork, Wiltshire 1968. *Proceedings of the University of Bristol Speleological Society.* **12**, 237–44.

CRAWFORD, C. B. 1961. Engineering studies of Leda Clay. In *Soils in Canada – Geological, Pedological and Engineering Studies*, (Ed. by R. F. Legget) Toronto.

CRAWFORD, C. B. 1963. Cohesion in an undisturbed sensitive clay. *Géotechnique*, **13**, 132–46.

CRAWFORD, C. B. 1968. Quick clays of eastern Canada. *Engineering Geology*, **2**, 239–65.

CRAWFORD, C. B. and W. EDEN, 1965. A comparison of laboratory results with in situ properties of Leda clay. *Proceedings of 6th International Conference on Soil Mechanics and Foundation Engineering*, **1**, 31–5.

CRAWFORD, C. B. and W. EDEN, 1967. Stability of natural slopes in sensitive clay. *Journal of Soil Mechanics and Foundation Division, American Society of Civil Engineers*, **93**, 419–36.

CULLING, W. E. H. 1963. Soil creep and the development of hillside slopes. *Journal of Geology*, **71**, 127–61.

CULLING, W. E. H. 1965. Theory of erosion of soil covered slopes. *Journal of Geology*, **73**, 230–54.

CULMANN, C. 1866. *Graphische Statik*. Zurich.

CURREY, D. R. 1964. A preliminary study of valley asymmetry in the Ogoturuk Creek area, N.W. Alaska. *Arctic*, **17**, 84–98.

CZUDEK, T. 1964. Periglacial slope development in the area of the Bohemian massif in northern Moravia. *Biuletyn Peryglacjalny*, **14**, 169–93.

DARCY, H. 1856. *Les Fontaines Publiques de la Ville de Dijon*. Dalmont, Paris.

DARWIN, C. 1882. *The Formation of Animal Mould through The Action of Worms with Observations on Their Habits*. John Murray. Edinburgh.

DAVIS, S. N. 1964. Silica in streams and ground-water. *American Journal of Science*, **262**, 870–91.

DAVIS, W. M. 1892. The convex profile of badland divides. *Science*, **20**, 245.

DAVIS, W. M. 1899. The geographical cycle. *Geographical Journal*, **14**, 481–504.

DAVIS, W. M. 1925. The basin range problem. *Proceedings of the United States National Academy of Science*, **11**, 387–92.

DAVIS, W. M. 1930. Rock floors in arid and in humid climates. *Journal of Geology*, **38**, 1–27 and 136–58.

DAVIS, W. M. 1932. Piedmont benchland and Primärrumpfe. *Bulletin of Geological Society of America*, **43**, 399–440.

DAVIS, W. M. 1954. *Geographical Essays*. (Ed. by D. W. Johnson). Dover. New York.

DAVISON, C. 1889. On the creeping of the soil-cap through the action of frost. *Geological Magazine*, **6**, 255.

DE FREITAS, M. H. and J. L. KNILL, 1967. The surface characteristics and strength of discontinuities in rock masses. *Paper presented to Geological Society of London, Engineering Group at Symposium on Rock Slopes, January 1967.*

DEAVEY, E. S. 1949. Biogeography of the Pleistocene. *Bulletin of the Geological Society of America*, **60**, 1315–416.

DEERE, D. U. 1967. Effect of pore pressures on the stability of slopes. *Paper presented to meeting of Geological Society of America at New Orleans, November 20–2.*

Bibliography

DEGE, W. 1941. Landformende vorgänge im eisnahen Gebier Spitzbergens. *Petermanns geographische Mitteilungen*, 87.

DEKLOTZ, E. J., BROWN, J. W. and O. A. STEMLER, 1966. Anisotropy of a schistose gneiss. *Proceedings of 1st Congress of International Society of Rock Mechanics*, **1**, 465–70.

DEMEK, J. 1964. Castle koppies and tors in the Bohemian highland (Czechoslovakia). *Biuletyn Peryglacjalny*, **14**, 195–216.

DENBIGH, K. G. 1951. *The Thermodynamics of the Steady State*. Methuen. London.

DENNY, C. S. 1956. Surficial geology and geomorphology of Potter County, Pennsylvania. *United States Geological Survey Professional Paper, 288*.

DENNY, C. S. 1967. Fans and pediments, *American Journal of Science*, **265**, 81–105.

DINGA, C. 1969. *A Quantitative Analysis Of The Effect Of Earth Rotation On Certain Parameters Of Meandering Alluvial Rivers*. Unpublished M.Sc. Thesis, Indiana State University.

DITTMER, M. J. 1938. A quantitative study of the subterranean members of three field grasses. *American Journal of Botany*, **25**, 654–7.

DOBBIE, C. H. and P. O. WOLF, 1953. The Lynmouth flood of August, 1952. *Proceedings of the Institution of Civil Engineers*, **2**, 522–88.

DOLE, R. B. and H. STABLER, 1909. Denudation. *United States Geological Survey Water Supply Paper 234*, 78–93.

DOUGLAS, I. 1967. Erosion of granite terrains under tropical rainforest in Australia, Malaysia and Singapore. *Proceedings of 14th General Assembly, International Union of Geodesy and Geophysics (Bern, Switzerland), Symposium on River Morphology*, 31–9.

DOUGLAS, I. 1968. Field methods of water hardness determination. *British Geomorphological Research Group, Technical Bulletin* No. 1, 35 pp.

DURY, G. H. 1959. *The Face of the Earth*. Penguin. London.

DYLIK, J. 1958. Periglacial investigations conducted in 1957 by the Lodz group of the Polish expedition to Spitsbergen. *Przrglad Geofizyczny*.

DYLIKOWA, A. 1964. Etat des recherches periglaciaires en Pologne. *Biuletyn Peryglacjalny*, **14**, 41–60.

EAKIN, H. M. 1961. The Yukon-Koyukuk region, Alaska. *United States Geological Survey Bulletin*, 631.

EDEN, W. J. and J. J. HAMILTON, 1957. The use of a field vane apparatus in sensitive clays. *ASTM Special Technical Publication 193*, 41–53.

EIDE, O. and L. BJERRUM, 1954. The slide at Bekkelaget. *Proceedings of the European Conference on the Stability of Earth Slopes, Stockholm*, **3**, 88–100.

EINSTEIN, H. A. 1950. The bed-load function for sediment transport in open channel flows. *United States Department of Agriculture Technical Bulletin*. 1026.

ELLISON, W. D. 1944. Studies of raindrop erosion. *Agricultural Engineering*, **25**, 131–6 and 181–2.

ELLISON, W. D. 1945. Some effects of raindrops and surface-flow on soil erosion and infiltration. *Transactions of American Geophysical Union*, **26**, 415–29.

ELLISON, W. D. 1948. Erosion by raindrop. *Scientific American*, Offprint 817.

EMBLETON, C. and C. A. M. KING, 1968. *Glacial and Periglacial Geomorphology*.

EMERY, K. O. 1947. Asymmetric valleys of San Diego County, California. *Bulletin of the South California Academy of Science, Pt. 2*. 61–71.

EMMETT, W. W. 1970. The hydraulics of overland flow on hillslopes. *United States Geological Survey Professional Paper 662-A*.

ENGELUND, F. and E. HANSEN, 1967. *A Monograph on Sediment Transport In Alluvial Streams*. Teknisk Forlag. Copenhagen.

EVANS, I. 1950. The measurement of the surface bearing capacity of soils in the study of earth-crossing machinery. *Géotechnique*, **2**, 46–57.

EVANS, I. S. 1970. General geomorphometry, derivatives of altitude, and descriptive statistics. Mimeo Paper read at *British Geomorphological Research Group, Symposium on Spatial Analysis in Geomorphology*, Cambridge, England.

EVERARD, C. E. 1963. Contrasts in the form and evolution of hillside slopes in central Cyprus. *Transactions of the Institute of British Geographers*, **32**, 31–42.

EVERARD, C. E. 1964. Climatic change and man as factors in the evolution of slopes. *Geographical Journal*, **130**, 65–9.

EVERETT, K. R. 1963a. Instruments for measuring mass-wasting. *Proceedings of International Permafrost Conference, Lafayette, Indiana*, 136–9.

EVERETT, K. R. 1963b. Slope movement, Neotoma Valley, Southern Ohio. *Ohio State University, Institute of Polar Studies*, Report No. 6.

EYRE, S. R. 1963. *Vegetation and Soils*. Arnold, London.

FABRE, L. A. 1903. La Dissymétrie des vallées et la loi dite de DeBaer, particulièrement en Gascogne. *Géographie*, **8**, 291–316.

FAHNESTOCK, R. K. 1963. Morphology and hydrology of a glacial stream, *United States Geological Survey, Professional Paper 422-A*.

FARMER, I. W. 1968. *Engineering Properties of Rocks*. Spon, London.

FEDERAL INTER-AGENCY RIVER BASIN COMMITTEE, 1953. Summary of reservoir sedimentation surveys for the United States through 1950. *Subcommittee on Sedimentation, Sedimentation Bulletin 5*.

FELLENIUS, W. 1927. *Erdstatische Berechnungen*. Berlin.

FELLENIUS, W. 1936. Calculation of the stability of earth dams. *Transactions of the 2nd Congress on Large Dams, 4*.

FELLER, W. 1950. *An Introduction to Probability Theory and Its Applications*. Vol. 1, New York.

FENNEMAN, N. F. 1908. Some features of erosion by unconcentrated wash. *Journal of Geology*, **16**, 746–54.

FISHER, O. 1866. On the disintegration of a chalk cliff. *Geological Magazine*, **3**, 354–6.

FLAXMAN, E. M. and R. D. HIGH, 1955. Sedimentation in drainage basins of the Pacific Coast States. *Soil Conservation Service*, Portland (Mimeographed).

FOURNIER, F. 1960. Débit solid des cours d'eau. Essai d'estimation de la perte en terre subie par l'ensemble du globe terrestre. *International Association for Scientific Hydrology*, Helsinki.

FREE, G. R., BROWNING, G. M. and G. W. MUSGRAVE, 1940. Relative infiltration and related physical characteristics of certain soils. *United States Department of Agriculture Technical Bulletin*, 729.

FREISE, F. W. 1933. Beobachtungen über erosion an urwaldgebirgsflüssen des brasilianischen staates Rio de Janeiro. *Zeitschrift für Geomorphologie*, 7.

GEIGER, R. 1965. *The Climate Near the Ground*. Harvard. Cambridge, Mass.

GEIKIE, S. A. 1880. Rock weathering as illustrated in Edinburgh churchyards. *Proceedings of Royal Society of Edinburgh*, **10**, 518–32.

GIBSON, R. E. and D. J. HENKEL, 1954. Influence of the duration of tests at constant rate of strain on the measured 'drained' strength. *Géotechnique*, **4**, 6–15.

GIFFORD, J. 1953. Landslides on Exmoor caused by the storm of August 15, 1952. *Geography*, **38**, 9–17.

GILBERT, G. K. 1884. The sufficiency of terrestrial rotation for the deflection of streams. *American Journal of Science*, **27**, 427–32.

GILBERT, G. K. 1877. *The Geology of the Henry Mountains.* United States Geographical and Geological Survey. Washington.

GILBERT, G. K. 1909. The convexity of hilltops. *Journal of Geology,* **17,** 344–51.

GILBERT, G. K. 1914. The transportation of debris by running water. *United States Geological Survey Professional Paper 86.*

GILLULY, J. 1949. Distribution of mountain-building in geologic time. *Bulletin of the Geological Society of America,* **60,** 561–90.

GLASSTONE, S., LAIDLER, K. and H. EYRING, 1941. *The Theory of Rate Processes.* McGraw-Hill. New York.

GOSSLING, F. 1935. The structure of Bower Hill, Nutfield. *Proceedings of the Geologists' Association,* **46,** 360–90.

GOSSLING, F. and A. J. BULL, 1948. The structure of Tilburstow Hill, Surrey. *Proceedings of the Geologists' Association,* **59,** 131–40.

GREGORY, K. J. 1966. Dry valleys and the composition of the drainage net. *Journal of Hydrology,* **4,** 327–40.

GREGORY, K. J. and D. E. WALLING, 1968. The variation of drainage density within a catchment. *Bulletin of International Association for Scientific Hydrology,* **13,** 61–8.

GREIG-SMITH, P. 1964. *Quantitative Plant Ecology.* Butterworth.

GRIGGS, D. T. 1936. The factor of fatigue in rock exfoliation. *Journal of Geology,* **44,** 781–96.

GRIM, R. E. 1953. *Clay Mineralogy.* New York.

GRIM, R. E. 1962. *Applied Clay Mineralogy.* New York.

GRISSINGER, E. H. 1966. Resistance of selected clay systems to erosion by water. *Water Resources Research,* **2,** 131–8.

GROVE, A. T. 1968. The last 20,000 years in the Tropics. *British Geomorphological Research Group, Occasional Paper* No. 5, 51–61.

GUILCHER, A. 1950. Nivation, cryoplanation et solifluction quaternaires dans les collines de Bretagne occidentale et du Nord de Devonshire. *Revue de Géomorphologie Dynamique,* **1,** 53–78.

GUILLION, Y. 1960. Monographie d'une paroi de Sablière, 1935–1959. *Zeitschrift für Geomorphologie,* Supp. 1.

GUNN, R. 1949. Isostasy – extended. *Journal of Geology,* **57,** 263–79.

GUTENBERG, B. 1941. Changes in sea level, post-glacial uplift and mobility of the earth's interior. *Bulletin of the Geological Society of America,* **52,** 721–72.

HACK, J. T. 1942. The changing physical environment of the Hopi Indians of Arizona. *Papers of The Peabody Museum, Harvard University,* **35,** 1–85.

HACK, J. T. 1957. Studies of longitudinal stream profiles in Virginia and Maryland. *United States Geological Survey Professional Paper 294-B.*

HACK, J. T. 1960. Interpretation of erosional topography in humid temperate regions. *American Journal of Science,* **258A,** 80–97.

HACK, J. T. and J. C. GOODLETT, 1960. Geomorphology and forest ecology of a mountain region in the Central Appalachians. *United States Geological Survey Professional Paper, 347.*

HADLEY, R. F. 1955. Development and significance of seepage steps in slope erosion. *Transactions of American Geophysical Union,* **36,** 792–804.

HADLEY, R. F. and G. C. LUSBY, 1967. Runoff and hillslope erosion resulting from a high-intensity thunderstorm near Mack, Western Colorado. *Water Resources Research,* **3,** 139–46.

HALL, E. B. and B. B. GORDON, 1963. Triaxial testing with large-scale pressure equipment. *Symposium on Laboratory Shear Testing of Soils, Ottawa, ASTM STP 361,* 315–28.

HALLSWORTH, E. G. 1963. An examination of some factors affecting the movement of clay in an artificial soil. *Journal of Soil Science,* **14,** 360–71.

HANSON-LOWE, J. 1935. The clinographic curve. *Geological Magazine,* **72,** 180–4.

HARVEY, D. 1969. *Explanation in Geography.* Arnold. London.

HEIM, A. 1932. Bergsturz und Menschenleben. *Vierteljahrschrift der Naturforschenden Gesellschaften in Zurich,* 77.

HENDRON, JT, A. J. 1968. Mechanical properties of rocks. In *Rock Mechanics,* Stagg and Zienkiewicz (ed.) New York.

HENKEL, D. J. 1957. Investigations of two long-term failures in London Clay slopes at Wood Green and Northolt. *Proceedings of the 4th International Conference on Soil Mechanics and Foundation Engineering,* **2,** 315–20.

HENKEL, D. J., KNILL, J. L., LLOYD, D. G. and A. W. SKEMPTON, 1964. Stability of the foundations of Monar Dam. *Proceedings of 8th International Congress on Large Dams,* **1,** 425–41.

HENKEL, D. J. and A. W. SKEMPTON, 1954. A landslide at Jackfield, Shropshire, in an over-consolidated clay. *Proceedings of the Conference on the Stability of Earth Slopes, Stockholm,* **1,** 90–101.

HEWLETT, J. D. 1961. Soil moisture as a source of base flow from steep mountain watersheds. *Southeastern Forest Experimental Station Paper 132. U. S. Department of Agriculture.*

HEWLETT, J. D. and A. R. HIBBERT, 1967. Factors affecting the response of small watersheds to precipitation in humid areas. *Proceedings of the International Symposium on Forest Hydrology, Pennsylvania State University (1969),* 275–90. New York.

HILGER, A. 1897. Uber verwitterungsvorgange bei krystallinischen und sedimentargesteinen, *Landwirkschaftliche Jahrbücher,* **8,** 1–11.

HIRSCHFELD, R. C. and S. J. POULOS, 1963. High pressure triaxial tests on a compacted sand and an undisturbed silt. *Symposium on Laboratory Shear Testing of Soils, Ottawa, ASTM STP 361,* 329–39.

HOLEMAN, J. N. 1968. The sediment yield of major rivers of the world. *Water Resources Research,* **4,** 737–47.

HOLLINGWORTH, S. E., J. H. TAYLOR, and G. A. KELLAWAY, 1944. Largescale superficial structures in the Northampton ironstone field. *Quarterly Journal of the Geological Society,* **100,** 1–44.

HOLMES, A. 1944. *Principles of Physical Geology.* Nelson. London

HOLTAN, H. N. and M. H. KIRKPATRICK, 1950. Rainfall, infiltration and hydraulics of flow in runoff computation. *Transactions of American Geophysical Union,* **31,** 771–9.

HOLTZ, W. G. and H. J. GIBBS, 1955. Shear characteristics of pervious gravelly soils as determined by triaxial shear tests. *Paper presented to Convention of American Society of Civil Engineers, San Diego, California, Feb. 9–11, 1955.*

HOLTZ, W. G. 1960. Effect of gravel particles on friction angle. *Proceedings American Society of Civil Engineers Research Conference on Shear Strength,* 1000–1.

HOOKE, R. LeB. 1967. Processes on arid-region alluvial fans. *Journal of Geology,* **75,** 438–60.

HOPKINS, D. M. and R. S. SIGAFOOS, 1952. Frost action and vegetation patterns on Seward Peninsula, Alaska. *United States Geological Survey Bulletin 974C,* 51–101.

HOPKINS, D. M. and B. TABER, 1962. Asymmetrical valleys in central Alaska. In *Abstracts for 1961 Geological Society of America Special Papers,* **68,** 116.

Bibliography

HOPKINS, D. M. and C. WAHRHAFTIG, 1959. Annotated bibliography of English-language papers on the evolution of slopes under periglacial climates. *Zeitschrift für Geomorphologie, Supp. Band* 1, 1–8.

HORN, H. M. and D. U. DEERE, 1962. Frictional characteristics of minerals. *Géotechnique*, **12**, 319–35.

HORTON, R. E. 1919. Rainfall interception. *Monthly Weather Review*, **47**, 603–23.

HORTON, R. E. 1933. The role of infiltration in the hydrologic cycle. *Transactions of American Geophysical Union*, **14**, 446–60.

HORTON, R. E. 1939. The analysis of runoff plot experiments. *Transactions of American Geophysical Union*, **20**, 693.

HORTON, R. E. 1945. Erosional development of streams and their drainage basins: hydrophysical approach to quantitative morphology. *Bulletin of the Geological Society of America*, **56**, 275–370.

HOUWINK, R. 1958. *Elasticity, Plasticity and Structure of Matter*. Dover. New York.

HUDSON, N. W. and D. C. JACKSON, 1959. Erosion Research. *Henderson Research Station. Report of progress, 1958-9.* Federation of Rhodesia and Nyasaland; Ministry of Agriculture.

HUNT, C. B., AVERITT, P. and R. L. MILLER, 1953. Geology and geography of the Henry Mountains region, Utah, *United States Geological Survey Professional Paper 288.*

HURTUBISE, J. E. and P. A. ROCHETTE, 1956. The Nicolet slide. *Proceedings of the 37th Canadian Good Roads Association*, 143–55.

HUTCHINSON, J. N. 1961. A landslide on a thin layer of quick clay at Furre, Central Norway. *Géotechnique*, **11**, 69–94.

HUTCHINSON, J. N. 1965. *The stability of cliffs composed of soft rocks, with particular reference to the coasts of south east England*. Unpublished Ph.D. thesis, University of Cambridge.

HUTCHINSON, J. N. 1967. The free degradation of London Clay cliffs. *Proceedings of the Geotechnical Conference, Oslo*, **1**, 113–18.

HUTCHINSON, J. N. 1968. Field meeting on the coastal landslides of Kent. *Proceedings of the Geologists' Association*, **79**, 227–37.

HUTCHINSON, J. N. 1969. A reconsideration of the coastal landslides at Folkstone Warren, Kent. *Géotechnique*, **19**, 6–38.

HUTCHINSON, J. N. and E. N. ROLFSEN, 1962. Large scale field shearbox tests on quick clay. *Geologie und Bauwesen, Heft I*, 31–42.

HVORSLEV, M. J. 1937. Uber die festigheitseigenschaften gestorter bindinger borden. *Ingeniorvidenskabelige Skrifter*, No. 45, Danmarks Naturvidenskabelige Samfund, København.

HVORSLEV, M. J. 1960. Physical components of the shear strength of saturated clay. *Proceedings of American Society of Civil Engineers Research Conference on Shear Strength of Cohesive Soils.*

INSLEY, A. E. and S. F. HILLIS, 1965. Triaxial shear characteristics of a compacted glacial till under unusually high confining pressures. *Proceedings of 6th International Conference on Soil Mechanics and Foundation Engineering*, **1**, 244–8.

INTERNATIONAL ATOMIC ENERGY AGENCY, 1967. Isotope and radiation techniques in soil physics and irrigation studies. *Proceedings of a symposium, Istanbul, June, 1967.* IAEA, Vienna.

JACKLI, H. 1957. Gegenwartsgeologie des bündnerischen Rheingebietes, *Beitrage zur Geologie der Schweitz, Geotechnical Series 36.*

JACKSON, M. L. 1958. *Soil Chemical Analysis*. Constable. London.

JAEGER, J. C. 1959. The frictional properties of joints in rock. *Geofisica Pura e Applicata*, **43**, 148–58.

JAEGER, J. C. and N. G. W. COOK, 1969. *Fundamentals of Rock Mechanics.*

JAHN, A., 1960. Some remarks on the evolution of slopes on Spitsbergen. *Zeitschrift für Geomorphologie, Supp. Band 1*, 49–58.

JENNINGS, J. N. and J. A. MABBUTT, (Editors), 1967. *Landform Studies from Australia and New Guinea.* Cambridge. London.

JENNY, H. 1941. *Factors of Soil Formation.* McGraw Hill, N.Y.

JEWELL, P. A. (Editor) 1963. The experimental earthwork on Overton Down, Wiltshire 1960. *British Association for the Advancement of Science.*

JOHNSON, D. W. 1932*a*. Rock planes of arid regions. *Geographical Review*, **22**, 656–65.

JOHNSON, D. W. 1932*b*. Rock fans of arid regions. *American Journal of Science*, **223**, 389–416.

JOHNSON, D. W. 1940. Contribution to symposium, 'Walther Penck's contribution to geomorphology'. *Annals of the Association of American Geographers*, **4**, 219–84.

JONES, O. T. 1924. The upper Towy drainage basin. *Quarterly Journal of the Geological Society*, **80**, 568–609.

JORRÉ, G. 1933. Le problème des terrasses goletz sibériennes. *Revue de Géographie Alpine*, **21**, 347–71.

JUDSON, S. 1949. Rock-fragment slopes caused by past frost action in the Jura Mountains, France. *Journal of Geology*, **57**, 137–42.

JUDSON, S. and D. F. RITTER, 1964. Rates of regional denudation in the United States. *Journal of Geophysical Research*, **69**, 3395–401.

KALINSKE, A. A. 1947. Movement of sediment as bed load in rivers. *Transactions of American Geophysical Union*, **28**, 615–20.

KELLER, W. D. 1957. *The Principles of Chemical Weathering.* Columbia, Missouri.

KELLER, W. D. 1963. *Chemistry in Introductory Geology.* Columbia, Missouri.

KELLOGG, C. E. 1941. *The Soils That Support Us.* New York.

KENNEDY, B. A. 1969. *Studies of Erosional Valley-Side Asymmetry.* Unpublished Ph.D. thesis, Cambridge.

KENNEY, T. C. 1967*a*. Shear strength of soft clay. *Proceedings of the Geotechnical Conference, Oslo*, 3–12.

KENNEY, T. C. 1967*b*. The influence of mineral composition on the residual strength of natural soils. *Proceedings of the Geotechnical Conference, Oslo*, 123–9.

KENNEY, T. C. 1967*c*. Slide behavior and the shear resistance of a quick clay determined from a study of the landslide at Selnes, Norway. *Proceedings of the Geotechnical Conference, Oslo*, 57–64.

KERPEN, W. and H. W. SCHARPENSEEL, 1967. Movement of ions and colloids in undisturbed soil and parent rock material columns. *Proceedings of International Atomic Energy Symposium, Istanbul*, 213–25.

KHOSLA, A. N. 1953. Silting of reservoirs. *Central Board of Irrigation and Power (India) Publication 51.*

KING, C. A. M. 1956. Scree profiles on Iceland. *Premier Rapport, Commission pour l'Etude des versants, International Geographical Union, Amsterdam*, 124–5.

KING, L. C. 1950. The study of the world's plainlands. *Quarterly Journal of the Geological Society*, **106**, 101–31.

KING, L. C. 1951. *South African Scenery.* Edinburgh.

KING, L. C. 1953. Canons of landscape evolution. *Bulletin of the Geological Society of America*, **64**, 721–52.

Bibliography

KING, L. C. 1955. Pediplanation and isostasy: an example from South Africa. *Quarterly Journal of the Geological Society*, **111**, 353–9.

KING, L. C. 1957. The uniformitarian nature of hillslopes. *Transactions of the Edinburgh Geological Society*, **17**, 81–102.

KING, L. C. 1962. *The Morphology of the Earth*. Edinburgh.

KIRKBY, M. J. 1963. *A Study of Rates of Erosion and Mass Movements on Slopes, with Special Reference to Galloway*. Unpublished Ph.D. thesis, Cambridge University.

KIRKBY, M. J. 1967. Measurement and theory of soil creep. *Journal of Geology*, **75**, 359–78.

KIRKBY, M. J. 1969. Erosion by water on hillslopes. In *Water, Earth and Man*. (Ed. by R. J. Chorley.) Methuen. London.

KIRKBY, M. J. 1971. Hillslope process–response models based on the continuity equation. *Transactions of Institute of British Geographers, Special Publication No. 3.*

KIRKBY, M. J. and R. J. CHORLEY, 1967. Throughflow, overland flow and erosion. *Bulletin of International Association for Scientific Hydrology*, **12**, 5–21.

KIRKBY, M. J. and A. V. T. KIRKBY, 1969. Erosion and deposition on a beach raised by the 1964 earthquake, Montague Island, Alaska. *United States Geological Survey Professional Paper 543-H.*

KIRKBY, M. J. and A. V. T. KIRKBY. In preparation. Slope processes and pediment forms in S.E. Arizona.

KIRKPATRICK, W. M. 1965. Effects of grain size and grading on the shearing behavior of granular materials. *Proceedings of 6th International Conference on Soil Mechanics and Foundation Engineering*, **1**, 273–7. University of Toronto Press.

KOJAN, E. 1967. Mechanics and rates of natural soil creep. *United States Forest Service Experiment Station (Berkeley, California) Report*, 233–53.

KOLOSEUS, H. J. and J. DAVIDIAN, 1966. Free-surface instability correlations. *United States Geological Survey Water Supply Paper 1592-C.*

KOONS, D. 1955. Cliff retreat in southwest United States. *American Journal of Science*, **253**, 44–52.

KRSMANOVIC, D. and Z. LANGOR, 1964. Large scale laboratory tests on the shear strength of rocky material. *Rock Mechanics and Engineering Geology, Supplement I*, 20–30.

LADANYI, B. and G. ARCHAMBAULT, 1969. Simulation of shear behavior of a jointed rock mass. *Paper presented at the 11th Symposium on Rock Mechanics, Berkeley.*

LAMARCHE, V. C. 1968. Rates of slope degradation as determined from botanical evidence, White Mountains, California. *United States Geological Survey Professional Paper 352-L.*

LAMBE, T. W. 1951. *Soil Testing for Engineers*. New York.

LAMBE, T. W. 1960. A mechanistic picture of shear strength in clay. *Proceedings of the American Society of Civil Engineers Research Conference on Shear Strength of Cohesive Soils.*

LAMBE, T. W. and R. V. WHITMAN, 1969. *Soil Mechanics*. New York.

LANGBEIN, W. B. 1949. Annual runoff in the United States. *United States Geological Survey Circular 52.*

LANGBEIN, W. B. and D. R. DAWDY, 1964. Occurrence of dissolved solids in surface waters in the United States. *United States Geological Survey Professional Paper 501-D*, 115–17.

LANGBEIN, W. B. and L. B. LEOPOLD, 1966. River meanders – theory of minimum variance. *United States Geological Survey Professional Paper 422-H.*

LANGBEIN, W. B. and S. A. SCHUMM, 1958. Yield of sediment in relation to mean annual precipitation. *Transactions of the American Geophysical Union*, **39**, 1076–84.

LAWS, J. O. and D. A. PARSONS, 1943. The relation of raindrop size to intensity. *Transactions of American Geophysical Union*, **24**, 452–9.

LAWSON, A. C. 1932. Rain-wash erosion in humid regions. *Bulletin of the Geological Society of America*, **43**, 703–24.

LEE, I. K. 1969 (Editor). *Soil Mechanics, Selected Topics*. Butterworths. London.

LEES, G. M. 1955. Recent earth movements in the Middle East. *Geologische Rundschau*, **43**, 221–6.

LEHMANN, O. 1933. Morphologische Theorie der Verwitterung von steinschlag wänden. *Vierteljahrsschrift der Naturforschende Gesellschaft in Zurich*, **87**, 83–126.

LEOPOLD, L. B. 1951. Rainfall frequency: an aspect of climatic variation. *Transactions of American Geophysical Union*, **32**, 347–57.

LEOPOLD, L. B. 1962. The Vigil Network. *Bulletin of International Association for Scientific Hydrology*, **7**, 5–9.

LEOPOLD, L. B., EMMETT, W. W. and R. M. MYRICK, 1966. Channel and hillslope processes in a semiarid area, New Mexico. *United States Geological Survey Professional Paper 352-G*.

LEOPOLD, L. B. and W. B. LANGBEIN, 1962. The concept of entropy in landscape evolution. *United States Geological Survey Professional Paper 500-A*.

LEOPOLD, L. B. and T. MADDOCK, JR., 1953. The hydraulic geometry of stream channels and some physiographic implications. *United States Geological Survey Professional Paper 252*.

LEOPOLD, L. B. and M. G. WOLMAN, 1957. River channel patterns; braided, meandering and straight. *United States Geological Survey Professional Paper 282-B*.

LEOPOLD, L. B., WOLMAN, M. G. and J. P. MILLER, 1964. *Fluvial Processes in Geomorphology*. Freeman. San Francisco.

LEWIS, G. L. 1956. Shear strength of rockfill. *Proceedings of 2nd Australian and New Zealand Conference on Soil Mechanics and Foundation Engineering*, 50–52.

LIAKOPOULOS, A. C. 1965a. Theoretical solution of the unsteady, unsaturated flow problems in soils. *Bulletin of the International Association of Scientific Hydrology*, **10**, 5–39.

LIAKOPOULOS, A. C. 1965b. Retention and distribution of moisture in soils after infiltration has ceased. *Bulletin of the International Association of Scientific Hydrology*, **10**, 58–69.

LINSLEY, R. K., KOHLER, M. A. and J. L. H. PAULHUS, 1949. *Applied Hydrology*. McGraw-Hill. New York.

LINTON, D. L. 1955. The problem of tors. *Geographical Journal*, **121**, 470–86.

LITTLE, A. L. and V. E. PRICE, 1958. The use of an electronic computer for slope stability analysis. *Géotechnique*, **8**, 113.

LOHNES, R. A. and R. L. HANDY, 1968. Slope angles in friable loess. *Journal of Geology*, **76**, 247–58.

LOW, P. F. and C. W. LOVELL, 1959. The factor of moisture in frost action. *Highway Research Board Bulletin*, **225**, 23–44.

LOWDERMILK, W. C. and H. L. SUNDLING, 1950. Erosion pavement formation and significance. *Transactions of American Geophysical Union*, **31**, 96–100.

LOWE, J. 1964. Shear strength of coarse embankment dam materials. *Proceedings 8th International Congress on Large Dams*, **3**, 745–61.

LOZINSKI, W. 1912. Die periglaziale Fazies der mechanischen Verwittening. *Report of 11th International Geological Congress*, **2**, 1039–53.

Bibliography

LOVERING, T. S. 1958. Significance of accumulator plants in rock weathering. *Bulletin of Geological Society of America*, **70**, 781–800.

LUBOWE, J. K. 1964. Stream junction angles in the dendritic drainage pattern. *American Journal of Science*, **262**, 325–39.

LUMB, P. 1962. The properties of decomposed granite. *Géotechnique*, **12**, 226–43.

LUMB, P. 1965. The residual soils of Hong Kong. *Géotechnique*, **15**, 180–94.

LUSTIG, L. K. 1969. Trend-surface analysis of the Basin and Range Province, and some geomorphic implications. *United States Geological Survey Professional Paper 500-D*.

LUTZ, J. F. 1934. The physicochemical properties of soils affecting erosion. *Missouri Agricultural Experimental Station Research Bulletin*, *212*.

MABBUTT, J. A. 1967. Denudation chronology in Central Australia: structure, climate and landform inheritance in the Alice Springs area. In *Landform Studies from Australia and New Guinea* (Ed. by Jennings, J. N., and J. A. Mabbutt). Cambridge. London.

MACKIN, J. H. 1948. The concept of the graded river. *Bulletin of the Geological Society of America*, **59**, 463–511.

MACDONALD, D. F. 1913. Some engineering problems of the Panama Canal in their relation to geology and topography. *United States Department of Interior, Bureau of Mines*, Bulletin 86, 1–88.

MALAURIE, J. N. 1952. Sur l'asymétrie des versants dans l'Ile de Disko, Groenland. *Académie des Sciences, Paris, Comptes Rendus*, **234**, 1461–2.

MANER, S. B. 1958. Factors affecting sediment delivery rates in the Red Hills physiographic area. *Transactions of American Geophysical Union*, **39**, 669-75.

MARSAL, R. J. 1967. Large scale testing of rockfill materials. *Proceedings of the American Society of Civil Engineers, Journal of the Soil Mechanics and Foundations Division SM2*, 27–43.

MARSHALL, T. J. 1958. A relation between permeability and size distribution of pores. *Journal of Soil Science*, 1–8.

MARSHALL, T. J. 1959. Relations between water and soil. Commonwealth Bureau of Soils, *Harpenden Technical Communication*, 50.

MAY, D. R. and J. H. A. BRAHTZ, 1936. Proposed methods of calculating the stability of earth dams. *Transactions of the Second Congress on Large Dams*, 4, 539.

MEGINNIS, H. G. 1935. Effect of cover on surface run-off and erosion in the Loessial Uplands of Mississippi. *U.S. Department of Agriculture, Circular* **347**, 15pp.

MEIGH, A. C. and K. R. EARLY, 1957. Some physical and engineering properties of chalk. *Proceedings of 4th International Conference on Soil Mechanics and Foundation Engineering*, **1**, 257–61.

MELTON, M. A. 1957. An analysis of the relations among elements of climate, surface properties, and geomorphology. *Technical Report 11, Project NR 389-042, Office of Naval Research*, Columbia University.

MELTON, M. A. 1965. Debris-covered hillslopes of the Southern Arizona Desert – consideration of their stability and sediment contribution. *Journal of Geology*, **73**, 715–29.

MENCL, V. 1965. Dilatancy of rocks. *Rock Mechanics and Engineering Geology*, **3**, 58–61.

METCALF, J. R. 1966. Angle of repose and internal friction. *International Journal of Rock Mechanics and Mining Sciences*, **3**, 155–62.

MEYER-PETER, E. and R. MÜLLER, 1948. Formulas for bed-load transport. *Proceedings of 3rd Meeting of International Association for Hydraulics Research, Stockholm*.

456

MEYERHOFF, H. A. 1940. Migration of erosional surfaces. *Annals of the Association of American Geographers*, **30**, 247–54.

MIDDLETON, H. E. 1930. Properties of soil which influence erosion. *United States Department of Agriculture Technical Bulletin 178.*

MILLER, J. P. 1961. Solutes in small streams draining single rock types, Sangre de Cristo Range, New Mexico. *United States Geological Survey Water Supply Paper 1535-F.*

MILNE, G. 1947. A soil reconnaissance journey through parts of Tanganyika Territory. *Journal of Ecology*, **27**, 192–265.

MITCHELL, J. K. 1956. The fabric of natural clays and its relation to engineering properties. *Proceedings of Highway Research Board*, **35**, 693–713.

MITCHELL, J. K. 1960. Fundamental aspects of thixotropy in soils. *Proceedings of American Society of Civil Engineers*, 783–810.

MITCHELL, J. K., CAMPANELLA R. G. and A. SINGH, 1968. Soil creep as a rate process. *Journal of Soil Mechanics and Foundations Division, Proceedings of the American Society of Civil Engineers*, **94**, SM1, 231–53.

MITCHELL, J. K. and W. N. HOUSTON, 1969. Causes of clay sensitivity. *Journal of Soil Mechanics and Foundations Division, Proceedings of the American Society of Civil Engineers*, SM3, 845–71.

MOHR, E. C. J. and F. A. VAN BAREN, 1954. *Tropical Soils*. Interscience Publishers. London.

MORGENSTERN, N. 1967. Mechanics of rock slope stability. In *Symposium on Rock Slopes, Geological Society of London, Engineering Group*. Mimeographed.

MORGENSTERN, N. R. and· V. E. PRICE, 1965. The analysis of stability of general slip surfaces. *Géotechnique*, **15**, 79–93.

MORISAWA, M. E. 1964. Development of drainage systems on an upraised lake floor. *American Journal of Science*, **262**, 340–54.

MORTENSEN, H. 1930. Einige oberflächenformen in Chile und auf Spitsbergen im rahmeneiner vergleichenden morphologie der klimazonen. *Petermanns geographische Mitteilungen*, 209.

MOSELEY, H. 1855. *The Mechanical Principles of Engineering and Architecture.*

MOSELEY, H. 1869. On the descent of a solid body on an inclined plane, when subjected to alternations of temperature. *Philosophical Magazine*, **38**, 99–118.

MÜLLER, L. 1964. The rock slide in the Vaiont valley. *Rock Mechanics and Engineering Geology*, **2**, 148–212.

MÜLLER, L. 1968. New considerations on the Vaiont side. *Rock Mechanics and Engineering Geology*, **6**, 1–91.

MUSGRAVE, G. W. 1947. Quantitative evaluation of factors in water erosion – a first approximation. *Journal of Soil and Water Conservation*, **2**, 133–8.

NEWLAND, P. L. and B. H. ALLELY, 1957. Volume changes in drained triaxial tests on granular materials. *Géotechnique*, **7**, 17–34.

NIKURADSE, J. 1932. Gesetzmässigkeiten der turbulenten strömung in glatten rohren. *Forschungshefte. Verein Deutscher Ingenieuve*, 356.

NITCHIPOROVITCH, A. A. 1964. Deformations and stability of rockfill dams. *Proceedings of 8th International Congress on Large Dams*, **3**, 879–94.

ØDUM, H. 1922. On 'Faarestiernes' natur. *Meddelelser fra Dansk geologisk Forening*, **6**, 1–29.

OLLIER, C. D. 1960. The inselbergs of Uganda. *Zeitschrift für Geomorphologie*, **4**, 43–52.

OLLIER, C. D. 1969. *Weathering*. Oliver and Boyd. Edinburgh.

457

Bibliography

OLLIER, C. D. and A. J. THOMASSON, 1957. Asymmetrical valleys of the Chiltern Hills. *Geographical Journal*, **123**, 71–80.

OOSTIG, H. J. 1958. *The Study of Plant Communities*. San Francisco.

OWENS, I. F. 1969. Causes and rates of soil creep in the Chilton Valley, Cass, New Zealand. *Arctic and Alpine Research*, **1**, 213–20.

PAIGE, S. 1912. Rock cut surfaces in desert ranges. *Journal of Geology*, **20**, 442–50.

PALMER, J. 1956. Tor formation at the Bridestones in N.E. Yorkshire, and its significance in relation to problems of valley-side development and regional glaciation. *Transactions of the Institute of British Geographers*, **22**.

PARIZEK, E. J. and J. F. WOODRUFF, 1957a. Mass wasting and the deformation of trees. *American Journal of Science*, **255**, 63–70.

PARIZEK, E. J. and J. F. WOODRUFF, 1957b. Description and origin of stone layers in the soils of the south-eastern States, *Journal of Geology*, **65**, 24–34.

PARKER, G. G. 1963. Piping, a geomorphic agent in landform development in the drylands. *International Association for Scientific Hydrology, Publication No. 65*, 103–13.

PATTON, F. D. 1966. Multiple modes of shear failure in rock. *Proceedings of 1st Congress of International Society of Rock Mechanics*, **1**, 509–13.

PECK, R. B. and W. V. KAUN, 1948. Description of a flow slide in loose sand. *Proceedings of the 2nd International Conference on Soil Mechanics and Foundation Engineering*, **3**, 296.

PECK, R. B., IRELAND, H. O. and T. S. FRY, 1951. Studies of soil characteristics, the earth flows of St Thuribe, Quebec. *Soil Mechanics Series No. 1*, University of Illinois.

PELLEGRINO, A. 1965. Geotechnical properties of coarse-grained soils. *Proceedings of 6th International Conference on Soil Mechanics and Foundation Engineering*, **1**, 87–92.

PELTIER, L. C. 1950. The geographic cycle in periglacial regions as it is related to climatic geomorphology. *Annals of the Association of American Geographers*, **40**, 214–36.

PENCK, A. 1894. *Morphologie der Erdeberfläche*. Stuttgart, 357–61.

PENCK, W. 1923. Uber die form andiner Krusten bewegungen und ihre beziehung zur sedimentation. *Geologische Rundschau*, **14**, 301–15.

PENCK, W. 1924. *Die Morphologische Analyse. Morphological Analysis of Landforms:* English translation by Czech. H. and K. C. Boswell. London, 1953.

PENCK, W. 1925. Die piedmontflächen des südlichen Schwarzwaldes. *Zeitschrift der Gesellschaft für Erdkunde zu Berlin*, 81–108.

PENCK, W. 1953. *Morphological Analysis of Landforms*, translated by Czech and Boswell. Macmillan, London.

PENNER, E. 1963a. Anisotropic thermal conduction in clay sediments. In *Proceedings of the International Clay Conference*, Pergamon, London.

PENNER, E. 1963b. Frost-heaving in soils, *Proceedings of International Permafrost Conference, Lafayette, Indiana*, 197–202.

PENNER, E. 1965. A study of sensitivity in Leda Clay. *Canadian Journal of Earth Science*, **2**, 425–41.

PENNER, E. 1967. Pressures developed during unidirectional freezing of water-saturated porous materials. *Proceedings of the International Conference on Low Temperature Science, 1966, Sapporo, Japan*, **1**, Pt. 2, 1401–12.

PHILIP, J. R. 1954. An infiltration equation with physical significance. *Soil Science*, **77**, 153–7.

458

PHILIP, J. R. 1957. The theory of infiltration: IV. Sorptivity and algebraic infiltration equations. *Soil Science*, **84**, 257–64.

PIGOTT, C. D. 1962. Soil formation and development on the Carboniferous Limestone of Derbyshire. *Journal of Ecology*, **50**, 145–56.

PITTY, A. 1968. The scale and significance of solutional loss from the limestone tract of the southern Pennines. *Proceedings of the Geologists' Association*, **79**, 153–77.

POLYNOV, P. B. 1937. *The Cycle of Weathering.* Murby. London.

POOROOSHASB, H. and K. H. ROSCOE, 1961. The correlation of the results of shear tests with varying degrees of dilation. *Proceedings of 5th International Conference on Soil Mechanics and Foundation Engineering*, **1**, 297–304.

POSER, H. 1954. Die Periglazial – erscheinungen in der urngebung der gletscher des zemmgrundes. *Gottinger Geographische Abhundlungen*, 15.

POWELL, J. W. 1874. Remarks on the structural geology of the valley of the Colorado of the west. *Washington Philosophical Society*, **1**, 48–51.

POWELL, J. W. 1875. *Exploration of the Colorado River of the West and its Tributaries.* Smithsonian Institution, Washington, D.C.

PRICE, N. J. 1958. A study of rock properties in conditions of triaxial stress. *Mechanical Properties of Non-Metallic Brittle Materials* (Ed. by W. H. Walton).

PRICE, N. J. 1960. The strength of coal-measure rocks in triaxial compression. *National Coal Board M.R.E. Report No. 2159.*

PRIGOGINE, I. and R. DEFAY, 1954. *Chemical Thermodynamics.* Longmans. London.

PUGH, J. C. 1955. Isostatic readjustment and the theory of pediplanation. *Quarterly Journal of the Geological Society*, **111**, 361–9.

PUGH, J. C. and L. C. KING, 1952. Outline of the geomorphology of Nigeria. *South African Geographical Journal*, **34**, 30–7.

QUEIROZ, L. 1964. Geotechnical properties of weathered rock and behavior of Furnas rockfill dam. *Proceedings of 8th International Congress on Large Dams*, **1**, 877–90.

QUIGLEY, R. M. and C. D. THOMPSON, 1966. The fabric of anisotropically consolidated sensitive marine clay. *Canadian Geotechnical Journal*, **3**, 61–73.

RAHN, P. H. 1966. Inselbergs and nickpoints in southwestern Arizona. *Zeitschrift für Geomorphologie*, **10**, 215–25.

RAINWATER, F. H. and L. L. THATCHER, 1960. Methods for collection and analysis of water samples. *United States Geological Survey, Water Supply Paper 1454.*

RANKINE, W. J. M. 1857. On the stability of loose earth. *Transactions of the Royal Society, London, 147.*

RAPP, A. 1960a. Talus slopes and mountain walls at Tempelfjorden, Spitsbergen. *Norsk Polarinstitutt Skrifter, 119.*

RAPP, A. 1960b. Recent developments of mountain slope in Kärkevagge and surroundings, northern Scandinavia. *Geografiska Annaler*, **42**, 71–200.

REICHE, P. 1950. *A Survey of Weathering Processes and Products.* University of New Mexico Publications in Geology, No. 3.

RENNER, F. G. 1936. Conditions influencing erosion on the Boise River Watershed. *United States Department of Agriculture Technical Bulletin*, **528**.

RINGHEIM, A. S. 1964. Experiences with Bearpaw shales at the South Saskatchewan River Dam. *Proceedings of 8th International Congress on Large Dams*, **1**, 529–50.

RIPLEY, C. F. and K. L. LEE, 1961. Sliding friction tests on sedimentary rock specimens. *Proceedings of 7th International Congress on Large Dams*, **4**, 657–71.

ROBERTSON, E. C. 1955. Experimental study of the strength of rocks. *Bulletin of Geological Society of America*, **66**, 1275–314.

ROBERTSON, J. M. and H. ROUSE, 1941. On the four regimes of open channel flow. *Civil Engineering*, **11**, 169–71.

459

ROBINSON, G. 1966. Some residual hillslopes in the Great Fish River Basin, S. Africa. *Geographical Journal,* **132,** 386–90.

RODDA, J. C. 1969. The flood hydrograph. In *Water, Earth and Man* (Ed. by R. J. Chorley).

ROS, M. and A. EICHINGER, 1928. Versuche zur klärung der frage der bruchgefahr. 11 Nichtmetallische Staffe. *Eidgenössische Materialprüfungsanstalt der Eigen Techn. Nochschule,* 27 June, 57 pp.

ROSCOE, K. H., SCHOFIELD, A. N. and C. P. WROTH, 1958. On the yielding of soils. *Géotechnique,* **8,** 22–53.

ROWE, P. W. 1962. The stress-dilatancy relation for static equilibrium of an assembly of particles in contact. *Proceedings of the Royal Society of London, Series A,* **269,** 500–27.

ROWE, P. W. 1963. Stress-dilatancy, earth pressure and slopes. *Proceedings of American Society of Civil Engineers, Soil Mechanics Division,* 37–61.

ROWE, P. W., BARDEN, L. and I. K. LEE, 1964. Energy components during the triaxial cell and direct shear tests. *Géotechnique,* **14,** 247–61.

ROZCYCKI, S. Z. 1957. Zones du model et phénomenès periglaciaires de la Terre de Torell. *Biuletyn Peryglacjalny,* **5,** Lodz.

RUBEY, W. W. 1938. The force required to move particles on the bed of a stream. *United States Geological Survey Professional Paper 189–E.*

RUDBERG, S. 1958. Some observations concerning mass movement on slopes in Sweden. *Meddelanden Från Uppsala Universitets, Geografiska Institution, Series A. No. 126,* 114–25.

RUDBERG, S. 1962. A report on some field observations concerning periglacial geomorphology and mass movements on slopes in Sweden. *Biuletyn Peryglacjalny,* **11,** 311–23.

RUHE, R. V. 1952. Topographic discontinuities of the Des Moines lobe. *American Journal of Science,* **250,** 46–56.

RUHE, R. V. 1959. Stone lines in soils. *Soil Science,* **87,** 223–31.

RUIZ, M. D. and F. P. DE CAMARGO, 1966. A large scale field shear test on rock. *Proceedings of 1st International Society for Rock Mechanics,* **1,** 257–61.

RUXTON, B. P. 1958. Weathering and subsurface erosion in granite at the Piedmont angle, Balos, Sudan. *Geological Magazine,* **95,** 353–77.

RUXTON, B. P. 1967. Slopewash under mature primary rainforest in northern Papua. In *Landforms Studies from Australia and New Guinea,* (Edited by Jennings, J. N. and J. A. Mabbutt). Cambridge. London.

RUXTON, B. P. 1968a. Measures of the degree of chemical weathering of rocks. *Journal of Geology,* **76,** 518–27.

RUXTON, B. P. 1968b. Ranges of weathering of Quaternary volcanic ash in north-east Papua. *Transactions of 9th International Congress of Soil Science (Adelaide),* 367–76.

RUXTON, B. P. and L. BERRY, 1957. The weathering of granite and associated erosional features in Hong Kong. *Bulletin of the Geological Society of America,* **68,** 1263–92.

SAVIGEAR, R. A. G. 1952. Some observations on slope development in South Wales. *Transactions of the Institute of British Geographers,* **18,** 31–52.

SAVIGEAR, R. A. G. 1956. Technique and terminology in the investigation of slope forms. *Union Géographical International; Premier Rapport de la Commission pour l'étude des versants, Rio de Janeiro,* 66–75.

SAVIGEAR, R. A. G. 1960. Slopes and hills in West Africa. *Zeitschrift für Geomorphologie Supp. 1,* 156–71.

SCHARPENSEEL, H. W. and W. KERPEN 1967. Studies on tagged clay migration due to water movement. *Proceedings of International Atomic Energy Agency Symposium.* Istanbul), 287–90.

SCHEIDEGGER, A. E. 1961*a*. Mathematical models of slope development. *Bulletin of the Geological Society of America,* **72,** 37–50.

SCHEIDEGGER, A. E. 1961*b*. *Theoretical Geomorphology.* Prentice Hall. Englewood Cliffs, N.J.

SCHEIDEGGER, A. E. 1966. Effect of map scale on stream orders. *Bulletin of International Association for Scientific Hydrology,* **11,** 56–61.

SCHICK, A. P. 1965. The effects of lineative factors on stream courses in homogeneous bedrock. *Bulletin of International Association for Scientific Hydrology,* **3,** 5–11.

SCHNEIDER, W. J. 1961. A note on the accuracy of drainage densities computed from topographic maps. *Journal of Geophysical Research,* **66,** 3617–18.

SCHULTZE, E. 1957. Large scale shear tests. *Proceedings of 4th International Conference on Soil Mechanics and Foundation Engineering,* **1,** 193–9.

SCHULTZE, E. and A. HORN, 1965. The shear strength of silt. *Proceedings of 6th International Conference on Soil Mechanics and Foundation Engineering,* **1,** 350–53.

SCHUMM, S. A. 1956*a*. Evolution of drainage systems and slopes on badlands at Perth Amboy, New Jersey. *Bulletin of the Geological Society of America,* **67,** 597–646.

SCHUMM, S. A. 1956*b*. The role of creep and rain-wash on the retreat of badland slopes. *American Journal of Science,* **254,** 693–706.

SCHUMM, S. A. 1963. The disparity between present rates of denudation and orogeny. *United States Geological Survey Professional Paper,* *454-H,* 13 pp.

SCHUMM, S. A. 1964. Seasonal variations of erosion rates and processes on hillslopes in western Colorado. *Zeitschrift für Geomorphologie. Supp. Band 5,* 215–38.

SCHUMM, S. A. 1965. Quaternary paleohydrology. In *The Quaternary of the United States* (Edited by Wright, H. E. and D. G. Frey), Princeton University.

SCHUMM, S. A. 1967. Rates of surficial rock creep on hillslopes in western Colorado. *Science,* **155,** 560–1.

SCHUMM, S. A. and R. J. CHORLEY, 1964. The fall of Threatening Rock. *American Journal of Science,* **262,** 1041–54.

SCHUMM, S. A. and R. J. CHORLEY, 1966. Talus weathering and scarp recession in the Colorado Plateaus. *Zeitschrift für Geomorphologie,* **10,** 11–36.

SCHUMM, S. A. and R. F. HADLEY, 1961. Progress in the application of landform analysis in studies of semi-arid erosion. *United States Geological Survey Circular 437.*

SCHUMM, S. A. and R. W. LICHTY, 1965. Time, space, and causality in geomorphology. *American Journal of Science,* **263,** 110–19.

SCHUMM, S. A. and G. C. LUSBY, 1963. Seasonal variations of infiltration and runoff on the Mancos Shale of Western Colorado. *Journal of Geophysical Research,* **68,** 3655–66.

SCHWARZENBACH, G. and H. FASCHKA, 1969. *Complexiometric Titrations.* Methuen. London.

SCOTT, R. F. 1963. *Principles of Soil Mechanics.* Reading, Mass.

SELBY, M. J. 1968. Cones for measuring soil creep. *Journal of Hydrology (New Zealand),* **7,** 136–7.

SEVALDSON, R. A. 1956. The Slide in Lodalen, October 6th, 1954. *Géotechnique,* **6,** 167–82.

SHARPE, C. F. S. 1938. *Landslides and Related Phenomena.* Columbia. New York.

461

Bibliography

SHERMAN, L. K. 1944. Infiltration and the physics of soil moisture. *Transactions of American Geophysical Union*, **25**, 57–71.

SHIELDS, I. A. 1936. Anwendung der Ahnlichkeitmechanik und der turbulenzforschung auf die geschiebebewegung. *Mitteilungen der Preussischen Versuchsanstalt für Wasserbau und Schiffbau*, 26.

SHOSTAKOVITCH, W. B. 1927. Der ewig gefrorne Boden Sibiriens. *Zeitschrift der Gesellschaft für Erdkunde zu Berlin*, 394–427.

SHREVE, R. L. 1966. Statistical laws of stream numbers. *Journal of Geology*, **74**, 17–37.

SHREVE, R. L. 1969. Stream lengths and basin areas in topologically random channel networks. *Journal of Geology*, **77**, 397–414.

SILVESTRI, T. 1961. Determinazione sperimentale de resistenza meccanica del materiale constituente il corpo di una diga del tipo 'Rockfill'. *Geotechnica*, **8**, 186–91.

SIMONETT, D. S. 1967. Landslide distribution and earthquakes in the Dewani and Torricelli Mountains, New Guinea. In *Landform Studies from Australia and New Guinea*, (Edited by Jennings, J. N. and J. A. Mabbutt). Methuen. London.

SIMONS, M. 1962. The Morphological Analysis of Landforms: a new review of the work of Walther Penck (1888–1923). *Transactions of the Institute of British Geographers*, **31**, 1–14.

SIMPSON, S. 1953. The development of the Lyn drainage system and its relation to the origin of the coast. *Proceedings of the Geologists' Association*, **64**, 14–28.

SITTER, L. H. DE, 1956. *Structural Geology*. New York.

SKEMPTON, A. W. 1945. Earth pressure and the stability of slopes. In *The Principles and Application of Soil Mechanics*, Institution of Civil Engineers, London.

SKEMPTON, A. W. 1953. Soil mechanics in relation to geology. *Proceedings of Yorkshire Geological Society*, **29**, 33–62.

SKEMPTON, A. W. 1960. Effective stress in soils, concrete and rocks. In *Pore Pressure and Suction in Soils*.

SKEMPTON, A. W. 1964. The long-term stability of clay slopes. (The Rankine Lecture.) *Géotechnique*, **14**, 75–102.

SKEMPTON, A. W. and A. W. BISHOP, 1950. The measurement of the shear strength of soils. *Géotechnique*, **2**, 90–108.

SKEMPTON, A. W. and J. D. BROWN, 1961. A landslide in boulder clay at Selset, Yorkshire. *Géotechnique*, **11**, 280–93.

SKEMPTON, A. W. and F. A. DELORY, 1957. Stability of natural slopes in London Clay. *Proceedings of 4th International Conference on Soil Mechanics and Foundation Engineering*, **2**, 378–81.

SKEMPTON, A. W. and J. N. HUTCHINSON, 1969. Stability of natural slopes and embankment sections. *Proceedings of 7th International Conference on Soil Mechanics and Foundation Engineering, State of the Art Volume*, 291–340.

SKEMPTON, A. W. and R. D. NORTHEY, 1952. The sensitivity of clays. *Géotechnique*, **3**, 30–53.

SLAYMAKER, H. O. 1968. *Patterns of Sub-aerial Erosion In Instrumented Catchments With Particular Reference To The Upper Wye Valley, Mid-Wales*. Unpublished Ph.D. thesis, Cambridge.

SMITH, D. I. 1965. Variations in the rate of geomorphological processes: a case study for limestone solutional erosion. In *Rates of Erosion and Weathering in the British Isles*, (Edited by D. I. Smith), British Geomorphological Research Group.

SMITH, D. I. 1969a. The solutional erosion of limestones in an arctic morphogenetic region. *Proceedings of Symposium on Karst Denudation, Brno, Yugoslavia*, 99–109.

Bibliography

SMITH, D. I. 1969*b*. The solutional erosion of limestone in the area around Maldon and Maroon Town, St James, Jamaica. *Journal of the British Speleological Association*, **6**, 120–35.

SMITH, H. T. U. 1949*a*. Periglacial features in the driftless area of southern Wisconsin. *Journal of Geology*, **57**, 196–215.

SMITH, H. T. U. 1949*b*. Physical effects of the Pleistocene climatic changes in non-glaciated areas; aeolian phenomena, frost action, and stream terracing. *Bulletin of the Geological Society of America*, **60**, 1485–516.

SOKOLOVSKI, V. V. 1956. *Statics of Soil Media*. Butterworths. London.

SPARKS, B. W. 1960. *Geomorphology*. Longmans. London.

SPARROW, G. W. A. 1966. Some environmental factors in the formation of slopes. *Geographical Journal*, **132**, 390–5.

STAGG, K. G. and O. C. ZIENKIEWICZ, 1968. *Rock Mechanics in Engineering Practice*.

STARKEL, L. 1959. Development of the relief of the Polish Carpathians in the Holocene. *Przeglad Geograficzny*, *31*.

STARKEL, L. 1967. *Guide to excursion of the symposium of the Commission on the Evolution of Slopes and of the Commission on Periglacial Geomorphology of the International Geographical Union, Poland*.

STEARN, W. H. 1935. Structure and creep. *Journal of Geology*, **43**, 323–7.

STEPHENS, C. G. 1961. The soil landscapes of Australia. *CSIRO Australian Soils Publication No. 18*.

STODDART, D. R. 1969. World erosion and sedimentation. In *Water, Earth and Man*, (Edited by R. J. Chorley). Methuen. London.

STONE, R. 1961. Geologic and engineering significance of changes in elevation revealed by precise levelling, Los Angeles area, California. *Geological Society of America Special Paper 68*, 57–8.

STRAHLER, A. N. 1950. Equilibrium theory of erosional slopes approached by frequency distribution analysis. *American Journal of Science*, **248**, 673–96 and 800–14.

STRAHLER, A. N. 1952*a*, Hypsometric analysis of erosional topography. *Bulletin of the Geological Society of America*, **63**, 1117–42.

STRAHLER, A. N. 1952*b*. Dynamic basis of geomorphology. *Bulletin of the Geological Society of America*, **63**, 923–38.

STRAHLER, A. N. 1966. *Physical Geography*. Wiley. New York.

STRAKHOV, N. M. 1967. *Principles of Lithogenesis*. vol 1. Translated by J. P. Fitzsimmons. Edited by Tomkeiff, S. I. and J. E. Hemingway. Consultants' Bureau, New York.

STRAUB, L. G. 1934. Effect of channel-contraction works upon regimen of movable bed streams. *Transactions of American Geophysical Union*, **15**, 455.

TABER, S. 1929. Frost heaving. *Journal of Geology*, **37**, 428–61.

TABER, S. 1930. The mechanics of frost heaving. *Journal of Geology*, **38**, 303–17.

TAYLOR, D. W. 1937. Stability of earth slopes. *Journal of the Boston Society of Civil Engineers*, **24**, 197–246.

TAYLOR, D. W. 1948. *Fundamentals of Soil Mechanics*. New York.

TE PUNGA, M. T. 1956. Altiplanation terraces in southern England. *Biuletyn Peryglacjalny*, **4**, 331–8.

TEIXEIRA DA CRUZ, P. 1963. Shear strength characteristics of some residual recompacted clays. *Proceedings 2nd Pan-American Conference on Soil Mechanics and Foundation Engineering*, **1**, 73–102.

TERZAGHI, K. 1936. The shearing resistance of saturated soils. *Proceedings of 1st International Conference on Soil Mechanics and Foundation Engineering*, **1**, 54–66.

463

Bibliography

TERZAGHI, K. 1943. *Theoretical Soil Mechanics*. New York.

TERZAGHI, K. 1944. Ends and means in soil mechanics. *Engineering Journal*, **27**, 608.

TERZAGHI, K. 1950. Mechanism of landslides. *Bulletin of the Geological Society of America, Berkey Volume*, 83–122.

TERZAGHI, K. 1953. Some miscellaneous notes on creep. *Proceedings of 3rd International Conference on Soil Mechanics and Foundation Engineering*, **3**, 205–6.

TERZAGHI, K. 1957. Varieties of submarine slope failures. *Norwegian Geotechnical Institute Publication No. 25*, Oslo.

TERZAGHI, K. 1958. Landforms and subsurface drainage in the Gaika region in Yugoslavia. *Zeitschrift für Geomorphologie, N.F. Band 2*, 76–100.

TERZAGHI, K. 1962a. Stability of steep slopes on hard unweathered rock. *Géotechnique*, **12**, 251–70.

TERZAGHI, K. 1962b. Dam foundation on sheeted granite. *Géotechnique*, **12**, 199–208.

TERZAGHI, K. 1962c. Discussion of a paper by A. W. Skempton. *Proceedings of 5th International Conference on Soil Mechanics and Foundation Engineering*, **3**, 144–5.

TERZAGHI, K. and R. B. PECK, 1948. *Soil Mechanics in Engineering Practice*. Wiley. New York.

THOMAS, M. F. 1965. Some aspects of the geomorphology of tors and domes in Nigeria. *Zeitschrift für Geomorphologie*, **9**, 63–81.

THOMAS, M. F. 1968. Some outstanding problems in the interpretation of the geomorphology of tropical shields. *Occasional Publication of British Geomorphological Research Group, No. 5*, 41–50.

THORNBURY, W. D. 1966. *Principles of Geomorphology*, Wiley. New York.

THORNTHWAITE, C. W., SHARPE, C. F. S., and E. F. DOSCH, 1942. Climate and accelerated erosion in the arid and semi-arid southwest, with special reference to the Polacca Wash drainage basin, Arizona. *United States Department of Agriculture Technical Bulletin No. 808.*

TOIT, A. L. DU, 1937. *Our Wandering Continents*. Edinburgh.

TINKLER, K. J. 1966. Slope profiles and scree in the Eglwyseg Valley, North Wales. *Geographical Journal*, **132**, 379–85.

TSUBOI, C. 1933. Investigation on the deformation of the Earth's crust found by precise geodetic means. *Japanese Journal of Astronomy and Geophysics*, **10**, 93–248.

TUAN, YI-FU, 1958. The misleading antithesis of Penckian and Davisian concepts of slope retreat in waning development. *Proceedings of the Indiana Academy of Science*, **67**, 212–14.

TUAN, YI-FU, 1959. Pediments in southeastern Arizona. *University of California Publications in Geography*, **13**.

TWENHOFEL, W. H. 1939. The cost of soil in rock and time. *American Journal of Science*, **237**, 771–80.

TWIDALE, C. R. 1967. Origin of the piedmont angle as evidenced in South Australia. *Journal of Geology*, **75**, 393–411.

TYLOR, A. 1875. Action of denuding agencies. *Geological Magazine*, **2**, 433–73.

UNITED STATES BUREAU OF RECLAMATION 1947. Laboratory tests on protective filters for hydraulic and static structures. *Earth Material Laboratory Report EM-132*, Denver.

UNITED STATES GEOLOGICAL SURVEY. Quality of surface waters of the United States. *United States Geological Survey Water Supply Papers.*

UNITED STATES GEOLOGICAL SURVEY, 1967. Mesa Verde National Park, Colorado, 1:24,000 topographic map.

VAN BURKALOW, A. 1945. Angle of repose and angle of sliding friction: an experimental study. *Bulletin of Geological Society of America*, **56**, 669–708.

VARGAS, M. 1953. Some engineering properties of residual clay soils occurring in S. Brazil. *Proceedings of 3rd International Conference on Soil Mechanics and Foundation Engineering*, **1**, 67–71.

VARGAS, M. and E. PICHLER, 1957. Residual soil and rock slides in Brazil. *Proceedings of 4th International Conference on Soil Mechanics and Foundation Engineering*, **2**, 394–8.

VARGAS, M., SILVA, F. P. and M. TUBIO, 1965. Residual clay dams in the state of São Paulo, Brazil. *Proceedings of 6th International Conference on Soil Mechanics and Foundation Engineering*, **2**, 579–82.

VEIHMEYER, F. J. and A. H. HENDRICKSON, 1931. The moisture equivalent as a measure of the field capacity of soils. *Soil Science*, **32**, 181–93.

VON BERTALANFFY, L. 1951. An outline of general systems theory. *British Journal of the Philosophy of Science*, **1**, 134–65.

VUCETIC, R. 1958. Determination of shear strength and other characteristics of coarse, clayey schist material compacted by pneumatic wheel roller. *Proceedings of 6th International Congress on Large Dams*, **4**, 465–73.

VYALOV, S. S., GORODETSKII, S. E., ERMAKOV, V. F., ZATSARNAYA, A. G. and N. K. PEKARSKAYA, 1966. Methods of determining creep, long-term strength and compressibility characteristics of frozen soils. *National Research Council of Canada, Technical Translation 1364*, 1969, 109 pp.

WAHRHAFTIG, C. and A. COX, 1959. Rock glaciers in the Alaska Range. *Bulletin of the Geological Society of America*, **70**, 383–436.

WALKER, E. H. 1948. Differential erosion on slopes of northern and southern exposure in western Wyoming. *Bulletin of Geological Society of America*, **59**, 1360 (Abstract).

WARD, R. C. 1967. *Principles of Hydrology*. McGraw Hill, London.

WARD, W. H. 1945. The stability of natural slopes. *Geographical Journal*, **111**, 170–91.

WARD, W. H. 1953. Soil movement and weather. *Proceedings of 3rd International Conference on Soil Mechanics and Foundation Engineering*, **1**, 477–82.

WASHBURN, A. L. 1960. Instrumentation for mass-wasting and patterned-ground studies in Northeast Greenland. *Biuletyn Peryglacjalny*, **8**, 59–64.

WATERS, R. S. and R. H. JOHNSON, 1958. The terraces of the Derbyshire Derwent. *East Midland Geographer*, **9**, 3–15.

WATERS, R. S. 1962. Altiplanation terraces and slope development in Vest-Spitsbergen and south-west England. *Biuletyn Peryglacjalny*, **11**, 89–101.

WATERS, R. S. 1964. The Pleistocene legacy to the geomorphology of Dartmoor. *Dartmoor Essays* (Edited by Simmons). Torquay.

WEAVER, J. E. 1937. Effects of roots of vegetation in erosion control. *United States Department of Agriculture Technical Bulletin, Soil Conservation Service*, Mimeo paper 2666.

WEGMAN, E. 1957. Tectonique vivante, dénudation et phénomènes connexes. *Revue de Géographie Physique et Géologie Dynamique*, Pt 2, **1**, 3–15.

WENTWORTH, C. K. 1928. Principles of stream erosion in Hawaii. *Journal of Geology*, **36**, 385–410.

WENTWORTH, C. K. 1943. Soil avalanches on Oahu, Hawaii. *Bulletin of the Geological Society of America*, **54**, 53–64.

WEYMAN, D. R. 1970. Throughflow on hillslopes and its relation to the stream hydrograph. *Bulletin of the International Association for Scientific Hydrology*, **15**, 25–33.

WHIPKEY, R. Z. 1965. Subsurface stormflow from forested slopes. *Bulletin of International Association for Scientific Hydrology*, **10**, 74–85.

Bibliography

WHITE, C. M. 1940. Equilibrium of grains on the bed of a stream. *Proceedings of the Royal Society of London, Series A*, **174**, 322–34.

WIEBOLS, G. A., JAEGER, J. C. and N. G. W. COOK, 1968. Rock property tests in a stiff testing machine. *Paper presented at 10th Rock Mechanics Symposium*, Rice University, Houston.

WILLIAMS, M. A. J. 1969. Prediction of rainsplash erosion in the seasonally wet tropics. *Nature*, **222**, 763–5.

WILLIAMS, P. J. 1957. Some investigations into solifluction features in Norway. *Geographical Journal*, **123**, 42–55.

WILLIAMS, P. J. 1959. An investigation into processes occurring in solifluction. *American Journal of Science*, **257**, 481–490.

WILLIAMS, P. J. 1962. An apparatus for investigation of the distribution of movement with depth in shallow soil layers. *Building Note 39, Division of Building Research, N.R.C., Canada.*

WILLIAMS, P. J. 1966a. Downslope movement at a subarctic location with regard to variations with depth. *Canadian Geotechnical Journal*, **3**, 191–203.

WILLIAMS, P. J. 1966b. Pore pressures at a penetrating frost line and their prediction. *Géotechnique*, **16**, 187–208.

WILLIAMS, P. W. 1965. The rate of limestone solution in western Ireland. In *Rates of Erosion and Weathering in the British Isles*, (Edited by D. I. Smith). British Geomorphological Research Group.

WILLIAMS, R. B. G. 1964. Fossil patterned ground in eastern England. *Biuletyn Peryglacjalny*, **14**, 337–49.

WISCHMEIER, W. H. and D. D. SMITH, 1958. Rainfall energy and its relationship to soil loss. *Transactions of American Geophysical Union*, **39**, 285–91.

WOLMAN, M. G. 1955. The natural channel of Brandywine Creek, Pennsylvania. *United States Geological Survey Professional Paper 271.*

WOLMAN, M. G. and J. P. MILLER, 1960. Magnitude and frequency of forces in geomorphic processes. *Journal of Geology*, **68**, 54–74.

WOOD, A. 1942. The development of hillside slopes. *Proceedings of the Geologists' Association*, **53**, 128–40.

WOODBURN, R. and J. KOZACHYN, 1956. A study of relative erodibility of a group of Mississippi gully soils. *Transactions of the American Geophysical Union*, **37**, 749–53.

WOOLNOUGH, W. G. 1927. The duricrust of Australia. *Proceedings of The Royal Society of New South Wales*, **61**, 24–53.

WOPFNER, H. and C. R. TWIDALE, 1967. Geomorphological history of the Lake Eyre Basin. In *Landform Studies From Australia and New Guinea.* (Edited by Jennings, J. N. and J. A. Mabbutt.) Cambridge. London.

WORRALL, W. E. 1968. *Clays: Their Nature, Origin and General Properties.* MacLaren. London.

WU, T. H. 1966. *Soil Mechanics.* Boston.

YATSU, E. 1955. On the longitudinal profile of the graded river. *Transactions of American Geophysical Union*, **36**, 655–65.

YATSU, E. 1966. *Rock Control and Geomorphology.* Sozosha.

YATSU, E. 1967. Some problems on mass-movement. *Geografiska Annaler*, **49A**, 396–401.

YONG, R. N. and B. P. WARKENTIN, 1966. *Introduction to Soil Behavior.* New York.

YOUNG, A. 1956. Scree profiles in West Norway. *Premier Rapport, Commission pour l'Etude des versants, International Geographical Union, Amsterdam*, 125.

YOUNG, A. 1958. *Some Considerations of Slope Form and Development, Regolith, and Denudational Processes.* Unpub. Ph.D. thesis, University of Sheffield.

YOUNG, A. 1960. Soil movement by denudational processes on slopes. *Nature*, **188**, 120–2.

YOUNG, A. 1961. Characteristic and limiting slope angles. *Zeitschrift für Geomorphologie, Band 5*, 126–31.

YOUNG, A. 1963*a*. Soil movement on slopes. *Nature*, **200**, 129–30.

YOUNG, A. 1963*b*. Some observations of slope form and regolith, and their relation to slope development. *Transactions of the Institute of British Geographers*, **32**, 1–29.

YOUNG, A. 1964. Slope profile analysis. *Zeitschrift für Geomorphologie Supp. band 5*, 17–27.

YOUNG, A. 1968. Slope form and the soil catena in savanna and rainforest environments. *Occasional Publication of British Geomorphological Research Group, No. 5*, 3–11.

YOUNGS, E. G. 1957. Moisture profiles during vertical infiltration. *Soil Science*, **84**, 283–90.

YOUNGS, E. G. 1958. Redistribution of moisture in porous materials after infiltration. *Soil Science*, **86**, 117–25 and 202–7.

ZINGG, A. W. 1940. Degree and length of land slope as it affects soil loss in runoff. *Agricultural Engineering*, **21**, 59–64.

SUBJECT INDEX